Varieties of Innovation Systems

Actors and Structures
Empirical Cross-national Research

Edited by Hans-Jürgen Andreß, Detlef Fetchenhauer, Karsten Hank, André Kaiser and Heiner Meulemann, University Cologne

Volume 3

Dr. Michael Ortiz studied sociology at the Universities of Bamberg and Salamanca. He works as a scientific assistant at the University of Mannheim.

Michael Ortiz

Varieties of Innovation Systems

The Governance of Knowledge Transfer in Europe

Campus Verlag
Frankfurt/New York

Printed with the kind assistance of "Johanna und Fritz Buch Gedächtnis-Stiftung" and "Axel Springer Stiftung".

Bibliographic Information published by the Deutsche Nationalbibliothek.
The Deutsche Nationalbibliothek lists this publication in the Deutsche Nationalbibliografie; detailed bibliographic data are available in the Internet at http://dnb.d-nb.de
ISBN 978-3-593-39898-3

Cover design: Campus Verlag GmbH, Frankfurt-on-Main
Printed on acid free paper.
Printed in the United States of America

This book is also available as an E-Book.
www.campus.de
www.press.uchicago.edu

Contents

Preface

Economic competition in global markets is increasingly determined by successful innovation performance, which is mainly driven by knowledge transfer and learning. Understanding the roots of the competitive advantages of national economies requires the analysis of their respective innovation and knowledge transfer systems. While approaches in the field of political economy mostly address comparative economic advantages in innovation and production systems, innovation studies predominantly point to absolute advantages in these systems. Both fields, however, frequently do not reflect the underlying structures and mechanisms of knowledge transfer in a sufficient manner. Moreover, the degree of coherence and complementarity of national innovation and production systems is often overestimated, while the regional variation within these systems is mostly insufficiently taken into account.

Consequently, in his book Michael Ortiz addresses the question on the degree of regional variation within the national production regimes postulated by the Varieties-of-Capitalism-Approach with a special focus on the field of knowledge transfer. This question is approached in a complex research design, which investigates three national production regimes and two regional case examples, respectively, on the basis of an internationally comparative qualitative case study. For this purpose he develops an own theoretical research approach which connects the two approaches of Varieties of Capitalism as well as National and Regional Innovation Systems. The author succeeds in adding the dimension of regional variation to the over-simplified assumptions of institutional coherence and complementarity in the debate on national production and innovation models by initiating an innovative "Regional-Varieties-of-Capitalism-Research-Agenda".

Furthermore, Michael Ortiz deals with the so far under-specified concept of knowledge transfer in innovation processes. He develops an own "heuristic model" of knowledge transfer systems, allowing the empirical analysis and modeling of national and regional knowledge transfer systems on the basis of distinct analytic dimensions of governance as well as central mechanisms and channels of knowledge transfer. He uses this heuristic model to develop a "Guide to the Comparative Analysis of National and Regional Innovation and Knowledge Transfer Systems", which he, as an essence of the analysis, plays back into the theoretical debate. This guide represents an important contribution to the reduction of existing deficits in the debate on innovation systems.

The book uses the empirical investigation of the three national production regimes of Germany, Spain and the UK, as well as the six extensive case studies on the regions of Baden-Württemberg, Metropolitan Region Hannover-Braunschweig-Göttingen-Wolfsburg, Catalonia, Madrid Community, East of England and Greater London. Michael Ortiz analyzes the degree and the nature of regional variation within national innovation and knowledge transfer models, and explains their major causes comprehensively. He focuses on the structural dimension of this variation by differentiating between historically grown industrial regions and metropolitan regions constructed in processes of regional planning, and, on the other hand, complements this through with the sectoral dimension by comparing the two sectors of automotive and biotechnology. As a result, the analysis draws a profound and differentiated picture of the processes of knowledge transfer on the national, regional, sectoral and structural level, as well as a detailed multiple embedded and differentiated comparative analysis of regional variation within national production and innovation models.

With this work, Michael Ortiz presents an important contribution not only to the debate on the politico-economic part of comparative economic sociology (Varieties-of-Capitalism-Approach), but also to the discussion on national and regional innovation systems. In empirical and methodological terms, the encompassing research design is sovereignly realized and convincingly presented. Because of its scientific and its high socio-political and politico-economic relevance, this book deserves to receive high international attention, and I wish it to be read by a broad expert audience in business, academia and society.

Prof. Dr. Knut Koschatzky
Fraunhofer Institute for Systems and Innovation Research ISI, Karlsruhe

1. Introduction

The first decade of the 21st century ended with a dramatic worldwide financial and economic crisis, culminating in the crash of *Lehman Brothers* in September 2008 and the debt crisis in the USA and the *Euro*-Zone in 2010 (Scharpf 2011), and questioning the market economy systems in almost all developed countries. With that, this crisis has also opened up a new chapter in the debate on the commonalities and varieties of capitalist systems, which dominated the field of political economy in the past decades (Streeck 2010). While this debate has in much of the postwar period been centered on the systemic antagonism of capitalism and socialism, the fall of communism in the late 1980s led the debate into an "end-of-history" (Fukuyama 1992) direction in the 1990s, presuming the global spread of the victorious capitalist system in its (neo-)liberal appearance, represented above all by the USA. This euphoria, however, even stronger inspired the field of political economy to develop a parallel research agenda focusing on the differences between national varieties of capitalism, not longer acknowledging the capitalist world as a unitary block, but, rather, as a diversified set of institutionally distinctly coordinated market economy systems (Chandler 1990; Albert 1993; Berger/Dore 1996; Crouch/Streeck 1997; Hollingsworth et al. 1994; Hollingsworth/Boyer 1997; Kitschelt et al. 1999; Whitley 1999). The rise of this varieties-of-capitalism-perspective in the 1990s and, above all, in the early 2000s through the influential contribution of Peter A. Hall and David Soskice (2001a), and its controversial reception in the light of the liberalization mainstream, now receive new vigor in the context of the actual crisis, in which whole national and international economic and financial systems are menaced in their stability by economic shockwaves, questioning their stability and their persistence as a whole.

While large parts of the debate on the varieties-of-capitalism-approach emphasized at least a certain trend (if not pressure) towards liberalization of coordination even in those market economies previously organized in a different way (Deeg 2005; Hay 2005; Morgan et al. 2005; Streeck/Thelen 2005; Cerny et al. 2006; Deeg/Jackson 2007; Hall/Thelen 2009; Mahoney/Thelen 2010; Thelen 2010), the actually observable differences in the impact of the recent crisis raise more strongly than before the question of long-term sustainability in globalized capitalism, efficient structural models and comparative institutional advantages, not only, but especially when looking at the deficiencies of the financialized (neo-)liberal Anglo-Saxon model or of the state driven Mediterranean type of capitalism (Streeck 2010). Latest contributions and debates, not only in the field of political economy, vindicate in this connection for example the German coordinated market economy model, which is celebrated for its technological leadership in export intensive consumer durables and equipment goods, its innovative capacity and international competitiveness (The Economist 2010), revitalizing the varieties perspective and providing new impetus to deeply understand the roots of these distinct success stories.

Even stronger than before, in the context of the current crisis also the consequences of economic globalization on national market economies are debated, in which economic action is increasingly determined by boundless globalized markets, flows of trade, goods and communication, as well as internationalized competition (Giddens 1990; Hirst/Thompson 1996; Held et al. 1999; Robertson 2001; McMichael 2005). In this debate, there is a broad consensus that in contrast to the 1980s and 1990s, for firms in the developed industrial nations the reorganization of firm structures as well as the rationalization and flexibilization of production processes no longer stand in the focus of their entrepreneurial activities. Rather, it is now acknowledged that in a situation of global competition, it is of increasing importance for these firms to come up with marketable product, process and service innovations faster and faster (a.o. Archibugi et al. 1999; Streeck 2004; Archibugi/Iammarino 2010). The *innovative ability* of firms and, thus, whole national economies becomes the decisive factor for the competitiveness of firms in internationalized markets, giving those types of market economies competitive edges which provide adequate and efficient institutional environments to their economic actors to be innovative.

In the centre of these innovation activities stand processes of producing and transferring knowledge and, thus, mutual learning. In the emergent "knowledge economy" (Cooke 2002), different forms of knowledge become quantitatively and qualitatively more and more important production factors, which is reflected by increasing knowledge related investments in

R&D, education and software and is continuously pushed on by the proceeding expansion of information and communication technologies (Al-Laham 2003; Cooke et al. 2007; Godin 2006; Heidenreich 2002; 2003). It is, consequently, indispensible for firms and whole regional and national economies to efficiently create structures and mechanisms for transferring these different types of knowledge between different kinds of actors to enable successful learning and innovation processes, since it is the ability to learn which finally determines their sustainable development and economic success. It will be these *structures and mechanisms of knowledge transfer* in distinct institutional market economy environments that will be the main concern of this study, contributing with that to the deeper understanding of the institutional roots of distinct models of learning and innovation systems.

While the current debates in the fields of political economy and innovation research mostly focus on the functioning and the performance of market economies or innovation systems as a whole, the underlying modes of learning and transferring knowledge are not yet investigated intensively, but, rather, regarded as basic but undifferentiated background processes (Audretsch/Lehmann 2005; Edquist 2005; Ortiz 2008; Acs et al. 1999). Especially the systemic character of knowledge transfer processes and their institutional embeddedness in macro-economic systems remains underexplored. It is one aim of this study to address this gap in the debate, to bring the structures and mechanisms of knowledge transfer underlying innovative activities stronger into the analytic focus, to develop a heuristic model of knowledge transfer systems at the macro level, and to investigate to which degree the institutional configurations coordinating knowledge transfer processes vary across distinct national and regional market economy environments.

In this regard, knowledge transfer is an often used but still insufficiently systemized concept in innovation research. In most studies on the issue the conceptualization of knowledge transfer in the field of innovation does not go far beyond stating that knowledge transfer occurs between many different types of actors and through distinct kinds of channels: It can take place between individuals, firms or other institutions and occurs throughout all stages of the innovation process, from the initial idea to the final commercialization of a specific product, process or service. Important channels of knowledge transfer can exist between the scientific world (universities, research institutes etc.) and the business sector (Mora-Valentín

2002; Etzkowitz et al. 2000; Etzkowitz/Leydesdorff 2000), but also within the business sector itself, when firms exchange knowledge between each other. Moreover, the scientific world and the business sector can maintain important channels of knowledge transfer to other actors, as for example political institutions or intermediary agents, like technology transfer agencies, chambers of industry and commerce or business associations. Furthermore, knowledge transfer between these actors can occur through a large variety of formal and informal mechanisms. Formal mechanisms can be, for example, the exchange of qualified staff, formal consultancy services, joint projects involving expert groups from different organizations, patenting, licensing, or contract and cooperative research. Informal mechanisms can include informal interactions between individuals in informal meetings, conferences, workshops, trade fairs, or publications (Abramson et al. 1997: 2f.; Varga 2009; Howlett 2010; Polt et al. 2010; Döring/Schellenbach 2006; Walter 2003; Bozeman 2000; Koschatzky/Hemer 2009; Franzoni/Lissoni 2009).

Despite these general features, however, a clearly systemized analytical framework on knowledge transfer processes in the context of innovation is still to be developed. The rationale for this need for further research on the topic of knowledge transfer is, from the point of view of the present study, quite simple: The distinct forms of knowledge transfer processes constitute the *heart of competitive edges* of firms and whole market economies. The intensity, quality, complexity, and efficiency of these processes are, therefore, important differentiating factors when analyzing distinctly successful firms, industries and national or regional production systems. In addition, also the obstacles to knowledge transfer, like structural inefficiencies, learning-averse business cultures, mental barriers, inadequate actors' constellations, insufficient knowledge supply from academia, lacking incentive structures or human capital related shortcomings must be investigated and understood in this connection, since they might be essential starting points for organizational and policy reform measures. A valuable contribution to the current debate on sustainable economic development and competitiveness requires, thus, both the explorative detecting of particular strengths and weaknesses of specific knowledge transfer systems, as well as the analytical systematization of the institutional structures and mechanisms underlying processes of knowledge transfer.

In this regard, it is broadly acknowledged that the innovative and learning ability of firms and individuals is decisively determined by their surrounding institutional environment into which they are embedded (Lundvall 1992; Granovetter 1985). Entrepreneurial action in general and innovation and knowledge transfer processes in particular are by no means taking place in an institutional vacuum. Rather, the *systemic character* of innovation processes requires specific, systemically condensed institutional arrangements e.g. at the national or regional level, which influence the innovation behavior of firms in a decisive way. To understand the strategies and the success or failure of firms in the global competition on innovation profoundly, it is, thus, necessary to focus the analysis on the specific national and regional institutional frameworks under which they operate. It is these different institutional frameworks which allow in the first place for a specialization in specific innovation strategies and which provide firms with the necessary resources to it. This relation of institutional frameworks and innovation strategies is particularly elaborated by the concepts of varieties of capitalism as well as national and regional innovation systems.

While the *national innovation system* (NIS) approach (Freeman 1987; Lundvall 1992; Metcalfe 1995) claims that the national level is decisive in providing a system of interconnected institutions to create, store and transfer knowledge and enable interactive learning processes, the *varieties-of-capitalism-approach* (VoC) (Hall/Soskice 2001a; Amable 2003; Schmidt 2002; Hancké et al. 2007a; Crouch/Streeck 1997) points to a wider set of complementary institutions to describe whole national production systems. Nevertheless, it also puts the innovation aspect into the centre of its analytical focus and claims that the comparative institutional advantages of specific national production models will particularly lead to different forms of technological specialization and innovation regimes: While the *liberal market economies* of the Anglophone countries have considerable strengths in technological sectors that require *radical* innovations, the *coordinated market economies* like Germany and other continental European countries perform particularly well in industries producing *incremental* innovations, and the *mixed market economies*, to which Europe's Mediterranean countries can be counted (Molina/Rhodes 2007), will have strengths in both radical and incremental innovations, but overall underperform the other two models. According to this approach, these different national production and innovation models arise from varying configurations of specific sets of institutions, as education and training systems, labor market regulation, corporate governance and financial systems, and the governance of inter-company relations in terms of market competition and technology transfer.

Moreover, the concept of *regional innovation systems* (RIS) rather emphasizes the region than the national level as central location for the coordination of innovation and learning processes. Here, according to a large part of the broad literature on the issue (a.o. Braczyk et al. 1998; Cooke et al. 1997; Cooke et al. 2004; Heidenreich 2001, 2004; Koschatzky 2001a), regional networks, local clusters of companies, universities and research and development facilities can establish relations of close interaction between each other. This interaction at the regional level is supported by geographic and socio-cultural proximity between the actors participating in innovation processes, and, in particular, by knowledge-supplying and -transferring institutions (Asheim/Isaksen 2002: 83) that provide the actors in a regional innovation system with specific "local collective competition goods." (Le Galès/Voelzkow 2001)

However, while these approaches have considerable strengths in describing and analyzing the specific actors' constellations, structural conditions, spatial aspects and cultural influences in innovation and knowledge transfer, they still lack to provide a conclusive theoretical framework which extensively and systematically captures the underlying mechanisms of learning and knowledge transfer, as well as the corresponding institutional configurations and causal relations between the distinct elements of national and regional knowledge transfer systems (Audretsch/Lehmann 2005; Edquist 2005; Acs et al. 1999; Ortiz 2008). This theoretical gap will be addressed by this study in elaborating an own *heuristic model* of knowledge transfer systems based on the implicit assumptions on knowledge transfer processes discussed in the approaches of VoC, NIS and RIS, summarizing and systemizing the basic structures and mechanisms of knowledge transfer in the different institutional dimensions of governance, i.e. networks, associations, organizations, the state, markets, and communities. With this model a "toolbox" of potentially observable mechanisms and institutional structures of knowledge transfer in innovation systems is presented, which systemizes the concrete mechanisms through which knowledge is transferred in the context of innovation clearly, and which can be used to reconstruct empirically observable knowledge transfer systems at the national as well as at the regional level.

Furthermore, in this connection a second, related gap in literature will be approached: Since all regional innovation and knowledge transfer systems are also embedded into wider institutional arrangements at the national level, the question emerges, to which degree these regional systems can vary from their respective national frameworks? Both NIS- and VoC-approaches claim that national institutional arrangements in the fields of production and innovation represent complementary and coherent systems that constitute almost homogeneous units with consistent logics of coordination (Crouch et al. 2009: 655ff.; Kenworthy 2009: 182f.; Voelzkow 2007; Höpner 2005). According to this, it is to be expected that all regional innovation and knowledge transfer systems within a specific national production system should show similar institutional configurations, structures and mechanisms in the fields of innovation and knowledge transfer. The question that is not sufficiently answered in the large body of literature on NIS and VoC so far is, to which degree these national systems can really be considered as coherent institutional configurations, and to which degree the regional innovation systems within these national models can differ from their respective national frameworks. In addition, and secondly, there remains considerable vagueness on how to explain this variation, if it exists. In this manner, also in the light of the actual revitalization of the debate on varieties of market economy systems, for policy and economic actors concerned with institutional adaptation and reform processes, deeper insights into the nature of regional variation in national systems appear to be absolutely useful and indispensible. Not only would a pointed overestimation of the coherence of national institutional configurations provoke an orientation towards incorrectly understood reference models, but it would also run the risk to oversimplify possible structural reform measures in the sense of "one-size-fits-all" solutions, disregarding regionally differentiated structural conditions and actors' constellations possibly requiring variegated and individually tailored adaptations.

It is, therefore, these two questions that will be addressed by this study and will stand in the centre of its analysis: The first is, in which way and to which degree the liberal, coordinated and/or "mixed" institutional arrangements at the level of national systems of innovation and production are also reflected on the regional level? In other words: Is it possible to identify in the regional innovation systems of national production models the corresponding liberal and coordinated (and mixed) structural mechanisms in the field of knowledge transfer, which the VoC-approach describes for the national level? Furthermore, the second question asks how exactly and why these structures and mechanisms are adapted at the regional level. Which are the specific institutional structures and mechanisms facilitating knowledge transfer at the regional level, and through which driving factors can these regional peculiarities be explained?

To address these research questions, the author conducted an international comparative qualitative case study of six regional innovation systems in three distinct national market economy settings. The aim of this case study is to test, to which degree, in which way and why the structures and mechanisms of knowledge transfer in regional innovation systems can differ within particular national production models. This case study is approached in several steps: In a first step, in chapter 2 some elementary concepts on the nature of knowledge and processes of knowledge transfer (2.1), as well as their institutional embeddedness (2.2) into social systems of innovation and production are discussed. For this purpose, the debates on VoC (2.3), NIS and RIS (2.4) are addressed with the aim to extract and elucidate the structures and mechanisms of knowledge transfer implicitly underlying these approaches, as well as the potential driving factors for regional variation within national systems provided by them. Above all, bringing together and integrating the varieties-centered perspective of comparative political economy with the innovation system approaches on the distinct spatial levels will be a central motivation of this discussion. In a second step, I develop a *heuristic model* of knowledge transfer systems (2.5) which addresses the conceptual shortcomings of the VoC-, NIS- and RIS-approaches by providing an explicit and systematic toolbox for the analysis of mechanisms and governance structures of knowledge transfer in social systems of innovation and production.

In a further step, this heuristic model will then be used to formulate research guiding hypotheses assuming which structures the knowledge transfer systems in the different national production models (liberal, coordinated, mixed) should take according to the approaches of VoC and NIS (chapter 3). These hypotheses will be used in the qualitative case studies as indicators to predict through which institutional mechanisms knowledge transfer should be organized in the investigated regions, and which modes of governance should dominate the knowledge transfer systems within the three distinct national market economy frameworks—testing with that if and to which degree the empirically observable regional knowledge transfer systems vary from the assumed national environment.

For the purpose of this analysis, in chapter 4, an internationally comparative, multiple embedded, qualitative case study design is developed: According to the VoC-typology by Hall and Soskice (2001b), three national innovation systems are selected: *Germany* as an example of the coordinated model, *the United Kingdom (UK)* for the liberal and *Spain* for the Mediterranean model. Within each of these national innovation systems, two regional innovation systems are sampled as units of analysis (4.1.2). In this, for every national model one *established industrial region* and one *metropolitan region* are investigated, to include *different types of regional innovation systems* (Tödtling/Trippl 2005) into the analysis, which are supposed to be characterized by markedly distinct structural conditions (2.4.2.5.3). The aim is to differentiate between the historically grown institutional structures in established industrial regions and the (mostly) newly created structures supporting innovation in the metropolitan regions. While the former are supposed to be characterized among other things by established regional practices, the experience of specific industrial trajectories and the entrepreneurial mentalities of a "commercial community" (Heidenreich 2004a; 2004b; Whitford 2001; Krauss 2005; Koschatzky 2001a), the latter represent central nodes in global networks of the emerging knowledge economy, where streams of information and knowledge agglomerate due to the presence of e.g. the headquarters of larger multinational companies, central financial places or national or international governmental institutions (Kujath 2005; Löw 2001). The aim of this differentiation is the following: Most contributions on VoC and the NIS-/RIS-concepts use the traditional industrial region type to develop their models of regional innovation systems within one country or whole national models, while other types of regions, and, above all, metropolitan regions remain under-reflected in this regard. However, taking the increased importance of metropolitan regions in the knowledge economy and the intensive diffusion of the concept in all developed countries in the last years, it appears to be absolutely necessary to stronger focus on these metropolitan regions as specific types of innovation and knowledge transfer systems, especially because they can be expected to be main drivers of regional variation within national institutional models.

In addition, in all six investigated regions a specific focus of analysis is put on the two sectors of *automotive* and *biotechnology*, to "control" the *sectoral specificities* as explanatory factors for regional variation (2.4.2.5.3). With the two sectors of automotive and biotechnology one traditional industrial sector with a "synthetic" and highly cumulative knowledge base and an incremental mode of innovation, and one younger sector with an "analytical" and lowly cumulative knowledge base and radical innovation processes are focused on (Cooke 2004; Asheim/Gertler 2005; Malerba 2004; Casper/Soskice 2004), which are assumed to be shaped by quite different structural and functional requirements in the fields of innovation and knowledge transfer. From the point of view of this study, also in this connection the concepts of VoC, NIS and RIS remain too imprecise with regard to distinct sectoral modes of knowledge transfer, their role as drivers of regional variation, as well the possibility of the parallel existence of distinct sectoral foci within one regional or national institutional model. The motivation for this sectoral differentiation is, therefore, to broaden the understanding of knowledge transfer processes and its variegated character also by the sectoral dimension.

Chapter 5 then discusses the results for Germany with the two investigated regions of *Baden-Württemberg* (5.2) and the *Metropolitan Region Hannover Braunschweig Göttingen Wolfsburg* (5.3), while in chapter 6 the results for the Spanish regions of *Catalonia* (6.2) and the *Madrid Community* (6.3) are presented, and chapter 7 focuses on the results for the case of the UK, with the regions of the *East of England* (7.2) and *Greater London* (7.3). Finally, chapter 8 summarizes the results and draws conclusions on the distinct analytical dimensions addressed by this study. Section 8.1 discusses the implications for a more systemic theory on knowledge transfer on the basis of the heuristic model developed in this study. Second, the necessity that this theory on knowledge transfer systems must strongly be aware of regional variation within national models is highlighted, giving an overview on the nature of this variation as well as the explanatory factors behind it in the three investigated market economy models (8.2). Third, the antagonism of traditional industrial and metropolitan knowledge transfer settings as driver of this regional variation is highlighted (8.3), while its sectoral determinants are focused on in section 8.4, differentiating between incrementally and radically innovating sectors and, in particular, between automotive and biotechnology. Fifth, the need for the development of a more varieties-centered research agenda on innovation and knowledge transfer through a stronger integration of the VoC-approach and the NIS-/RIS-concepts is discussed (8.5), to achieve a more innovation- and learning-oriented foundation of our understanding of capitalist diversity, as well as a more varieties-oriented perspective on innovation and knowledge transfer processes. Finally, the implications for the VoC-approach and the resulting policy recommendations are summarized (8.6).

2. Discussing Central Concepts and Definitions

2.1 Knowledge Transfer: Economic Action and Knowledge Economy

2.1.1 The Emergence of Knowledge Economy

This study tries to contribute to a profound understanding of processes of knowledge transfer in regional innovation systems and their dependence on their respective overarching national institutional configurations. In this, it is closely connected to the actual debate on the emergence of *knowledge society*, which is one of the most cited concepts in management science, economics and social sciences (Heidenreich 2003a). It describes the rising importance of knowledge in almost all parts of daily life, pointing to the change from industrial to knowledge societies (Willke 2001; Drucker 1994; Toffler 1990). Willke (2001) defines a knowledge society or "knowledge based society" as follows:

> "It is possible to talk of a knowledge society or knowledge based society, if, on the one hand, the structures and processes of the material and symbolic reproduction of a society are penetrated by knowledge based operations in a way that the processing of information, the symbolic analysis and expert systems become prior compared to other factors of reproduction. A decisive additional precondition of the knowledge society is that knowledge and expertise are subject to a process of *continuous revision* and that, thus, innovations become an everyday part of the knowledge work." (Willke 2001: 291; translation: MO; italics in original)

Especially in the economic sphere the globalization of markets assigns a decisive role to innovation and the creation of new knowledge, since they are key prerequisites for the success in competitive markets and economic development (Audretsch/Thurik 2001; Heidenreich 2000; Tushman/Nelson 1990). An increasing part of economic activities in highly developed economies takes place in the commercialization of information and knowledge based products and services. *Knowledge economies*, thus, rely less on the abundance of natural resources than on the capacity to improve the quality of human capital and to create new ideas and apply them to equipment and people (Cooke et al. 2007). As the OECD (2001) observes, knowledge as an input is becoming quantitatively and qualitatively more important, which is reflected by increasing knowledge related investments in R&D, education, software and information technologies. Finally, these developments are continuously pushed on by the proceeding expansion of information and communication technologies (Cooke et al. 2007: 26f.; Al-Laham 2003: 1f.; Heidenreich 2002; 2003).

For this study, the debate on the emerging knowledge economy provides an important framework, since it points to the fact that innovation and learning processes, as well as the production, transfer and diffusion of different kinds of knowledge are crucial elements for understanding social and, in particular, economic action and structures in contemporary societies, and, moreover, represent indispensible factors of success in the economic competition on global markets. It not only demonstrates the importance of addressing the research questions outlined above, but also the relevance of a deeper understanding of the processes, structures and mechanisms underlying innovation and knowledge transfer at the national and regional level for the scientific debate as well as for political and economic decision makers.

2.1.2 What Is "Knowledge" and How Can It Be "Transferred"?

A study addressing research questions that point to structures and mechanisms of knowledge transfer in the context of innovation has to define in a first step what economically useful knowledge is. Thus, in the following, some brief theoretical considerations on the nature of knowledge will be presented, to then discuss the resulting implications for the transfer of knowledge and give a respective definition of the notion of knowledge transfer and, finally, to provide first evidence for the non-linear and systemic character of knowledge transfer processes.

There is broad consensus that *knowledge* has to be differentiated from *information* (Schmoch 2003: 138ff.). Although knowledge is based substantially on information, it is only the purposeful networking of information that creates knowledge. Following Lambooy (2002), who distinguishes three levels of complexity, *data* are unstructured facts, while *information* represents structured facts. As *knowledge*, he defines the competences of individuals to evaluate and assess data and information, as well as to resolve problems (Lambooy 2002: 1019). Thus, knowledge can be described as resource of action of individual and collective actors. However, an *economic value* only emerges from the transformation of this resource into *marketable products and services* (Fischer/Fröhlich 2001: 8).

Moreover, Lundvall and Johnson (1994) distinguish four types of economically useful knowledge: *Know-what*—knowledge of facts, being similar to the notion of information; *know-why*—scientific knowledge on principles and laws that play a decisive role in the development of technological innovations; *know-who*—knowledge on social relations, on who holds relevant knowledge and information when and where; and, finally, *know-how*, describing skills and capabilities to do something. According to the authors, in contrast to the two former types of knowledge, know-who and know-how are much more difficult to codify, transfer and learn (Lundvall/Johnson 1994: 23ff.).

This leads to another important distinction: The distinction between implicit and explicit knowledge (Polanyi 1985). While information and *explicit* knowledge can be processed, transferred and saved through codification, this is not possible with *implicit* (tacit) knowledge. This type of knowledge is hardly formalizable, communicable and transferrable, since it is bound to the specific backgrounds, experiences, interpretations, ideas and convictions of persons. Implicit knowledge consists of technical and cognitive elements. The technical aspect consists of the informal and hardly documentable skills and capabilities. The cognitive element refers to schemes, mental models, convictions, views on the world and perceptions that are taken for granted by the individuals and reflect an idea of reality (Nonaka/Takeuchi 1997; Berger/Luckmann 1966).

This implicit dimension of knowledge causes the fact that knowledge is bound to specific contexts, social relations and concrete persons. It is, thus, spatial and social proximity, personal communication and interaction that basically enable the exchange and transfer of implicit knowledge. However, it has to be stated at this point that implicit and explicit knowledge are not dichotomous or mutually exclusive, but the extreme poles of a continuum. There is always some explicit and codified knowledge needed to understand implicit knowledge, and without a certain stock of implicit knowledge it is not or only hardly possible to decode, interpret, understand and use complex explicit knowledge (Strambach 2004; Schmoch 2003: 142ff.).

In contrast to information, which can be collected, the emergence, application and transfer of knowledge is connected to the social process of *learning*. In earlier economics and innovation research, these learning processes were mostly described as *learning by searching*, meaning the purposeful search of new knowledge through R&D-activities. Moreover, many scholars have shown that learning also occurs as a "by-product" in routine activities. These learning processes are labelled *learning by doing, learning by using* and *learning by interacting* (Arrow 1962; Lundvall 1988; Kline/Rosenberg 1986; Johnson 1992). In the present, there is no elaborated theory on knowledge economy (Strambach 2004: 5), but recent works suggest that knowledge as a factor of production or as a commodity is subject to another logic of action than material goods (Dosi 1996; Foray/Lundvall 1996; Willke 2001). One central difference between immaterial or symbolic goods like knowledge and material goods is that the use and the market value of the former are strongly dependent on the respective context and the specific situation of action. Moreover, knowledge does not exhaust in its use, but increases through its application. Thus, investments in knowledge are rather connected to *increasing* than to *decreasing* returns (Dosi 1996; Foray 1998).

Concerning property rights, knowledge, like a material good, turns into the customer's possession when it is sold, but, nevertheless, it remains the property of the producer and can still be used by him. Consequently, in many cases knowledge at least partly becomes a public good, since it can diffuse unintentionally to other actors and can be used by them as positive externality. In economic terms, therefore, knowledge as an immaterial good is both *non-rivalrous* and *non-excludable* (Karlsson/Manduchi 2001: 104f.; Polt et al. 2010: 94; Döring/Schellenbach 2006: 378), implying that the generation and the transfer of knowledge is always accompanied by uncertainties and risks with regard to the results of these activities (Gresse 2010: 17). These uncertainties and risks can constitute one major barrier to knowledge transfer, in particular when looking at knowledge transfer within and from the business sector (Döring/Schnellenbach 2006: 380). The regulation of IPRs, thus, becomes a crucial point not only when looking at knowledge transfer, but also for the whole debate on the emerging knowledge society (Strambach 2004).

In addition, and following Malerba (2004), knowledge can be more or less *cumulative*, i.e. the degree to which the creation of new knowledge builds upon already existing stocks of knowledge. This cumulativeness of knowledge can emerge from three different sources, from which the first one is *cognitive* in nature. Learning processes and past knowledge not only constrain current research, but also lead to new questions and new knowledge. The second source of cumulativeness is related to the firms and their organizational *capabilities*. Since these capabilities are highly firm specific, they generate knowledge that is highly path-dependent. Finally, also the *feedback* from the markets can be a source of cumulativeness of knowledge, in the sense of "success breeds success". This means that successful processes of knowledge transfer and innovation create profits that can be reinvested in further innovation and knowledge transfer activities (Malerba 2004: 20). With regard to the research interest pursued in this study, this cumulative character of knowledge is particularly relevant, since it not only leads to specific stocks of knowledge at the firm level, but also to particular economic trajectories at the technological, sectoral, regional and national level, as will be discussed in detail below.

Taken together, these qualities of knowledge as an immaterial good also have consequences for the individual and collective actors involved in the production and transfer of knowledge. The interactions between suppliers and users of specific knowledge take place under conditions of uncertainty, since the results are not exactly predictable. The reason for this is that the exchange of knowledge requires an active social process of communication realized under specific social and cognitive conditions. Consequently, in dealing with knowledge, the actors have to actively form and interpret these conditions (Stehr 1994). In this, knowledge has a *dynamic* character; it emerges and changes in the interaction process through the actors' expectations and patterns of interpretation. That is why the quality of knowledge based products and services results from cumulative learning processes, depending as well on the experience and competence of the producers as on the abilities of the user to integrate external knowledge into the own knowledge base. In these interaction processes informal institutions like trust play a decisive role for a successful exchange of knowledge. Moreover, social networks, references and reputation exercise key functions in this connection, since they are substitutes for the missing quantifiable quality criteria of knowledge (Heidenreich 2011a; Strambach 2004).

Based on this, according to Howlett (2010) the process of *knowledge transfer* can be characterized as follows:

"Knowledge transfer describes how knowledge and ideas move between the knowledge source to the potential users of that knowledge. [...][It consists of; MO] activities to transfer good ideas, research results and skills between, for example, universities and other research organizations, business, the third sector, public sector and/or the wider community." (Howlett 2010: 2)

Closely connected to the notion of knowledge transfer is the notion of *technology transfer* (Walter 2003: 14ff.). In this context, technology can be defined as a specific form of knowledge that includes individual and collective knowledge in implicit and explicit forms to use the findings of natural, social and engineering sciences in a product- and process-oriented way. With that, technology not only contains knowledge about the properties of particular materials, but also about their production, interaction and combination in terms of solutions for specific tasks or problems. Technology transfer, thus, points to the transfer of a specific type of knowledge, namely knowledge about a technology, including artefacts like specific tools, machines or software and skills (Gresse 2010: 22ff.). For this study, which focuses on knowledge transfer in the field of innovation, the notion of knowledge transfer will be used in a wider sense including processes of transferring technology.

Furthermore, knowledge transfer can take place between individuals, firms or other institutions and occurs throughout all stages of the innovation process, from the initial idea to the final commercialization of a specific product, process or service. Intensive channels of knowledge transfer can exist between the scientific world (universities, research institutes etc.) and the business sector (Varga 2009; Mora-Valentín 2002; Bozeman 2000; Etzkowitz et al. 2000; Etzkowitz Leydesdorff 2000), but also within the business sector itself, when firms exchange knowledge among each other. Moreover, the scientific world and the business sector can both maintain important channels of knowledge transfer to other actors, as for example political institutions or intermediary agents, like technology transfer agencies, chambers of industry and commerce or business associations. Knowledge transfer can either take place *directly* between the holder of specific knowledge and the receiver, when for example a firms' R&D-department interacts with a research institute at a university, or *indirectly*, when the process of knowledge transfer is mediated by these intermediary actors (Walter 2003: 18f.; Polt et al. 2010: 96; Döring/Schnellenbach 2006: 387f.).

In this, an important factor for the emergence and the success of knowledge transfer processes is the *absorptive capacity* (Cohen/Levinthal 1989; Cohen/Levinthal 1990) of the receiving knowledge transfer partner. This includes the necessary cognitive capacities as well as the willingness to incur the costs of learning new knowledge, i.e. it must appear economically rational to the recipient to use the knowledge that is to transfer (Döring/Schnellenbach 2006: 380; Gresse 2010: 103). Moreover, as Abreu et al. (2008) point out, the ability of the actors involved in knowledge transfer, and in particular of firms, to assimilate and exploit external knowledge relies not only on their R&D-expenditures, but also on prior knowledge embodied and accumulated in human capital and individual skills, the organisational structure, management practices and the type and intensity of interaction and cooperation with external partners, as could be other firms, universities or public institutions (Abreu et al. 2008: 10). Insufficiencies in these conditions for knowledge transfer could, therefore, operate as major barriers to knowledge transfer. Since these insufficiencies might be addressed by specifically tailored institutional support to being overcome, as, for example, through the involvement of intermediary actors or the provision of funding, with regard to my research question it appears to be plausible to assume that not only individual actors and organizations on the micro-level differ in their willingness and capacity to absorb knowledge, but also that these characteristics are "quite likely to differ between regions depending on their actual institutional framework" (Döring/Schnellenbach 2006: 380), as will be shown below.

Like the proper innovation process, also processes of knowledge transfer usually are iterative and involve multiple steps of transfer. In general, they can take place through a large variety of formal and informal mechanisms. *Formal* mechanisms, defined as building on contractual relationships, could be for example the exchange of qualified staff, formal consultancy services, joint projects involving expert groups from different organizations, patenting, licensing, or contract and cooperative research. *Informal* mechanisms without contractual basis could include interactions between individuals in informal meetings, conferences, workshops, trade fairs, or publications (Varga 2009: 1f.; Abramson et al. 1997: 2f.; Howlett 2010: 4ff.; Polt et al. 2010: 92f.; Döring/Schellenbach 2006: 379; Walter 2003: 20ff.; Bozeman 2000: 628ff.). Since formal and informal mechanisms of knowledge transfer are closely interconnected to each other (Grimpe/Hussinger 2008: 2f.; Grimpe/Fier 2010: 638), and, as will be demonstrated, arc both important for successful innovation activities, for this study, both formal and informal mechanisms will be included into the analysis. Moreover, and following Walter (2003), also academic spin-offs and business start-ups will be regarded as important mechanisms of knowledge transfer within the business sector and between the business sector and the scientific world. With that, also the respective intermediary actors involved in the process of new business creation will be in the focus of this study, as for example business incubators, technology centres, science parks, venture- and seed-capital providers and business angels (Walter 2003: 18f.; Koschatzky/Hemer 2009: 212f.; Franzoni/Lissoni 2009).

These general characteristics of knowledge transfer processes already point to their predominantly *non-linear* character. While only in a limited number of cases a linear model of knowledge transfer can be observed, in which knowledge is transmitted directly and in a planned, unidirectional and routinized way from one knowledge transfer partner to the other, and in which it is directly and immediately received and applied there (Gresse 2010: 36f.), most knowledge transfer processes are of a non-linear kind. Rather, these processes are mostly characterized by and dependent of multiple feedback loops, overlaps, callbacks and the involvement of manifold intermediary actors, as described above (Walter 2003: 19f.). Moreover, as Fischer and Fröhlich (2001) state, knowledge is not only generated and transferred by individuals and single organizations, but also by their often complex patterns of interaction (Fischer/Fröhlich 2001: 6) which are shaped and coordinated by particular institutional set-ups (Tappeiner et al. 2008: 863), making it possible to speak of knowledge and technology transfer as *systemically embedded processes*, as for example Abramson et al. (1997) do.

In sum, it has been shown that implicit and explicit forms of economically useful knowledge can be distinguished from each other, and that the transfer of these different types of knowledge is object to specific cumulative social processes of learning by searching, learning by doing, learning by using and learning by interacting. These learning processes are embedded in and shaped by formal and informal institutions that create specific systemic environments, allowing the application of specific formal and informal, as well as direct and indirect mechanisms of knowledge transfer. For the investigation of structures and mechanisms of knowledge transfer in this study, this implies that the analysis has to focus on exactly these systemic environments to come to a deeper understanding of these processes. The most prominent theoretical approaches dealing with this systemic institutional embeddedness of knowledge transfer and innovation are the concepts of innovation systems and production systems that will be outlined in the subsequent sections. Since it is the aim of this study to investigate structures and mechanisms of knowledge transfer in the context of innovation processes, in a next step the notion of innovation will be defined and the role of knowledge transfer in innovation processes will be discussed.

2.2 The Institutional Embeddedness of Innovation and Learning

2.2.1 Defining Innovation

In most instances it can be assumed that the transfer of economically useful knowledge as defined above does not occur because of its intrinsic value, but rather on the base of *economic interest.* That is the reason why processes of transferring economically useful knowledge are discussed in the literature in the context of *innovation.* In this study, innovations are defined as "[…] new creations of *economic significance.* They may be brand new but are more often new combinations of existing elements. Innovations may be of various kinds (e.g., technological and organizational)." (Edquist 1997: 1; italics added) Thus, an innovation's economic usability is important. This is also the decisive difference to the notion of "inventions": While an invention describes the first appearance of an idea, like for example for a new product or a new production process, innovations are newnesses, for which a solvent economic demand exists or is to expect. Therefore, innovations can be described as the first attempts to realize new ideas or processes in practice and to use them economically. In this, not all inventions can be used economically by far, whereas a multitude of innovations are not inevitably the result of inventions, i.e. of scientific-technical newnesses (Edquist 1997:1f.; Fagerberg 2005: 4f.).

Apart from their economic significance, the second quality of innovations that can be derived from the definition above is that they are mostly new combinations of already existing elements. To realize an innovation, a firm or organization must be able to combine different kinds of knowledge, resources, competences and qualifications with each other. It is, thus, essential for many innovations to possess extensive knowledge about, for example, production processes and markets, to have sufficient financial resources and production plants at one's disposal and to employ qualified staff. Here, according to *Joseph Schumpeter*, the "entrepreneur" exercises the important role of combining the decisive factors for the innovation process with each other, from his point of view labor, capital, and also knowledge. With that, Schumpeter equals innovations with new combinations of these already existing factors (of production) (Fagerberg 2005: 5).

Schumpeter's definition of innovations is a very broad one, since they are described by setting up a new production function. "[…] this function describes the way in which quantity of product varies if quantities of factors vary. If, instead of quantities of factors, we vary the *form* of the function, we have an innovation. […] we will simply define innovation as the setting up of a new production function. This covers the case of a new commodity as well as those of a new form of organization such as a merger, of the opening up of new markets, and so on." (Schumpeter 1939: 87, cited by Edquist 1997: 9) Also Lundvall (1992) states in this connection: "Almost all innovations reflect already existing knowledge, combined in new ways." (Lundvall 1992: 8) Knowledge transfer can thus be regarded as a necessary condition for innovation in the sense of an input factor, although it might not be a sufficient condition for successful innovation, since other factors as, for example, financial resources might be necessary as well.

Building on this, different kinds of innovation can be differentiated from each other. Schumpeter, e.g., differentiates between new products, new methods of production, new sources of supply, the exploitation of new markets and new forms to organize business. However, and according to Fagerberg (2005), in the study at hand it will only be differentiated between two types of innovation: Between product innovation on the one hand and process innovation on the other. The former describes new or improved products and services, the latter an improvement in the production of these products and processes (Fagerberg 2005: 6f.).

Moreover, innovation is to be regarded as a dynamic process and represents as such the result of a series of developments and innovations that build on, complement and relate to each other (Fagerberg 2005: 5ff.). With that, the notion of innovation used in this study is also delimited from "linear" models of innovation. These models assume that innovation is primarily a product of applied science. According to this, innovation processes get their *linear character* through a clearly defined and *cascade-like course* of stages they have to pass through. Here, *scientific research*, which is always the starting point of any innovation, is followed by *product or process development* as a second stage and, finally, by the stages of *production* and *marketing*. Consequently, in these models scientific research plays a decisive role for the success of innovation processes, and all further stages are connected to each other by processes of *knowledge transfer* (Kline/Rosenberg 1986: 285f.; Fagerberg 2005: 8).

These linear models of innovation are to be criticized now in a twofold manner: On the one hand, they assume a generally occurring chain of causation and knowledge transfer between the single stages of the innovation process, what in reality only holds for a minority of innovations. Although it is possible indeed, that certain innovations stem directly from the results of scientific research, following Kline and Rosenberg (1986) this is not true for a majority of the realized innovations. Rather, as already indicated by Schumpeter, firms normally first fall back on already existing knowledge and combine it in new ways. Also experiences and routines of consumers, customers, suppliers and other partners are often integrated into innovation processes or even initiate them (Kline/Rosenberg 1986: 291; Fagerberg 2005: 9).

On the other hand, in the linear models of innovation, the numerous *feedbacks* and *loops* between the single stages of innovation processes which are characteristic for many innovations are not sufficiently taken into account. In fact, errors and backlashes in single stages of the innovation process can lead to a rethinking of anteceding steps and, thus, fundamentally change the innovation process as a whole. Furthermore, it cannot generally be assumed that single factors do only play a role in specific stages of innovation processes. Thus, the results of basic scientific research, for example, could not only be integrated in the stage of initiation, but also in any other stages of innovation processes to the final marketing (Kline/Rosenberg 1986: 286ff.; Fagerberg 2005: 9).

However, recent debates on the "linear model" also point out that the idea of linearity in innovation processes should not per se and absolutely be rejected. It is argued that certain innovation processes might also be characterized by at least some linear and, thus, less systemic elements (Balconi et al. 2010). In this connection, for example, Heidenreich (2011b) shows for the specific case of multinational companies, that there is always a tension for this kind of innovating companies to choose, on the one hand, between the advantages of the *social and systemic embeddedness* of innovation processes and, on the other hand, the *organizational and corporate control* of these processes. Heidenreich (2011b) demonstrates that under certain conditions it might be quite useful for some companies to focus on the internalization, routinization and standardization of innovation processes, to secure competitiveness and efficiency in the field of innovation, but also to control relevant knowledge that is not wished to be shared and to avoid having to share intellectual property rights (IPRs) or other benefits with other actors (Heidenreich 2011b; Heidenreich et al. 2011; Mattes 2010). Contrariwise, the risk of this strategy would be to get disconnected from diversified stocks of knowledge and external innovative impulses. Nevertheless, and in particular for small and medium sized enterprises, it can be assumed that this more linear and standardized strategy of organizing innovation processes is not very likely to be realized, if not impossible at all, not least because they lack the necessary resources and organizational capabilities for it. Therefore, in this study the view on innovation as a predominantly non-linear process will be shared.

In sum, it can be stated that the transfer of economically useful knowledge should be analyzed in the context of innovation, where it has a major facilitating function. In this, knowledge transfer as an input factor can be conceptualized as a necessary condition for successful innovation, but not as a sufficient condition, since other factors might be necessary as well. As has been shown, both processes of knowledge transfer and innovation, under specific circumstances might include certain linear features, but overall show non-linear characteristics. This non-linear nature of innovation and knowledge transfer also refers to the systemic character of these processes that will be discussed subsequently.

2.2.2 The *Systemic* Character of Innovation and Knowledge Transfer

The brief considerations on the predominantly non-linear character of innovation processes in the previous section already point to the fact that firms normally do not innovate in isolation from other firms or organizations. Rather, they keep complex relationships of interaction, collaboration and interdependence with them. Often, these relationships are characterized by their reciprocity and a multitude of feedback mechanisms (Edquist 2005: 185). Thus, in the following, a *systemic* character of innovation and knowledge transfer processes will be assumed. According to Lundvall (1992), in a general sense, systems are constituted by a number of elements and the relationships between them. Consequently, also systems of innovation are constituted by different elements and the specific relationships and interactions between them in the production, diffusion and use of new and economically useful knowledge. According to that, systems of innovation must always be regarded as *social* systems. In the heart of systems of innovation stand learning and knowledge transfer, and these processes are, as has been shown before, social activities, which require the interaction between different individual, institutional and social actors (Lundvall 1992: 2).

Moreover, Edquist (2005) states, that the elements of a system and the relations between them should always form a *coherent* whole. Such a system must have properties different of those of the single constituents. Furthermore, a system must always fulfill a specific function, i.e. it must perform or achieve something. Finally, it must be possible to discriminate between the system and its environment, i.e. it must be possible to identify the boundaries of the system. This is especially important if specific systems are investigated empirically, like in this study, since in this case their precise extent has to be indicated (Edquist 2005: 187).

The main components of these larger social systems are institutions and organizations. Given that in the debate on innovation systems it is not always clear, what these terms exactly mean (Edquist 2005: 186), in the following some brief remarks to specify them will be made. According to Edquist and Johnson (1997) institutions can be defined as „sets of common habits, routines, established practices, rules or laws, that regulate the relations and interactions between individuals and groups." (Edquist/Johnson 1997: 46) This definition allows for looking at institutions in their basic function of shaping people's cognitions, views, visions and actions, as well as at their function of providing a frame of orientation for the decision making processes of rational actors (Edquist/Johnson 1997: 46). In this regard, institutions reduce the uncertainty in individual and collective actions and, thus, also the amount of information that must be collected for an action. Consequently, they can be interpreted as "rules of the game" of a society that structure social interaction and, thus, reduce the transaction costs connected to them (North 1990: 3).

Analytically, this notion of institutions must be differentiated from the notion of "organizations". North (1990) states in this connection: „Conceptually, what must be differentiated clearly are the rules from the players. The purpose of the rules is to define the way the game is played. But the objective of the team within that set of rules is to win the game […]. Organizations are created with purposive intend in consequence of the opportunity set resulting from the existing set of constraints (institutional ones as well as the traditional ones of economic theory) and in the course of attempts to accomplish their objectives are a major agent of institutional change." (North 1990: 5) According to that, organizations are the „players", which are partly formed by their institutional environments, but influence them at the same time, and are, hence, vehicles of their change. They are formal structures, which have an explicit purpose and are consciously created to its fulfillment (Edquist/Johnson 1997: 47).

However, for the analysis at hand, this strict analytical differentiation between institutions and organizations will not be applied. Following Lundvall (1992), also banks, universities and firms will be labeled as institutions here, as in everyday language use. They can be characterized as "formal institutions" (Lundvall 1992: 26). Like this, a bank for example (so to say a "financial institution") can be regarded as an already formalized and organized form of institutionalized actions of borrowing and lending money (Lundvall 1992: 26).

In sum, it can be stated that processes of innovation, learning and knowledge transfer are embedded into systemic environments. For the study at hand this implies that the systemic character of knowledge transfer and innovation has to be taken into account and that the analysis of these processes should, thus, focus on particular *systems of innovation and production*. Moreover, these processes are structured, shaped and mediated by specific formal and informal institutions. This institutional set-up can, however, vary to a considerable degree among innovation and production systems. Therefore, differences in the institutional set-up of specific innovation and production systems should be supposed to constitute important explanatory factors for differences in the concrete structures and mechanisms of knowledge transfer observable in these systems. Since it can be derived from all this that innovation and production systems should be regarded as complex social systems (Blättel-Mink 2006), in the following, a specific sociological view on innovation and production systems as institutional orders or social and organizational fields will be developed, that allows to apply some analytical tools of neoinstitutionalist theory to the analysis, which are assumed to be helpful when addressing the research questions outlined above.

2.2.3 A Sociological Perspective on Innovation Systems

Taking these considerations on the systemic character of innovation and knowledge transfer, based on a sociological perspective, thus, innovation and production systems can be analyzed as specific *institutional orders*. In general, an institutional order can be defined as an arrangement of *rules of the game* that interact in the regulation of social processes like e.g. production or innovation processes. These orders are characterized by four attributes: First, by *institutions*, i.e. cognitively, normatively and strategically embedded patterns of behaviour, interpretation and social relations that are taken for granted and shape the actions and the reasoning of the actors in a social field (Jepperson 1991). Second, they are characterized by *institutional complementarities*, i.e. structural similarities and the interaction of different institutional spheres (Amable 2000). Third, institutional orders are not only determined by actual constellations and power relations, they also develop an own dynamic which is reflected by specific *path dependent* patterns of change. This means, that also inferior solutions are continued, if the change towards a completely new path would be too risky or too expensive. A fourth characteristic of institutional orders are their immanent differences in interests and perceptions, often leading to *conflicts*. These conflicts can only provisionally be regulated. Consequently, institutional orders are by no means homogeneous, coherent and consensually shared regulatory structures. Rather, they provide regulation structures for the actions as well as for the integration of the divergent and sometimes conflicting strategies of the participating actors (Friedberg 1995). Thus, referring to innovation systems, institutional orders can be described as arenas in which conflicts and cooperations are realized, knowledge is exchanged and learning processes take place.

Moreover, based on a neoinstitutionalist perspective, innovation systems can also be described as *organizational* or *social fields* (Heidenreich 2006; Krauss 2005). This notion was introduced by the neoinstitutionalist organizational research to catch the institutionalized, i.e. regulated environments of organizations (Scott 1994) and refers to both the totality of organizations and actors being involved in an arena of social and cultural production and the dynamic relationships among them (Bourdieu 1971: 161; DiMaggio 1979: 1462f.). Social or organizational fields only exist to the extent of their institutional definition or "structuration" (DiMaggio/ Powell 1983), including both connectedness (the existence of transactions tying organizations to one another) and structural equivalence (the similarity of position in a network structure) (Scott 2008: 185; DiMaggio/Powell 1983: 148ff.): "By organizational field, we mean those organizations that, in the aggregate, constitute a recognized area of institutional life: key suppliers, resource and product consumers, regulatory agencies, and other organizations that produce similar services or products" (DiMaggio/Powell 1983: 148). Referring to markets, Fligstein (2001) states, that the "theory of fields implies that the search for stable interactions with competitors, suppliers, and workers is the main cause of social structures in markets." (Fligstein 2001: 18) This makes clear, that the notion of organizational or social fields corresponds essentially to the concept of innovation systems, focussing on the same actors and general structural features.

Furthermore, the field perspective can help to answer the question, what makes an innovation system distinct from its respective environment. Here, the concept of institutional *isomorphism* plays a central role, describing the tendency that the organizations in a specific field become more similar and homogeneous to one another than to organizations outside the field. "[…] isomorphism is a constraining process that forces one unit in a population to resemble other units that face the same set of environmental conditions. At the population level, such an approach suggests that organizational characteristics are modified in the direction of increasing compatibility with environmental characteristics […]." (DiMaggio/Powell 1983: 149) According to the general definition of systems above, thus, isomorphism in an organizational field might help to discriminate between the system and its environment, i.e. to identify the boundaries of the system. There are two types of isomorphism: Competitive and institutional. The former points to a system rationality emphasizing market competition, niche change and fitness measures. The latter widens this perspective by recognizing that organizations not only compete for resources and consumers, but also for political power and legitimacy, for social as well as for economic fitness (DiMaggio/Powell 1983: 148ff.; Scott 2008: 152). This institutional isomorphism can be induced by particular *coercive*, *mimetic* or *normative* mechanisms and pressures (DiMaggio/Powell 1983: 150ff.; Meyer/Rowan 1991: 49).

In sum, investigating the occurrence and the success of innovation and learning processes based on a sociological perspective means bringing social actors and institutional structures into the centre of attention. This can be achieved by looking at innovation systems as specific institutional orders or social and organizational fields. This sociological approach to the study of economic behavior highlights the mechanisms regulating the manner in which formal elements of institutional structures in combination with informal social organization of networks and norms facilitate, motivate, and *govern* economic action in general and innovation processes in particular. Thus, both distal and proximate causal mechanisms are addressed and incorporated, allowing for a *comparative* institutional analysis of economic life (Nee 2005: 49). For this study, this approach gives the opportunity to investigate differences and similarities in the formal and informal structures and mechanisms of regional innovation systems with regard to their respective national institutional arrangements.

In particular, the concept of institutional isomorphism claimed by neoinstitutionalist theory appears to be helpful when investigating the assumption of institutional coherence of national systems of innovation and production, and should, above all, be a useful tool to explain possible variation within them. For instance, some regional innovation systems within national production and innovation systems should show stronger tendencies of institutional isomorphism with regard to their respective national institutional framework than others, as could be assumed e.g. for traditional industrial regions that have experienced long historical paths of industrial and institutional development parallel to their national environment. Contrariwise, other types of regions, as e.g. metropolitan regions with their newly created institutional frameworks and particular structural conditions in the fields of innovation and knowledge transfer can be assumed to be much less coherent with regard to their national environment. However, since the specific form of institutional order shaping the character of social systems of innovation and production is always the result of a complex interplay of its different constituting institutional subspheres, in a next step the "governance approach" will be briefly discussed to differentiate between the different (ideal-) types of institutions that constitute particular forms of institutional order in the fields of innovation and knowledge transfer.

2.2.4 The "Governance" of Social Systems of Innovation and Production

As becomes clear from the previous sections, innovation and knowledge transfer are systemic processes that are embedded into complex social systems of innovation and production. These innovation and production systems are characterized by particular forms of collective institutional order, that shape the perspectives and interactions of the actors involved in innovation and knowledge transfer, and provide them with specific institutional structures and mechanisms through which they can transfer knowledge. In particular, these different forms of institutional orders provide the actors within innovation and production systems with specific "collective competition goods" (Le Galès/Voelzkow 2001), which may contribute to their innovativeness and stimulate knowledge transfer. This provision of specific collective competition goods in innovation and production systems is realized by many different types of institutions, which in their specific interplay result in particular governance models of knowledge transfer, as will be shown subsequently.

To describe the differences between these types of institutions and their different modes of providing collective competition goods, the "governance approach" appears to be helpful. This approach basically describes the concept of institutional orders that has already been introduced above, but, according to Hollingsworth, Schmitter and Streeck (1994), allows for a more differentiated analysis of distinct *ideal types* of institutions that coordinate and regulate activities within a specific social or economic system. These authors define a *governance system* as "the totality of institutional arrangements—including rules and rule-making agents—that regulate transactions inside and across the boundaries of an economic system." (Hollingsworth et al. 1994: 5) Also for the pretended comparative analysis of institutional structures and mechanisms of knowledge transfer in innovation and production systems the governance approach seems to be useful, since it allows for differentiating between distinct *national and regional governance models* (i.e. combinations of ideal types of institutions), that might lead to different outcomes in the mechanisms and structures of knowledge transfer within these models. These ideal types of institutions constitute the basic components of a governance system and find their expression in six forms of social order: The market, the organization, the state, the network, the community and the association (Hollingsworth et al. 1994: 5ff.; Voelzkow 2007: 41). In the following, these six ideal types of institutions will be briefly discussed.

First, the *market* is a type of institution that tries to coordinate the interactions of rational and utility maximizing actors through the mechanism of atomistic competition. This model assumes that in the exchange of private goods and services individual and collectivities (particularly firms) are primarily concerned with profit. "Markets are arenas in which individual or corporate actors holding separate property rights in different resources voluntarily engage in free, legally enforceable contractual exchanges. Markets are the places for buying and selling, with prices providing more or less distorted information for the allocation of goods and services (Hollingsworth et al. 1994: 5). However, since the market mechanism is" always confronted with the risk of opportunistic and "free rider" behavior, it frequently fails to provide collective competition goods—although this is not completely impossible. One could for example imagine a situation in which one single company alone pays for the provision of a collective competition good—even though other agents will act as free rider—because the utility gained from the risen competitiveness of the whole innovation system exceeds its individual costs (Le Galès/Voelzkow 2001: 7).

In contrast to this, in an *organization* the interactions between the different actors are coordinated through hierarchies that are deduced from specific formalized as well as informal rights and duties. "Corporate hierarchies [...] are institutions that permit some actors to wield authority over others on the basis of property rights vested in a formal organization to which both categories of actors belong." (Hollingsworth et al. 1994: 5) In this model of social order, the production of collective competition goods will be realized by a vertical integration, in the sense of a merger of several firms of a nation or region. However, in the process of vertical integration the collective competition good will of course lose its collective characteristics and will become a private good of the vertically integrated large enterprise (Le Galès/Voelzkow 2001: 7).

Third, also the *state* is based on hierarchical control to coordinate the interactions of the actors. In this, the state relies on its monopoly of power that allows him to implement decisions, if necessary by force. In this model the provision of collective competition goods is mostly realized through cooperation between private companies and public institutions. These specialized public institutions and administrative units on the national or regional level, as, for example, national, regional and local governments, public initiatives, regional institutions for the promotion of economic development and trade, universities or public research facilities, provide economic actors with specific information, consulting and other services (Le Galès/Voelzkow 2001: 7f.).

The fourth ideal model of institutions is the *network*. A network describes stable relations and interactions of persons or organizations (firms), without that the actors lose their autonomy, as it would be the case in an integration into an organizational hierarchy. In a network, the coordination of economic transactions is neither realized through simple market based exchange processes, nor through instruction-based hierarchies. Rather, the actors in a network have trust in the fact that their own input into the network will be compensated at a later point with a more or less equivalent *quid pro quo*. This allows for a coordination of economic activities that opens up the dualism of market and organization, since not all inputs have to be compensated immediately with an own price. Good examples for the provision of collective competition goods through networks are R&D-cooperations, joint ventures, strategic alliances or stable business relations between industrial firms and their highly specialized suppliers (Voelzkow 2007: 45f.; Schmoch 2003: 113ff.).

Fifth, in the *community* the central mechanism of coordination between the actors is informal, voluntary, and sometimes spontaneous solidarity among the members of a social unit. The motivation for the members in this ideal type of institutions is not profit, but the appreciation of the other members of the community or the desire to belong to a group. This solidarity can also serve as the basis for cooperations between several small and medium-sized companies to jointly fulfill specific operational functions and, thus, provide certain collective competition goods. In this, the prevailing collective identity in this model helps to overcome the individually rational orientation which under the market mechanism would have caused market failure (Le Galès/Voelzkow 2001: 8).

Finally, the (corporatist) *association* is the sixth form of institutions that is to mention in this context. In this model the central actors are associations and organizations of collective representation of interests, like business and employer associations or trade unions (Hollingsworth et al. 1994: 7). The interactions within these associations, as well as the interactions between them are coordinated by the *concertation* of interests. These associations can represent e.g. their members' interests towards the state or towards other associations. In this, the associations assume considerable regulatory potential, so that they can enforce token decisions through effective organization structures, internally as well as in external collective agreements. These associations of collective interest representation also provide services like information and consulting to their members and other actors, so that they also contribute to the production of collective competition goods (Le Galès/Voelzkow 2001: 8).

In conclusion, it has been shown that the production of collective competition goods in innovation and production systems can be realized through a wide range of institutional variants creating specific models of social order and national and regional governance systems. Referring to the research question, this means that also structures and mechanisms of knowledge transfer should be influenced by a broad variety of institutions and will be embedded within different governance models. This allows for a differentiated view on specific regional and national solutions, and also displays the existence of *functional equivalents* within these models. Hence, in can be supposed that certain elements of the six ideal types of institutions can be found in all national and regional innovation and production systems, but their respective *weights* can be assumed to be distributed *unevenly*. The governance approach, thus, emphasizes strongly the "contingency of orders", i.e. the variety of ways in which certain tasks, as the production of collective competition goods or the promotion of knowledge transfer, can be assigned to the market, the organization (firm), the state, the network, the community or the (corporatist) associations. Therefore, this approach allows for analyzing a broad range of theoretical possibilities and leaves it to empirical research, in particular to *horizontal comparisons* through case studies, to identify specific governance systems and investigate their functioning. In this study, the governance approach will be used to develop a "heuristic model" of knowledge transfer systems (section 2.5) in the context of innovation, which will then be the basis for the empirical qualitative case studies in the three countries and six investigated regions.

2.3 Production Regimes: Institutional Frameworks for Knowledge Transfer

As becomes clear from the discussion of the previous sections, learning and knowledge transfer are, like innovation, systemic processes that are structured, shaped and mediated by specific formal and informal institutions creating specific forms of social order and governance systems. A study pretending to investigate similarities and differences of specific structures and mechanisms of knowledge transfer, thus, has to look at the relevant systemic institutional environments, into which processes of learning and knowledge transfer are embedded. In this regard, my point of view is that the specific institutional environments of economically relevant learning and knowledge transfer processes are best described by the theoretical concepts of VoC as well as NIS and RIS. These concepts are institution-based and focus on the specific institutional orders and governance models of innovation and production systems. In this, they seem to fit particularly well for addressing the research questions outlined above, which attempt to analyze to which degree structures and mechanisms of knowledge transfer in regional innovation systems can differ from their respective national institutional framework. Although other approaches, as, e.g., the concept of *learning regions*, might have a more explicit analytical focus on learning and knowledge transfer, they mostly remain too unspecific when describing the institutional embeddedness of these processes and the concrete governance models shaping them. While the concept of learning regions (Koschatzky 2001a) focuses on spatial proximity (Maskell/Malmberg 1999a: 9), specific localized capabilities (Hudson 1999: 64), untraded interdependencies (Morgan 1998: 151) and trust (Maskell/Malmberg 1999a: 19) as decisive factors for successful learning and knowledge transfer processes, it lacks to elaborate clearly on the institutional framework from which these factors emanate (MacKinnon et al. 2002: 294; Markusen 1999: 870f.). The position taken by the study at hand is that all these factors indeed play a crucial role for processes of innovation and knowledge transfer, but that the respective institutional framework these processes are embedded in is decisive in determining the concrete structures and mechanisms that are applied to realize them. Therefore, in the following sections the concepts of VoC as well as NIS and RIS will be discussed in detail to illustrate the institutional frameworks of innovation and knowledge transfer at the national and regional level.

As this discussion will show both the concepts of NIS and RIS as well as the VoC-approach only point to learning and knowledge transfer as explaining factors and central preconditions for successful innovation activities. However, they still lack a clear analytical framework for defining the specific roles single institutions or particular institutional configurations play for learning and knowledge transfer, and, thus, to systematically describe the concrete mechanisms applied, as well as their impact on innovation processes (Edquist 2005; Audretsch/Lehmann 2005). It is a central aim of this study to address this gap and to contribute to the specification of the understanding of the particular structures and mechanisms of learning and knowledge transfer especially in regional innovation systems and to suggest a more precise analytical framework of their institutional determinants. Moreover, also the question of regional variation within national knowledge transfer systems will be approached, first, by identifying possible theoretical explanatory factors for regional specialisms, and, second, by confronting these theoretical considerations with empirical evidence based on six regional case studies.

2.3.1 National Production Regimes—The Varieties-of-Capitalism-Approach

For investigating the institutional frameworks of innovation and knowledge transfer this study uses the theoretical concepts of VoC as well as NIS and RIS. Looking at these concepts, it is certainly the VoC-approach that takes the widest perspective on innovation and knowledge transfer using a large set of institutional spheres to describe differences between national production regimes, their technological specialization and their specific governance models. Even if it also includes the institutions of the narrowly defined national "innovation sub-system" (see below) into its analysis, it goes far beyond this to describe the specific economic performance of *whole national production systems*. Nevertheless, as will be shown, it also puts the innovation aspect into the centre of its concept and claims that the comparative institutional advantages of the respective production models will particularly lead to different forms of innovation regimes. Both approaches are, thus, closely connected to each other. However, the VoC-approach is not an explicit theoretical concept on innovation and knowledge transfer. Even so, it draws a precise picture of the wider institutional frameworks of national production systems into which all processes of innovation and knowledge transfer in a market economy are embedded. Therefore, in this study it will be read as a theoretical approach *implicitly* describing institutional structures and mechanisms of innovation and knowledge transfer at the national level, based on which in further sections theoretical assumptions on the specific institutional configurations of particular national knowledge transfer systems will be derived. In the following, the basic elements of the VoC-approach will be discussed.

2.3.2 Liberal and Coordinated Market Economies

The VoC-approach focuses on different governance models of specific national production regimes. As an internationally comparative approach it tries to identify institutional differences in the coordination of economic activities between developed capitalist economies and attempts to explain their specific economic (and social) performance. The body of contributions to this topic is large, and the discussion on it controversial. Early works in the field of comparative political economy were presented e.g. by Alfred Chandler (1990), who differentiated managerial, personal and cooperative capitalism, and Michel Albert (1993), who introduced the distinction between Rhenish and Atlantic capitalism. Today, his analysis is mainly noted for its differentiation between shareholder and stakeholder capitalism. Other earlier contributions have been the edited volumes by Berger and Dore (1996), Crouch and Streeck (1997), Hollingsworth et al. (1994), Hollingsworth and Boyer (1997), Kitschelt et al. (1999), and Whitley (1999). But the most systematic and influential one has probably been that of Peter A. Hall and David Soskice (2001a).

In their VoC-approach, these authors present an actor-centred approach to the political economy, which sees the political economy as a field populated by multiple actors seeking to advance their interests in a rational way in strategic interaction with others. Although these actors might be individuals, firms, producer groups or governments, the VoC-approach is a *firm-centred* approach looking at firms as the crucial actors of capitalist economies (Hall 2006: 183; Hall/Soskice 2001b: 6). The concept of firms in this view is *relational*, and firms are seen as actors seeking to develop and exploit core competencies or dynamic capabilities, that can be interpreted as capacities for developing, producing and distributing goods and services profitably. Since these competencies are ultimately relational, the firms in a political economy face a wide range of *coordination problems*. Their economic success, thus, depends substantially on its capacity to coordinate effectively with a wide range of actors, as, for example, customers, suppliers, stakeholders, collaborators, trade unions, business associations and governments (Hall 2006: 183; Hall/Soskice 2001b: 6).

Hall and Soskice (2001b) identify four institutional spheres of political economies in which firms must develop relationships to these actors to resolve their coordination problems. The first one is the sphere of *industrial relations*, in which the firms have to face the problem of coordinating the bargaining over wages and working conditions with their labor force, representative organizations of labor and other employers. The second sphere is the one of *vocational training and education*, where firms have to confront the challenge of securing a workforce with adequate skills, while the workers have to decide how much to invest in what kind of skills. Third, firms are also facing coordination problems in the sphere of *corporate governance*, in which they have to secure their access to specific forms of finance and where investors seek to request assurances of returns on their investments. Finally, the fourth institutional sphere in which firms have to face coordination problems is the one of *inter-company relations*. In this sphere firms have to establish relations with other firms, as customers, suppliers and competitors, to secure adequate supplies of inputs, access to technology and knowledge, and a stable demand for its products, among others in the fields of technology transfer, joint R&D-activities and standard-setting. Coordination problems in this sphere may arise from the sharing of proprietary information and knowledge and the risk of exploitation in joint ventures (Hall/Soskice 2001b: 6f.).

Building on this, national political economies can be compared according to the specific way in which firms approach the coordination problems they are confronted with in these four institutional spheres. According to that, Hall and Soskice differentiate between two basic types of political economies, liberal market economies and coordinated market economies, "which constitute ideal types at the poles of a spectrum along which many nations can be arrayed." (Hall/Soskice 2001b: 8) In the first type, the *liberal market economies* (LMEs), firms coordinate their activities primarily through hierarchies, competitive market arrangements and investments in transferable assets. These market based relations between the actors of a political economy are mostly characterized by an *arm's-length* exchange of goods and services under circumstances of competition and formal contracting. The firms and other actors in this type of market economy adjust their disposition to demand and supply goods or services in response to the price signals generated by such markets. Moreover, this market orientation of LMEs is also reflected by short-term oriented company finance, general education, deregulated labor markets and strong inter-company competition (Hall/Soskice 2001b: 8; Jackson/Deeg 2006: 22).

In contrast to this, in *coordinated market economies* (CMEs) the firms and other actors depend much more on non-market relationships to coordinate their activities with other actors and to create their core competencies. These non-market coordination mechanisms are basically characterized by "more extensive relational and incomplete contracting, network monitoring based on the exchange of private information inside networks, and more reliance on collaborative, as opposed to competitive, relationships to build the competencies of the firm." (Hall/Soskice 2001b: 8) In these coordinated market economies, the equilibrium outcomes of firm behavior are usually the result of *strategic* interaction among firms and other actors. CMEs are, thus, characterized by strategic investments in specific assets, that manifest themselves in long-term industrial finance, high levels of vocational training, cooperative industrial relations and cooperation in R&D, technology and standard setting across companies (Hall/Soskice2001b: 8; Jackson/Deeg 2006: 22; Hall 2006: 182ff.).

These different types of coordination are also supported by the respective roles exercised by the state in the different market economy models: In LMEs the state is defined as "liberal", because it takes an arms' length approach to business and labor and plays, therefore, only a limited role in setting rules and moderating conflicts. The administration of the rules is often delegated to specific self-regulating bodies and regulatory agencies. In this model the state acts as an agent of market preservation by providing framework legislation with the aim to locate decision-making power in companies and to limit the bargaining power of organized labor (Hall/ Soskice 2001b: 45ff.; Wood 2001; Schmidt 2006: 5). In CMEs, by contrast, the state can be defined as "enabling" since it is active not just to arbitrate among economic actors but rather to facilitate their activities. The state in CMEs often leaves the rules to be administered by the corporatist actors themselves, while the state acts as a "co-equal" or "bystander" with business associations, management and trade unions in the fields of labor regulation and wage bargaining. In this model the state has the function to protect the non-market coordinating institutions characterizing the production system by establishing and periodically renewing the framework laws that concede regulatory authority in private bodies, including employers' associations, chambers of industry and commerce as well as trade unions (Schmidt 2006: 5; Wood 2001; Hall/Soskice2001b: 45ff.).

In constructing this analytical framework on the varieties of political economies, Hall and Soskice (2001b) reverse one fundamental theoretical assumption of institutional economics: While conventionally it is argued that the institutional structures of the economy (e.g. markets, hierarchies, networks etc.) follow firm strategy, i.e. firms seeking the most efficient institutions for performing certain tasks, the authors emphasize that in many important respects the strategy of firms follows the institutional structures. This can be explained by the fact that the firms acting in any political economy will always face a set of coordinating institutions whose character is not fully under their control. Therefore, systematically different firm behavior across institutional settings can be observed, which parallels the overarching institutional structures of the political economy (Hall/Soskice 2001b: 14f.; Jackson/Deeg 2006: 22f.).

2.3.3 Institutional Coherence, Complementarity and Modes of Innovation

These differences between LMEs and CMEs are reinforced by *institutional complementarities* between the four institutional spheres of a political economy. The notion of institutional complementarity can be defined as follows: "Institutional complementarity is present when the existence of the particular form taken by an institution in one area reinforces the presence, functioning or efficiency of another institution in another area." (Amable et al. 2005: 316) This implies that political economies with a particular type of coordination in one institutional sphere of the economy should be likely to develop complementary practices in other institutional spheres as well. Thus, in LMEs, for example, short-term industrial finance requires the possibility to quickly enter and exit from business activities, what fits best with a system of industrial relations allowing inexpensive and non-bureaucratic hiring and firing of labor (Jackson/Deeg 2006: 23; Hall/Soskice 2001b: 18).

Moreover, in the VoC-approach this assumption of institutional complementarity is closely connected to the concept of *institutional coherence* (Kenworthy 2009: 182f.; Höpner 2005). The notion of institutional coherence refers to the fact that the different institutional spheres of a specific national economy show analogous modes of coordination: "Coherence [...] refers to structures. Institutions are coherent if they are designed according to identical principles." (Höpner 2005: 333) Following this assumption, as Hall and Soskice (2001b) do, for a political economy it should be beneficial in terms of economic performance, if all institutional spheres follow the same logic of coordination: Either, all institutional spheres function according to market based coordination, or all spheres follow the non-market, strategic logic of coordination. Consequently, only if the institutional spheres of a market economy are completely coherent, this will increase its economic performance, while incoherent institutional regimes in a market economy (one sphere coordinated in a market based, one sphere in a strategic way) will have negative consequences on its economic performance. Thus, at this point, institutional coherence (structural consistency) and institutional complementarity (economic advantage and functionality) go hand in hand and mutually reinforce each other (Hall/ Soskice 2001b: 17ff.; Voelzkow 2007: 56ff.; Hall/Gingerich 2009: 460; Jackson/Deeg 2006: 23f.; Kenworthy 2009: 181ff.).

Consequently, the concepts of institutional complementarity and coherence indicate that institutional practices of various types should not be distributed randomly across nations. Rather, national economies should cluster along the dimensions dividing liberal and coordinated market economies described above, since specific political economies converge on complementary practices across different institutional spheres (Hall/ Soskice 2001b: 18). To support these assumptions, the authors present some empirical evidence that demonstrates that within the OECD countries the Anglophone countries like the USA, UK, Ireland, Canada, Australia and New Zealand can be classified as LMEs, while many continental European countries such as Germany, Austria, Netherlands, Belgium, Sweden and Denmark represent CMEs (Hall/Soskice 2001b: 19f.).

In between the two poles of liberal and coordinated market economies lie the more ambiguous cases showing *incoherent* institutional regimes (Kenworthy 2009: 181ff.). To this group of countries belong for example Spain, France, Italy, Portugal and Greece. However, also these *mixed market economies* show some common signs of institutional clustering as well, a fact that has not only brought Hall and Soskice (2001b), but also many other authors (Molina/Rhodes 2007; Schmidt 2002; Amable 2003; Ebbinghaus 1999) to acknowledge at least the existence of a third, "Mediterranean" type of capitalism. For this study, this view will be shared, and a third type of Mediterranean market economies (MMEs) will be included into the analysis, that is particularly characterized by large levels of *state intervention* in the coordination of economic activities (Hall/Gingerich 2009: 59f.).

In general, these MMEs show strengths and weaknesses in both market-based and strategic coordination in the different institutional spheres of the political economy. The shortcomings of the coordination by markets and corporatist actors are *compensated by an active role of the state* (Molina/Rhodes 2007; Royo 2007; Ebbinghaus 1999). While for LMEs the state has been described as *liberal*, and for CMEs as *enabling*, for the MMEs it can be defined as *influencing*, in a sense that it tends to intervene wherever necessary. In this, the state in MMEs can, on the one hand, exercise an *enhancing* role for business and labor activity because public intervention might have a positive impact on the productive capabilities and interactions of the economic actors. On the other hand, the active role of the state can also be *hindering* economic activities, whenever public intervention shows a negative impact on productive capabilities and interactions of the economic actors (Schmidt 2006: 5f.).

However, the fundamental assumption behind the VoC-approach is that these specific institutional configurations of particular political economies provide firms with advantages for engaging in specific types of economic and entrepreneurial activities. This relation is captured by the concept of *comparative institutional advantage*: "Firms can perform some types of activities, which allow them to produce some kinds of goods [or services; MO] more efficiently than others because of the institutional support they receive for those activities in the political economy, and the institutions relevant to these activities are not distributed evenly across nations." (Hall/Soskice 2001b: 37) These comparative institutional advantages manifest themselves in specific *modes of innovation* in which the firms within a particular political economy have strengths in.

According to this, LMEs have considerable strengths in supporting *radical innovations*, i.e. completely new products, processes and services, while CMEs should be highly supportive in promoting *incremental innovations*, i.e. the stepwise further development and improvement of already existing products, processes and services. Since different technological sectors are characterized by different modes of innovation, this implies that the firms in LMEs focus on technological sectors that are fast-moving and require rapid product development based on scientific research and analytical knowledge, such as biotechnology, software development, semiconductors, telecommunication, financial services, or entertainment. In contrast to this, firms in CMEs like Germany find particularly favorable conditions to engage in technological sectors in which "diversified quality production" (Streeck 1991; 1997) and a synthetical knowledge base is required, as e.g. consumer durables, mechanical engineering, factory equipment, automotive, chemicals, or electrical engineering (Voelzkow 2007: 59ff.).

In conclusion, thus, the VoC-approach traces back the comparative institutional advantage of specific national production regimes to particular institutional strengths in promoting specific modes of innovation. With that, it ultimately describes whole political economies as national innovation systems (Hübner 2009: 144; Ebner 2009: 125). As a political-economic approach it emphasizes the *comparative* advantages of national economies, and, therefore, cross-national patterns of product or process specialization. These comparative advantages in the innovation activities of national market economy models are strongly connected to the complementarity of their respective institutional sub-spheres and the coherence of their coordination mechanisms. With regard to the pretended analysis of the relation of national and regional knowledge transfer systems this implies that nationally specific and coherent patterns of innovation and knowledge transfer processes should exist, making use of particular national institutional set-ups promoting specific modes of innovation. However, it is the position taken in this study that exactly these assumptions of institutional complementarity and coherence of national production systems are overemphasized in the VoC-approach and that there should be at least some kind of regional variation in the institutional structures and mechanisms of innovation and knowledge transfer within these models. Therefore, in the next section, a detailed critique of these assumptions will be discussed.

2.3.4 Criticizing the Varieties-of-Capitalism-Approach

Much critique has been formulated on the VoC-approach. Among the most important points of criticism are contributions claiming that it is too static and focuses too much on institutional permanency and path-dependence, not emphasizing enough the dynamic elements of economic and institutional change (Crouch/Farrel 2004; Crouch 2007; Deeg/Jackson 2007; Streeck/Thelen 2005; Hancké/Goyer 2005). Other authors have argued that it is too functionalist (Hancké et al. 2007b; Allen 2004; Boyer 2005; Howell 2003), or that it has a tendency to a certain institutional determinism and an almost mechanistic conception of institutional complementarities, neglecting the underlying power structures, including social class (Coates 2005; Crouch/Farrel 2004; Amable/Palombarini 2009; Thelen 2003; Pontusson 2005; Jackson/Deeg 2006). Moreover, it has also been criticized that it divides the capitalist world into only two archetypical models of market economy and lacks the tools for diversifying this bifurcation (Schmidt 2002, Watson 2003; Pontussen 2005; Boyer 2005), for example by highlighting different roles of the state in coordinating economic activities (Schmidt 2006), and that it creates these types of market economies by reading back empirical information on the countries it seeks to make its paradigmatic cases, namely Germany and the USA (Hancké et al. 2007b; Crouch 2005).

This study criticizes in particular one central argument of the VoC-approach: That it overemphasizes the institutional coherence and complementarity of national market economy models and that it is, thus, not sufficiently aware of the (particularly regional) diversity within national systems (Voelzkow 2007; Crouch/Voelzkow 2009; Crouch et al. 2009; Kenworthy 2009). The implicit argument of the VoC-approach is that only if all institutional subspheres of a political economy show *coherent* types of coordination, this would lead to effects of complementarity and, thus, improve the economic performance of the respective market economy model as a whole. In contrast, if a political economy is characterized by incoherent types of coordination in its different institutional subspheres, the institutional complementarity will be less pronounced, showing negative effects on the economic performance of the respective market economy model (Voelzkow 2007: 62ff.; Kenworthy 2009: 181ff.). The assumption is that the coordination mechanisms in the different institutional spheres of a political economy are linked together so tightly that neither institutional change in single institutional subspheres, nor regional variation within national market economies are likely to occur (Crouch/Voelzkow 2009: 2).

Although Hall and Soskice (2001b) admit, that certain "difference in corporate strategy can be conditioned by the institutional support available to firms at the regional or sectoral levels" (Hall/Soskice 2001b: 15), they clearly emphasize variations in corporate strategy evident at the national level: "[…] our approach predicts systematic differences in corporate strategy across nations, and differences that parallel the overarching institutional structures of the political economy." (Hall/Soskice 2001b: 15) They argue this to be "justified by the fact that so many of the institutional factors conditioning the behavior of firms remain nation specific. There are good reasons why that should be the case. Some of the relevant institutions were deeply conditioned by nationally specific processes of development, as are most trade unions and employers' associations. In others, the relevant institutions depend heavily on statutes or regulations promulgated by national states, as do many institutions in the financial arena and labor market, not to mention the sphere of contract law." (Hall/Soskice 2001b: 16)

This clear focus on national production and innovation systems as coherent institutional regimes appears to be highly questionable and insufficiently examined in an empirical manner (Kenworthy 2009: 196ff.). One reason for this is that many authors of the VoC-approach base their conceptual and empirical analyses strongly on particular regions within national economies to then draw conclusions on and describe the national model as a whole. Thus, for example, Baden-Württemberg for the German coordinated economy, or Silicon Valley for the liberal economy of the USA are introduced and investigated as paradigmatic cases for the respective production models. However, this implies that all other regions within these national models are assumed to show similar institutional configurations, an assumption that appears to be highly debatable (Amable 2003: 13). The point of view of this study is, therefore, that these regions should be regarded as specific rather than paradigmatic cases.

Moreover, and referring to the discussion on regional innovation systems following below, it seems to be plausible to assume that regional systems might have enough regulatory competences to create institutional environments diverging clearly from their respective national frameworks. Specific regional business cultures, mentalities and identities might contribute to these differences in the institutional structures. And many of the institutions, as trade unions, business associations and chambers of industry and commerce that the authors mention as central for the national level might, thus, have important regional branches following quite different strategies. Furthermore, regions might have experienced different historical paths of development and might have divergent economic and business structures as well as sectoral focuses, requiring deviant institutional support. Especially in the fields of innovation and knowledge transfer these regional peculiarities can be supposed to be particularly pronounced, since innovation and learning processes depend strongly on specific locational factors, as spatial proximity, *local* collective competition goods and trust based relationships, for example. Therefore, the governance of innovation and knowledge transfer at the regional level might be quite different from the overarching governance structures at the national level.

While the assumption that national market economy models are coherent and complementary institutional orders indeed appears to be useful to describe path-dependencies, lock-in effects and institutional stability within national political economies, it can be supposed to be less helpful to explain processes of institutional change and adaptation. In the narrow terms of the VoC-approach these processes of change are unlikely to happen or would only be possible if the whole market economy model including all institutional subspheres would move in a *symmetrical* way. This study takes a quite different perspective: It tries to stress the diversity within national economic models, assuming less institutional coherence and complementarity of the different institutional subspheres and regional subsystems of national market economies. Although the existence of typical and dominant national models is not neglected, the determinism behind it is questioned, focusing on instances on the regional level in which economic actors might find institutional support for strategies that diverge from these national models. Even stronger, and following Crouch and Voelzkow (2009), it is supposed that there might exist "creative incoherencies" within national innovation and production systems, "when internal diversity produces a loose coupling of different institutional spheres, leading to the creation of autonomous subsystems and governance structures on the sectoral and/or local [and regional; MO] level." (Crouch/Voelzkow 2009: 6)

When discussing the institutional coherence and complementarity implicated by the approaches of NIS and VoC, one could also ask the question, why regional or sectoral concentrations of business activities or clusters can be found in countries whose national model of innovation is not particularly favorable to their performance, as e.g. Trigilia (2004) does. Although these approaches do not completely exclude the possibility of further concentrations of governance at the sub-national level, these regional peculiarities should nevertheless still be consistent with the dominant national institutional framework (Trigilia 2004: 230ff.). In concrete terms, the fact that certain regions in Germany specialize in the automotive industry, some in mechanical engineering and others in the chemical industry, would not necessarily contradict the VoC-approach, since all these industrial sectors show incremental innovation processes which fit into the dominant coordinated model. However, a region specializing in biotechnology as a high-tech industry basing on synthetic stocks of knowledge and being characterized by radical innovation processes should require institutional support different from the national model to be successful. Therefore, the VoC-approach would have difficulties to explain the emergence and persistence of biotech-industries in regions like Baden-Württemberg, Lower Saxony, Central Franconia or the Munich Region. These cases at least indicate, that regional variation of institutional business support exists and that the national level must not necessarily be decisive in determining economic and innovation activities (Crouch/Voelzkow 2009: 10).

Altogether, this critical discussion demonstrates that in most of the literature on the VoC-approach the assumpton of institutional coherence and complementarity appears to be overemphasized and that the important question remains unanswered, to which degree regional subsystems within national economies can deviate from their respective national institutional frameworks. This study tries to address exactly this gap, by investigating structures and mechanisms of knowledge transfer in regional innovation systems within different national market economy models. On the one hand, the aim is to show if and to which degree these structures and mechanisms in the field of knowledge transfer on the regional level can deviate from their respective national institutional frameworks. On the other hand, also the explaining factors of these diverging regional patterns will be investigated. For this purpose, an own theoretical approach will be developed, which, in a first step presents a systematic heuristic model on knowledge transfer systems in national and regional innovation and production systems, and then, in a second step, tries to derive hypotheses on which institutional structures and mechanisms of this heuristic model should be relevant in specific national market economy models according to the VoC-literature.

Prior to this, in the next sections the second strand of theory describing the institutional frameworks of innovation and knowledge transfer will be discussed, namely the *innovation systems approach*. In this regard, a particular focus of analysis will be directed on the different *territorial dimensions* of innovation systems, the national as well as the regional. The aim is to provide theoretical evidence for the specific conditions shaping processes of innovation and knowledge transfer in regional innovation systems and to discuss to which degree these specific regional conditions can contribute to the creation of structures and mechanisms of knowledge transfer that diverge from their respective national institutional framework as it is described by the VoC- and NIS-approaches.

2.4 The Innovation Systems Approach—Spatial Dimensions of Innovation

Compared to the VoC-approach that differentiates distinct ideal types of market economy models and, thus, whole national production systems, the "systems of innovation" approach has a more focused perspective on processes of innovation and learning. This approach basically builds on two fundamental assumptions that have already been introduced above: The predominantly non-linear (Kline/Rosenberg 1986; Fagerberg 2005) and systemic character (Lundvall 1992; Edquist 2005) of innovation and learning processes. From the systems of innovation perspective innovation and learning are conceptualized as complex and interactive processes (Fischer et al. 2001: 2ff.) that are thought as relationships of collaboration and interdependence characterized by their reciprocity and a multitude of feedback mechanisms (Edquist 2005: 185). These processes are embedded into specific set-ups of formal and informal institutions that govern, structure and shape these interactions between the different actors involved in processes of innovation, learning and knowledge transfer (Fischer et al. 2001: 8f.; Edquist/Johnson 1997; Blättel-Mink 2006).

According to Lundvall (1992) a system of innovation is "constituted by elements and relationships which interact in the production, diffusion and use of new, and economically useful, knowledge" (Lundvall 1992: 2), while Edquist (1997) includes in his definition of an innovation system "all important economic, social, political, organizational, institutional, and other factors that influence the development, diffusion, and use of innovations." (Edquist 1997: 14) The use of the term "systems" is, therefore, a pragmatic one, and should not be understood in a systems theoretical way as done by Bathelt (2003) (Asheim Coenen 2006: 172). Moreover, at the current stage of theory-building, there is no general agreement on which elements and relations are essential to the core of the conceptual framework and what their precise content is (Fischer et al. 2001: 8).

Of particular interest for the study at hand are the different *territorial dimensions* of innovation systems, that have been discussed intensively in the literature (a.o. Lundvall 1992; Nelson 1993; Braczyk et al. 1998; Malecki/Oinas 1999), and which base on the assumption that most innovation systems are build around some kind of *spatial proximity*, might it be at the local, regional or national level (Fischer et al. 2001: 12ff.). In this, national innovation systems are, like national production systems in the VoC-approach, conceptualized as coherent institutional frameworks for innovation and knowledge transfer. This leaves the question unanswered to which degree regional innovation systems within these national systems can vary from their respective national frameworks. To address this question in detail, in the following sections the national and regional dimensions of innovation systems will be discussed. In a first step, the NIS-concept will briefly be outlined, especially with the aim to illustrate its relation to the VoC-approach. In a further step, the RIS-approach will be discussed in detail focusing on the specific regional factors determining structures and mechanisms of knowledge transfer, which might account for possible regional variation within national institutional frameworks.

2.4.1 National Innovation Systems

The origins of the NIS-approach trace back to studies presented by Freeman (1987; 1988; 1995), Lundvall (1988; 1992) and Nelson (1993), who initially developed and evolved the "systems of innovation" approach. Christopher Freeman (1987) was among the first to apply the concept to describe and interpret the economic performance of Japan in the post-war period. In his analysis, he identified a number of central and distinctive elements in Japans' national system of innovation to which could be attributed its success in terms of innovation and economic growth (Freeman 1988: 338). Thus, the first studies using the systems of innovation approach focused on the national level, although it has subsequently also been applied to contexts outside the original national setting, above all to the regional level, as will be shown below.

According to Freeman (1987), in general terms a *national innovation system* can be defined as "[…] the network of institutions in the public and private sectors whose activities and interactions initiate, import, modify and diffuse new technologies." (Freeman 1987: 1) Similarly, also Metcalfe (1995) defines a national innovation system as a "[…] set of distinct institutions which jointly and individually contribute to the development and diffusion of new technologies and which provides the framework within which governments form and implement policies to influence the innovations process. As such it is a system of interconnected institutions to create, store and transfer the knowledge, skills and artefacts which define new technologies." (Metcalfe 1995: 462f.)

Moreover, Lundvall (1992) especially emphasizes the role of interactive learning processes as well as the production and exchange of knowledge in innovation systems, highlighting with that important factors concerning the research question investigated in this study. In this, he defines national innovation systems in a *narrow* and in a *broad* sense: "The narrow definition would include organizations and institutions involved in searching and exploring—such as R&D departments, technological institutes and universities. The broad definition [...] includes all parts and aspects of the economic structure and the institutional set-up affecting learning as well as searching and exploring—the production system, the marketing system and the system of finance present themselves as subsystems in which learning takes place." (Lundvall 1992: 12) Herewith, culturally and institutionally stabilized network relationships between different firms and organizations are attributed a central role for the success of innovation processes.

Especially this broader definition points to the specific focus the NIS-approach has on science, research, development and education, and their particular connectedness to their respective national production regime to explain the innovative capacities of particular national innovation systems. As Block and Keller (2009) claim, this emphasis on the interaction of academia and business sector constitutes a major strength of the NIS-approach: "Many of these studies of innovation systems focus on the interface between the public and private sectors, looking particularly at public funding of research and higher education, the growth of the scientific and technical labour force, the systems for establishing and protecting IPRs for innovators and the mechanisms that facilitate the movement of ideas from the research laboratory to the market. The great strength of this literature is that it looks simultaneously at the role of government and the role of business and raises important questions about the interaction between the two." (Block/Keller 2009: 462) It is this interface between private firms and universities, other higher education facilities, and research institutes, their concrete forms and their particular framework conditions which stand in the centre of a high number of contributions to the NIS-concept (Lundvall 1992: 12ff.; Nelson/Rosenberg 1993: 53ff.; Edquist 1997: 2ff.; Padmore et al. 1998: 615; Porter 1998a: 186; Mason/Wagner 1999: 85f.; Weingart 2001: 185; Fagerberg 2005: 12; Blättel-Mink 2006: 149ff.; Block/Keller 2009: 462). A special emphasis in these contributions is mostly put on knowledge transfer activities between academia and the business sector, investigating the role of intermediary institutions and public innovation policy measures. Even if in these contributions also distinct possible channels and mechanisms of knowledge transfer are named (Padmore et al. 1998: 606ff.; Mason/Wagner 1999: 86f.), they mostly lack an exhaustive discussion of these mechanisms as well as a systematic conceptualization of their concrete impact on knowledge transfer processes and the innovation system as a whole. Instead, the focus is rather on describing the respective sets of relevant institutions and measures, their main missions and orientations, their interplay, as well as the formulation of policy recommendations. This study considers this a major conceptual shortcoming in the NIS-approach and derives from it the need to introduce a more elaborated and systematic concept of structures and mechanisms of knowledge transfer into the debate, which will be presented in section 2.5.

In addition, the broader definition of national innovation systems presented above already refers to the fact that the NIS-concept is based on the fundamental assumption that innovation activities differ between nations. This means both that national economies are specialized in terms of technological fields, and that the mode of innovation has national specificities. Andersen and Lundvall (1997) formulate this as follows: "Since innovation is rooted in processes of learning by doing, learning by using, and learning by interacting, the specialization of the production system is crucial in characterizing a national system of innovation in terms of technological fields. The mode of innovation will reflect, among other factors, the predominating patterns of governance." (Andersen/Lundvall 1997: 254)

At this point, also the relation of the concepts of VoC and NIS becomes clear: While the latter focuses more on the knowledge producing, transmitting and processing parts of the national economy, the former includes a considerably wider set of institutions into the analyses to, ultimately, describe whole political economies as national innovation systems (Hübner 2009: 144; Ebner 2009: 125). The decisive difference appears to be that most of the literature on NIS as a socio-economic approach (Blättel-Mink 2006: 52) focuses on non-market relations, as trust, networks, mutual understanding and culture and, thus, identifies factors of *absolute* competitive advantage, i.e. factors that would improve the performance of *any* economy. In contrast to this, the VoC-approach as a political-economic approach emphasizes *comparative* advantages and, therefore, cross-national patterns of product or process specialization.

However, in particular the broader definition of national innovation systems presented by Lundvall (1992) makes clear that to understand the structures and processes of knowledge transfer in depth, it is indispensable to also include the wider institutional arrangements of the national market economy into the analysis, as they are described above all in the VoC-approach that has been discussed in detail before. Therefore, it will be the position taken in this study that both approaches are *complementary* in describing the institutional conditions of innovation and knowledge transfer processes at the national level, considering a national innovation system as an "innovation sub-system" of national production systems. With this perspective a mutual shortcoming in both approaches is addressed: While most of the contributions on NIS have a too narrow focus on the scientific and higher education sector and partly disregard the productive environment (Nelson/Rosenberg 2009: 61), the VoC-approach draws only insufficient attention to the academic knowledge production and its connection to the business sector. The combined perspective suggested in this study, thus, appears to be well suited for addressing the research questions outlined above, not least because it allows to include, apart from the absolute competitive advantages highlighted by the NIS-approach, also the comparative advantages emphasized by the VoC-approach into the analysis, making it possible to investigate the occurrence of regional knowledge transfer systems within distinct national market economy models in a comparative perspective.

Overall, these considerations on national innovation and production systems now raise the important question, to which degree the regional innovation systems within these systems can vary from their respective national institutional environment. As has already been pointed out, both the NIS- and, in particular, the VoC-approach claim the homogeneity and coherence of national institutional arrangements. To critically investigate this assumption of institutional coherence of national innovation and production systems, in the following the regional innovation systems approach will be discussed in detail, demonstrating the particular conditions of innovation and knowledge transfer at the regional level and providing theoretical and conceptual evidence why these specific regional conditions could create regional variation in the institutional structures and mechanisms of innovation and knowledge transfer within national systems.

2.4.2 Regional Innovation Systems—Approaching a Fuzzy Concept

2.4.2.1 Regions

After its first development and application at the national level, the systems of innovation approach has subsequently also been applied to other spatial dimensions, and in particular to the regional level. To precisely describe the regional dimension of innovation and knowledge transfer within the framework of the regional innovation systems (RIS-) approach, in a first step the notion of "regions" will be defined, before the basic elements of the approach will be discussed in detail.

In general, regions can be differentiated with regard to their administrative, geographic, socio-cultural and institutional characteristics. For the planned analysis of regional innovation systems this might be important, because the specific occurrences of these dimensions might influence the success of firms' innovation processes in a region. The notion "regions" stems from the Latin word "regere", which could be translated with "to govern". This is also one possible and important meaning that will be attributed to the notion of "region" in the context of this study: The region is an administrative unit of a nation state. In this, especially competences in the field of the promotion of regional economic development stand in the centre of the analysis (Cooke/Leydesdorff 2006: 6; Ohmae 1993). Moreover, and following Cooke et al. (1997), according to a geographic notion, regions can also be defined as territories, which emerged along specific political, cultural and economic paths and developed their own identities. These regions are territories, which are smaller than the nation-states they belong to, but they possess significant political rooms for maneuver and competences, which contribute to their internal cohesion and delimit them externally from their respective nation-states and from other regions (Cooke et al. 1997: 480; Ohmae 1993).

Finally, institutional and socio-cultural notions of regions are closely interconnected with each other, since the emergence of many regional orders relies on two central processes: On the one hand, regions can emerge from *regionalization* processes, i.e. the delimitation of a supralocal territory by higher politico-administrative authorities, as a nation-state for instance—often regardless of the cultural history of the affected territory (Lagendijk 2001: 87ff.). While the borders of these regions are, thus, fixed "from above", as in the case of the German *Länder* or the English *Government Office Regions*, the emergence of many regions can, on the other hand, be traced back to regionalist flows. Here, socio-cultural aspects, as e.g. a common language or commonly shared historical experiences, possess a much more important meaning, like in some of the Spanish *Comunidades Autónomas*. In these cases, a perceived disregard and discrimination by the nation-state, as well as inefficiencies of the central government are often the breeding ground for *regionalist* tendencies that could lead to the creation of a new regional institutional order (Lagendijk 2001: 87ff.). "This institutional ordering creates new norms, routines and habits as it yields up a new governance structure. It can be the expression in institutional terms of a region's 'social capital'." (Cooke et al. 1997: 480)

Thus, in both cases, regions can also be described as institutionalized spaces with a negotiated and evolving collective social *order*. It is this collective social order that is underlying the action of the regional actors and creates trust between them, by institutionalizing specific routines of behavior, values and norms. This might happen in the form of specific implicit conventions, as well as in the form of clearly defined und codified rules of the game and laws. Altogether, it can be summarized that this social and cultural capital determines collective action in the respective regions, and, thus, exercises a decisive influence on a region's specific development path and shapes its capacity to generate and promote innovation processes and knowledge transfer (Cooke et al. 1997: 480).

2.4.2.2 Regional Innovation Systems—Defining the Theoretical Concept

Bringing the different elements outlined in the previous sections together, the RIS-concept looks at regions as the central location for the coordination of economic activities on the geographic meso level, between the local level on the one hand and the national and supranational level on the other. In this concept, the regional level is ascribed a decisive role for the production of innovation and successful processes of learning and knowledge transfer, because here regional institutional networks, local clusters of companies and research and development facilities can establish relations of close interaction between each other. With Asheim and Isaksen (2002) regional innovation systems can be defined as "places where close interfirm communication, socio-cultural structures and institutional environment may stimulate socially and territorially embedded collective learning and continuous innovation. […] Basically, regional innovation systems consist of two main types of actors and the interaction between them […]. The first actors are the firms in the main industrial cluster in a region including their support industries [with their networks of customers and suppliers; MO]. Secondly, an institutional infrastructure must be present, i.e., research and higher education institutes, technology transfer agencies, vocational training organizations, business associations, finance institutions etc., which hold important competence to support regional innovation." (Asheim/Isaksen 2002: 83)

Parting from this definition, and following Koschatzky (2001a), the basic elements of a regional innovation system are the economic structure (sectoral specialization patterns), the business structure (SMEs, large firms, MNEs, business demography), the inter-firm relationships (customers, suppliers, competitors, contractors, cooperation partners), the business R&D-activities (R&D-investments, qualification and training of staff), the knowledge and technology supply infrastructure (universities, other higher education institutions, research institutes, scientific parks, technological centers), the financial support and infrastructure (banks, public financial bodies, seed- and venture-capital suppliers, business angels), service providers and intermediary actors (business associations, chambers, technology transfer agencies, consultancies, technology oriented information services, vocational training), policy actors (national and supranational policy measures, regional policy measures and actors, financial transfers), and, finally, the regional environment (regional milieu and innovation culture, living conditions, qualification levels of the workforce) (Koschatzky 2001a: 175f.; Cooke 1998: 10). As Cooke et al. (1997) highlight, the interaction among these different institutional and non-institutional elements can be conceptualized as "flows of knowledge and information, flows of investment and funding, flows of authority and even more informal arrangements such as networks, clubs, fora and partnership. It can further be hypothesized that there will be strong and weak, regular and irregular, intense or relaxed kinds of interaction which shape the system." (Cooke et al. 1997: 478)

These conceptual definitions make clear that one major building block in explaining a region's innovative capacity in the RIS-approach is its ability to enable intensive and adequate processes of knowledge transfer. In this connection, the RIS-approach acknowledges as the decisive point the geographic, social and cultural *proximity* (Boschma 2005) between the actors participating in innovation processes, making it possible for them to accumulate, acquire and recombine the knowledge available in the respective region in a faster way. With that, regional inter-firm networking, inter-personal connections, local learning processes and implicit knowledge embedded in specific regions become central locational factors of regions. Especially the *regional embeddedness* of specific forms of knowledge contributes to the fact that these location factors cannot be found everywhere and cannot be transferred to other places (Asheim/Isaksen 2002: 82f.; Howells 1999: 84).

With that, the RIS-concept is closely connected to the notion of regional *clusters*. This notion, especially coined by Michael E. Porter (1998b), describes geographically concentrated networks between firms and institutions along a specific value added chain (Porter 1998b: 78). A cluster includes upstream and downstream production and service activities as well as a specialized infrastructure supporting these activities (Koschatzky 2001a: 197). In this, and in contrast to regional innovation systems, a cluster must not be innovative *per se*, since apart from young and innovative industries also firms in traditional industrial sectors characterized by momentums of persistency can cluster together. However, the decisive difference compared to a regional innovation system is, that the latter mostly includes one or several clusters and that these are not only integrated by processes of production and value added chains, but also and particularly by regional institutional orders and cultures (Tödtling/Trippl 2005: 1206). As Asheim and Coenen (2006) point it, "a regional innovation system can be conceptualized as regional clusters surrounded by supporting knowledge organizations." (Asheim/Coenen 2006: 167) Consequently, the RIS-concept draws a more complete picture of innovation processes than the concept of regional clusters. Moreover, Asheim and Isaksen (2002) point out that in comparison to clusters regional innovation systems are characterized by closer inter-firm cooperations and a stronger integration of knowledge-supplying and -transferring institutions into these cooperations (Asheim/Isaksen 2002: 83).

Other authors as e.g. Rosenfeld (1997), Swan and Prevezer (1996), Simmie and Sennett (1999), Enright (1996) and van den Berg et al. (2001) define clusters in a much more restricted way than Porter as spatially bounded concentrations of independent, specialized and interconnected firms, without pointing to supportive institutional structures. In this sense, clusters emerge spontaneously and in an unplanned way. Since a specific supporting institutional infrastructure would involve concrete planning activities, for Asheim and Isaksen (2002) regional clusters in the notion of Porter more likely already represent regional innovation systems. However, for this study regional clusters will be regarded as further concentrations of interactive firm activities within regional innovation systems. There might exist several of these clusters in different sectors and industries within one regional innovation system. And these clusters might be stronger or weaker connected to the respective regional (and national) institutional infrastructure, as e.g. universities, research institutes, standard setting agencies or trade associations. Overall, clusters within regional innovation systems can then also be considered as important entities with regard to knowledge transfer, since, apart from competition, they are above all characterized by intensive frequency and impact of interaction as well as cooperation (Porter 1998a: 198ff.). Here, knowledge transfer is likely to occur in both directions, vertically within the chains of customers and suppliers and horizontally with other firms and institutions (Martin/Sunley 2003: 10).

It is these institutions that exercise the particularly important function to provide the actors in a regional innovation system with specific "*local* collective competition goods" (Le Galès/Voelzkow 2001; see also section 2.2.4), which may contribute to their innovativeness or stimulate cooperation and knowledge transfer between firms, schools, universities, technology transfer agencies, research and development facilities as well as political and administrative actors. Examples for these goods are the access to specific technological knowledge, functioning sales markets, vocational training of a qualified workforce specifically tailored to the needs of regional industry, specific consultancy and information services, access to regional research and development facilities, collective representation of interests, the acquisition of and access to expensive equipment, funding of R&D projects, financial support for cooperations, as well as joint marketing activities. These goods are provided by institutions, as e.g. universities, research institutes, technology transfer agencies, and regional and local support institutions, which Cooke et al. (2000) names knowledge generation and diffusion sub-systems. Altogether, the specific combination of local collective competition goods provided in a particular geographical space create specific *regional infrastructures*, that offer highly specialized services from which particularly SMEs can profit, since they mostly lack the necessary resources to access or provide them themselves (Le Galès/ Voelzkow 2001; Voelzkow 2007; Ortiz 2008).

Such as in other types of complex social systems of innovation and production, the provision of these local collective competition goods in regional innovation systems can be realized by many different types of institutions. As already pointed out in the discussion on the governance approach in section 2.2.4, six ideal types of institutions can be differentiated in their contribution to the provision of these goods: The market, the organization, the state, the network, the community and the association (Hollingsworth et al. 1994: 5ff.; Voelzkow 2007: 41). These ideal types of institutions constitute the basic components of *regional governance systems*, and from their specific combinations and roles within these systems emanate particular forms of regional institutional order. With regard to the question of regional variation within national systems of innovation and production, in this study it will, thus, be assumed that the specific form of regional institutional order in regional innovation systems can deviate from the overarching national institutional framework and can, therefore, lead to different outcomes in the concrete institutional structures and mechanisms of knowledge transfer at the regional level. The assumption at this point is that particular regional developments and experiences in terms of cultural identity, historical practices, actors' constellations, economic structures and sectoral specializations might be important explanatory factors for these differences. Also in the theoretical debate on the RIS-approach these factors have been discussed intensively to describe the nature, the basic structures and the functioning of regional innovation systems. In the following sections, these parts of the debate on the RIS-approach will be recapitulated to detect explanatory factors potentially accounting for regional variation of knowledge transfer systems within national institutional frameworks.

2.4.2.3 Regional Innovation Networks, Trust and Learning Institutions

From the perspective of the RIS-approach for the firms integrated into a regional innovation system the intensive exchange of knowledge with other firms and economic actors in the region—enabled by complementary and cooperative network relations between them—constitutes a decisive competitive advantage (Heidenreich 2011a: 175ff.). To address the research questions guiding this study it is, therefore, central to understand the nature of these knowledge transfer processes in the way they are (implicitly) conceptualized in the RIS-approach to then identify respective determining factors which could explain regional variance within national systems. In this, the RIS-approach emphasizes distinct forms of network based cooperations in regional innovation systems, ranging from horizontal business networks between competitors and vertical networks of firms and their customers and suppliers along specific value added chains, to networks including other regional and intermediary actors as well as formalized and informal networks.

With regard to business networks, the RIS-approach stresses that, on the one hand, these inter-firm relations can be structured *heterarchically*.

According to that, the network-like relations between the firms of a regional innovation system do not base on unilateral power and relationships of dependence, but on trust, reliability, reputation, custom, reciprocity, the openness to learn, as well as an inclusive and mutual empowering attitude. On the other hand, these relations can as well contain *hierarchical* elements. This would be the case, if a cooperation network is dominated and strategically led by one or several focal firms, as one can observe for example in some specific supplier networks of the automotive industry. In this case, economic objectives of efficiency are combined through a "marketlike" coordination of economic activities with hierarchical forms of interaction, which are derived by asymmetries of power in the network relationships (Cooke 1998: 8f.; Heidling 2000: 70; Sydow 2010: 379ff.; Heidenreich 2011a: 168f.). However, in both cases the *socio-cultural* environment surrounding these intercompany networks is seen as decisive. Here, this environment is understood in the sense of established practices and the entrepreneurial mentalities of a "commercial community" (Cooke 1998: 9). In this kind of environment, entrepreneurial successes can be assessed in a qualified way and the advantages of innovations and improvements can be discussed immediately. New ideas are picked up fastly by other companies and are modified by them, so that they can become the basis for more new ideas (Cooke 1998: 8f.).

For firms in a regional innovation system entering into cooperations with other firms and actors, Heidenreich (2011a; 2006) recognizes particularly four advantages: Apart from the opportunity to use local resources and to reduce transaction costs, like, e.g. transport and communication costs, regional innovation systems especially offer the chance to achieve advantages of *specialization*. Like that, in a regional cluster often a specialized regional labor market as well as a broad range of specialized suppliers and service-firms develop. This makes it possible for firms to use scale effects to a higher degree and to produce specialized outputs cheaper and in a higher number of pieces. Finally, firms in a region have an increased chance to profit from the experiences, competencies and the knowledge of other regional actors, when they are included in their networks (Heidenreich 2011a: 175ff.). However, all this requires specific organizational capabilities of the respective firms making them able to transfer knowledge and to learn. In the following, with regard to the research questions addressed in this study, this last point will be looked at in detail.

In a knowledge based economy especially the capability to react fastly to changes on global markets is of high importance for the competitiveness of companies (see 2.1.1). Thus, learning becomes a core competence for firms. Possessing the capability to absorb and process knowledge from different fields as well as to establish network relationships with other firms and actors is essential for firms to be successful in the competition with their competitors on the world markets (Maskell/Malmberg 1999b: 180f.). In this, the management has the task to create an institutional environment that is conducive to interactive learning processes. Through network and cooperative relationships with other firms and actors, but also through the mobility of staff, for example, firms become "learning organizations." (Cooke 1998: 13)

In this regard, the RIS-approach accentuates that regional innovation systems provide firms with the framework conditions they need to develop and improve their learning and innovation aptitude and to exchange knowledge with other actors. Here, from an organizational perspective, Cooke (1998) particularly refers to three conditions underlying successful learning and innovation processes in RIS: First, he points to a sufficient degree of *reflexivity*, i.e. the ability to continuous self-monitoring and reflection of one's own central moral concepts, in consideration of the points of view of the other participating actors in the learning process. Second, he states the ability to *interdisciplinary* thinking and action, overcoming established professional categories and making it possible to produce innovations out of a specific context of application. Finally, *heterogeneity* is the third condition for successful learning and innovation processes, i.e. the ability of bringing together the most different talents and experiences for dealing with specific challenges (Cooke 1998: 15).

Apart from these purely business oriented networks, large parts of the literature on RIS also emphasize innovation networks including firms and other regional and intermediary actors. In this view it is a decisive factor for the success of knowledge transfer, learning and innovation processes at the regional level that firms can also establish dense communication and cooperation *networks* not only to other companies, but also to research institutions and public administrations, as well as to intermediary actors as e.g. technology transfer agencies, employers-, business- and trade-associations or chambers of industry and commerce (Heidenreich 2000; 2011a; Weyer 2000; Koschatzky 2001a; 2001b; Sydow 2010: 379ff.). The RIS-approach especially highlights the lasting, trust-based and mutual nature of these cooperations. In this connection, Heidenreich (2000) emphasizes above all the opportunity to establish low-level interaction. Regional networks offer the chance to establish noncommittal and informal contacts without bigger effort and without explicit company interests. From this low-level, often unplanned and lowly selective contacts, for the participating actors often arises the advantage to receive more "stimulations beyond their established patterns of thought and behavior." (Heidenreich 2000: 96; translation: MO) In this also lies the "strength of weak ties" mentioned by Granovetter (1985). Moreover, as an approach being strongly coined by works on economic geography, RIS stresses that geographical proximity to other actors increases the frequency of interaction, which can lead—together with specific regional identities and traditional regional forms of socialization—to the rise of common interpretations of reality and trust-based relationships. It is especially these common worldviews and mutual trust among the actors which convert these regional networks into platforms in which implicit, context-bound and non-codifiable knowledge can be transferred (Heidenreich 2000: 95ff.).

In these trust-based relationships all participating actors must be willing to advance performance, although the content of the later *quid pro quo* is not exactly predictable: "Trust is necessary, to bear the uncertainties, risks, openness and ambiguities, which are always linked to network relationships." (Heidenreich 2000: 102) In this connection it is possible to distinguish three forms of development and stabilization of regional trust-relationships: First, one could mention *characteristic-based* trust. In this case, regional trust emerges on the base of common values and traditions. With embedding the different actors from business, science and politics into common and everyday contexts of experience, as well as through commonly shared matter of courses and background assumptions, the obstacles in communication are reduced and mutual understanding for each other increases (Lane/Bachmann 1996: 371f.; Heidenreich 2000: 102f.; Heidenreich 2011: 173f.; Maskell/Malmberg 1999b: 180).

Secondly, regional trust can emerge on the basis of *regional institutions.* These institutions set bindingly expectable patterns of relations, behavior and interpretation and help to overcome the collective problems of the regional actors. They create collectively usable outputs and prevent the rise of "free-rider-behavior" through binding rules of access. Especially regional governments and administrations, regional education and research facilities, regional economic and industry-policies, regional associations and chambers, as well as regional banks contribute to this stabilization of innovations networks (Lane/Bachmann 1996: 371f.; Heidenreich 2000: 103f.; Heidenreich 2011: 173f.; Ortiz 2008).

Finally, trust can emerge from already made positive, stable and long-lasting cooperation-experiences. Consequently, in this case trust is based on the assumption that the other actors will continue their behavior in a way they did before. However, this *experience-based* trust is not suited for establishing a collective regional order. Following Heidenreich (2000), such kind of order could only be created by the combination of *experience-based* and *institution-based* trust. According to that, the specific disadvantages of both types of trust, low scope on the one hand and the relative rigidity on the other, could be averted by tying back the regional institutions to the action of the cooperating regional actors. In this, regional institutions would less be the guarantors of unchangeable rules, but rather the arenas for the regional actors' communication-, coordination- and negotiation-processes. Here, the regional order would not only base on political decisions, but especially on the compromises and agreements of the regional actors. These "learning" institutions offer to the actors the possibility to exchange their different experiences and perspectives. This way, the actors learn to coordinate discursively and to redefine their goals and projects by themselves and within the interaction with other actors (Heidenreich 2000: 104f.; Heidenreich 2011: 174; Gunasekara 2006: 110f.).

Summarizing this, it can be stated that regional innovation systems constitute particularly well suited institutional frameworks for firms to realize successful knowledge transfer, learning and innovation processes. Through horizontal or vertical cooperations with other firms, specific socio-cultural environments and the opportunity to focus on particular patterns of specialization firms in regional innovation systems can become learning organizations that develop considerable competences to deal with the challenges of reflexivity, interdisciplinarity and heterogeneity that underlie all processes of learning and transferring knowledge. Nevertheless, it can also be assumed that the specific occurrences of these factors, as the degree of hierarchy or heterarchy, the degree of dominance of one or more focal firms (as well as the absence of such firms), the strength of the cohesion of the commercial community, and the degree of sectoral and sub-sectoral specialization within one region will influence a region's institutional structures and mechanisms of innovation and knowledge transfer and could, thus, be explanatory factors of regional innovation systems varying from their respective national institutional framework.

In addition, it has been shown that apart from offering a wide range of opportunities to establish formal and institutionally stabilized cooperations, regional innovation systems represent particularly well suited frameworks for firms to enter into low-level and trust-based network relationships with other firms and actors. This points to the very important informal dimension of learning and knowledge transfer mentioned above. Knowledge transfer always implicates both formal and informal institutional structures and mechanisms to be successful and to lead into mutual learning and innovation. It can, hence, be identified as a decisive strength of the RIS-approach to put a particular focus on this informal dimension. For this study, this allows to also investigate differences and similarities of informal mechanisms of knowledge transfer as regional platforms, informal meetings, conferences, workshops, trade fairs, or publications etc. It should be assumed that these differences in the ability of regions to establish and promote formal and informal regional networks and cooperations as well as trust-based relationships among regional actors should be important further differentiating factors of regional innovation systems within national systems of innovation and production.

2.4.2.4 Regional Innovation Systems and their Environment

For an in-depth understanding of the structures and functions of regional innovation systems, it is indispensible to also include their contacts to their extra-regional environment into the analysis, since these contacts might work as important differentiating factors between distinct regional innovation systems. Regional innovation systems are by no means closed boxes whose economic performance would not be influenced by their respective environment, might it be national or supranational (Löw 2001: 63ff.). Rather, in the debate on the approach, they are characterized as open and globally networked spaces operating in an intensive global competitive surrounding (a.o. Heidenreich 2006: 5). These external contacts of regional innovation systems are expected to adopt a broad variety of forms. Especially with regard to the transfer of knowledge, the actors in a regional innovation system find themselves in an intensive and permanent process of exchange with firms and R&D-institutions or other knowledge suppliers outside the region (Sternberg 2002: 147f.).

In this context, Tödtling and Trippl (2005), for example, differentiate two important dimensions of extra-regional contacts between regional innovation systems and their environment. First, from a firms' perspective, *complementary* contacts to firms outside the region are crucial in the face of global competition and accelerating technological change. These external contacts provide the firms with the access to technologies, knowledge and ideas, which they cannot access within the limited context of the region (Tödtling/Trippl 2005: 1206).

Moreover, in this connection Heidenreich (2006; 2011) refers to the role of *multinational companies* (MNCs) and their respective subsidiaries in regional innovation systems. These MNCs play an important role in embedding a region into global flows information, trade and investments and contribute decisively to the coordination of extra-regional contacts. MNCs, which can also be described as cross-border, inter-organizational networks (Ghoshal/Bartlett 1990: 603f.) or "arenas of collective learning" (Heidenreich 2006: 6), thus, establish cross-border transfers of competences, expertise and knowledge. With that, they contribute to the innovativeness of the respective regions, since regional institutions and companies can profit from the efforts of these multinational enterprises and their local subsidiaries in recruiting, procurement, research, development, and training. Consequently, the success of regional innovation systems does also depend on their ability to establish themselves as "neo-marshallian" nodes in national and global flows of information, communication, production and investments (Amin/Thrift 1992: 577).

The second dimension of external contacts is the one of *public interventions*, particularly by national and supranational political institutions. Although the distribution of competencies of these distinct levels can differ with varying degrees of regional political autonomy (Perry/May 2007: 1041f.; Koschatzky/Kroll 2007: 1116ff.; Koschatzky 2009: 2ff.), it is possible to identify a general pattern in the distribution of competencies among the three levels: While the regional level often assumes responsibility for lower and medium levels of education and training, incubation and innovation centers, technology transfer agencies and cluster policies, the national level has competencies for universities, specific R&D-institutions, as well as the funding for R&D and innovation. Finally, the supranational European level contributes to regional innovation systems with its structural funds, specific programs to promote innovation processes, as e.g. the RIS and RITTS Program, or the Framework Programs for R&D and technological development (Tödtling/Trippl 2005: 1207ff.). Regional governance of innovation and knowledge transfer (Perry/May 2007: 1041f.; Perry 2007: 1053ff.) has, thus, to be understood as a *multi-level process*: "Regional innovation systems are not determined by regional governance alone, but firmly integrated in multi-level governance structures." (Koschatzky/Kroll 2007: 1116ff.)

To sum up, external contacts of regional innovation systems to national and supranational institutions, the international scientific community, as well as to global flows of information, trade and investments via MNCs, are an important factor for the success of innovation processes. They also play a role for the analysis of regional structures and mechanisms of knowledge transfer in this study, since these national and supranational institutions among other things create programs and provide funding to promote innovation and knowledge transfer on the regional level. Furthermore, specific regional institutions help the actors to access and use the specific resources, competencies and knowledge of MNCs and national and international science networks, and contribute, therefore, to diffusing this global knowledge into the regional networks. The effects of these extra-regional contacts of regional innovation systems can now be supposed to work in two different directions with regard to the fit of regional institutional structures and mechanisms of knowledge transfer to their respective national frameworks: On the one hand, it could be assumed that since the national level has the responsibility for universities, specific R&D-institutions and for the funding of R&D and innovation etc., a predominant national institutional model could be distributed equally across all regional innovation systems of a national economy. On the other hand, differences in the ability of a region to establish extra-regional contacts and cooperations, the number of MNCs operating in a region and their role in connecting a region to global markets, as well as the specific division of labor between the regional, national and supranational level in structuring and promoting innovation and knowledge transfer can be supposed to be important explanatory factors for regional variation within national systems of innovation and production.

2.4.2.5 *Types of Regional Innovation Systems: Governance, Sectoral and Structural Dimensions of Regional Knoweldge Transfer*

2.4.2.5.1 *Governance Dimensions of Regional Knowledge Transfer*

Taking these general characteristics, different types of RIS have been distinguished in the debate on the concept, which can be assumed to be characterized by major differences in their institutional structures and mechanisms of knowledge transfer and which will be, thus, briefly discussed with regard to my research question. In this connection, one important distinction is made by Asheim (1998) and Cooke (1998; 2004), who suggest to differentiate between three types of RIS to reflect the conceptual variety and empirical diversity of the relationship linking the regional production structure to specific governance modes and institutional set-ups. The first type are *territorially embedded regional innovation systems*, in which the innovation activities of firms are based on localized learning processes stimulated by geographical, social and cultural proximity. These firms mainly employ synthetic knowledge and have only a few interactions with scientific organizations. The most prominent examples of this type of RIS are the networks of SMEs in industrial districts, as they can be found e.g. in Italy's Emilia-Romagna. These territorially embedded systems actively promote adaptive technological and organizational learning in a territorial context, by providing bottom-up, network-based support, e.g. through technology centers, innovation networks, or industry centers providing market research and intelligence services (Asheim/Gertler 2005: 300f.).

The second type of RIS is the *regionally networked innovation system*, in which firms and organizations are still embedded in a specific region and are characterized by localized and interactive learning processes. However, these systems have a more planned character by policy interventions, which intentionally try to strengthen the region's institutional infrastructure, innovation capacity and collaboration. The main advantage for the firms in this type of region is that these measures provide them with access to information and competence supplementing their locally derived competence. The firms may have to supplement their informal knowledge characterized by a high tacit component with expertise coming from more systematic and basic R&D, to carry out more radical innovations. To avoid situations of institutional lock-in, in the long run firms must also gain access to wider pools of both analytical and synthetic knowledge on a national and global basis. Hence, these policy measures are typically directed towards e.g. developing and strengthening vocational training organizations, regionally based R&D-institutes, technology transfer agencies and intermediary institutions, as chambers and business associations. RIS of this type can mostly be found in Germany, Austria and the Nordic Countries (Cooke 2004: 10ff.; Asheim/Gertler 2005: 301ff.).

The third important type of RIS is the *regionalized national innovation system*. In contrast to the two preceding types, in this type of RIS parts of the business sector and the institutional infrastructure are more functionally integrated into national and international systems. Thus, innovation activity takes place primarily in cooperations with exogenous actors and mainly incorporates the R&D-functions of universities, research institutes and firms. Collaboration between actors within this type of RIS comes closer to the linear model of innovation, since it is mostly realized through specific projects to develop more radical innovations based on formal analytical-scientific knowledge. Examples for this kind of RIS would be regions in which planned science parks and technological centers ("technopoles") predominate the innovative action. Here, the R&D-laboratories of large firms and governmental research institutes cluster together, sometimes in close proximity to universities and technical colleges, but, as has been shown e.g. by Asheim and Cooke (1998), in most cases have only limited links to local industry. Thus, although these planned innovative environments might help firms with a high level of internal resources and expertise that are situated in a weak regional cooperative environment, the lack of their regional embeddedness makes it questionable if science parks and technological centers might be able to promote innovation activities in the respective regions as a whole (Asheim/Gertler 2005: 302f.; Asheim/Cooke 1998).

With regard to my research question this distinction of three different (ideal-) types of RIS and governance models of knowledge transfer raises the question, if several of them can coexist within the institutional framework of the same national production model? My assumption is that to a certain degree in single regional innovation systems within national production models governance structures in the field of knowledge transfer can be created diverging from those at the national level. Among other things, I assume that not only regional regulatory capacities are strong enough to create regionally specific governance structures, but also that specific historical development paths and experiences as well as particular sectoral and structural conditions will lead to models of governance diverging from that at the national level. For the *detailed* analysis of the six regional case studies presented below the distinction of the three (ideal-) models of regional innovation systems, is, however, not fine-graded and precise enough. From my point of view this categorization misses to stronger differentiate regional governance structures by the six ideal types of social order identified in section 2.2.4, what would give the comparative anaylsis of regional innovation and knowledge transfer a conclusive analytical framework. I will address this weakness in the comparative analyses of regional innovation systems by developing an own heuristic governance model of knowledge transfer processes in innovation systems in section 2.5, allowing to reconstruct particular regional and national knowledge transfer systems, to analyze their predominating governance structures and modes of coordination, and to investigate their comparative institutional advantages as well as the degree of regional variation within distinct national innovation and production models.

2.4.2.5.2 Sectoral Dimensions of Regional Knowledge Transfer

Another major differentiation of regional innovation systems has been introduced by Cooke (2004), who distinguishes institutional regional innovation systems (IRIS) and entrepreneurial regional innovation systems (ERIS). The former is seen to describe the traditional regional innovation systems, as they can be found in the German regions as Baden-Württemberg and in the regions of the Nordic countries, whose leading industries basically rely on *synthetic* knowledge bases. As Asheim and Gertler (2005) point out, "its effectiveness flows from the positive effects of systemic relationships between the production structure and the knowledge infrastructure, embedded networking governance structures and supporting regulatory and institutional frameworks." (Asheim/Gertler 2005: 304) IRIS are particularly successful if technology generation and innovation occur in a path dependent rather than disruptive way, "where institutions have grown incrementally to meet needs in an evolving but well-understood sectoral innovation system." (Cooke 2003: 57)

In contrast to this, ERIS rely much less on the strong systemic elements of the IRIS type discussed above, and instead derive their competitiveness and dynamism from local venture capital, entrepreneurs, market demands, scientists, and incubators to foster innovation that draws primarily from an *analytical* knowledge base. Cooke (2004) refers to this as a "new economy innovation system" or a "venture capital driven innovation system", which is supposed to be more flexible and adjustable and, thus, less likely to end up in institutional or technological lock-in situations. On the other hand, ERIS lack the same long-term stability characterizing IRIS and provide less systemic support for historical technological trajectories, what has to be taken into account when referring to their long-term economic sustainability (Asheim/Gertler 2005: 304). Prominent examples for ERIS would be Silicon Valley, Route 128 in the USA and the East of England region in the UK.

This distinction of different types of RIS shows clearly that the specific institutional order can differ decisively between distinct regional innovation systems, and that these different institutional orders can promote quite different production models, sectoral specializations, modes of innovation and learning, and knowledge transfer strategies. In this, the RIS-concept shows important similarities to the VoC-approach (Hall/Soskice 2001a) discussed above, since it shares the assumption that institutional structure shapes firms' strategy. In this sense, for example the IRIS type or the regionally networked innovation system would fit best into the coordinated market economy model described by the VoC-approach, while the ERIS type would correspond more to the liberal market economy model. However, if these different types of RIS exist, and if they also imply different structures and mechanisms of knowledge transfer, then the question emerges, what their relation to their respective national institutional framework is, i.e., if within one particular national innovation and production system only one of these types of RIS can exist, or if different types of them can exist parallel to each other within the same national institutional context. Unfortunately, this question on the relation of regional innovation systems to their respective national institutional frameworks remains mainly unanswered in the literature so far, which can be considered as a central weakness of the concept. Also the answers to the question, if particular *adaptations* or *hybrid forms* of these different types of RIS can exist, remain considerably vague. This study has the aim to contribute to this debate by addressing these questions.

For that, it is necessary to point closer to the debate on the sectoral dimensions of innovation systems (Malerba 2004). One major reason for differences between sectoral systems are the different knowledge bases underlying distinct sectors: While, for example, some traditional industrial sectors like automotive or mechanical engineering rely strongly on synthetic stocks of knowledge that are characterized by a high degree of cumulativeness, others, as the young sectors of biotechnology and microelectronics base much more on analytical knowledge having a much less cumulative character (Coriat/Weinstein 2004: 342ff.; Malerba 2004: 9ff.; Casper/Soskice 2004: 349ff.). In this, firms in the former sectors should predominantly be found in ERIS, while firms the latter sectors should mostly operate in IRIS.

Basically, the question that emerges from these sectoral dimensions of innovation with regard to my research interest is if the institutional structure in specific national innovation systems and their respective regional innovation systems shapes firm's strategy in the field of knowledge transfer in all sectors operating within them in the same way? The affirmative position would mostly correspond to the authors of VoC, which claim that specific institutional infrastructures at the market economy level might attract firms operating in particular sectoral fields that can take advantage of this particular institutional support (Casper/Soskice 2004: 349ff.). Conversely, another reasonable position would be that specific sectors might create their own, particularly tailored institutional support, turning the strategy follows structure paradigm into a structure follows strategy view, as some authors of the sectoral innovation system approach do: "[…] sectoral systems may greatly differ with respect to their typical institutions. […] Sectoral institutions may emerge either as a result of deliberate planned decisions by firms or other organizations, or as the unpredicted consequence of agents' interaction." (Malerba 2004: 27)

While the VoC-position would imply in this regard a high degree of institutional coherence and a strong homogeneity of institutional structures and mechanisms of knowledge transfer across different sectors and regions, the sectoral systems perspective would imply a higher institutional variation across different sectors and regions within the same national production model (Casper/Soskice 2004: 362). However, also the sectoral systems view acknowledges the strong role exercised by national institutional frameworks: "National institutions may have major effects on sectoral systems, as, for example, the patent system, property rights or antitrust regulations. These effects may differ from sector to sector, as a consequence of the different features of these systems." (Malerba 2004: 27) In addition, also a certain sectoral specialization of national innovation and production models according to their dominant institutional structures is not neglected.

Another important question that arises from this sectoral systems debate is then, which structures and mechanisms exactly do coordinate processes of knowledge transfer in industrial sectors that—according to the VoC-approach—should not fit into the overall institutional production model? In concrete terms this would mean for example: Which institutional structures and mechanisms are predominantly coordinating knowledge transfer in the German biotechnology sector? According to the VoC-approach, firms in this sector with their radical mode of innovation, their low technological cumulativeness and high dependence on analytical knowledge bases should not find adequate institutional support for their economic and entrepreneurial activities within the framework of a CME like Germany and, thus, either not operate there or underperform their competitors in LMES. Nevertheless, the empirical evidence shows that there are considerable strengths in the biotechnology sector in Germany. This makes it necessary to ask, if knowledge transfer in this sector adapts to the structures and mechanisms of the overall coordinated model, or, if other, more liberal structures and mechanisms of knowledge transfer are created and used within the coordinated framework (Casper/Soskice 2004: 358f.). If the latter should be the case, this would imply that different sectoral conditions have to be considered as important explanatory factors also for regional variation within national models: Regional innovation systems within Germany with considerable strengths in biotechnology should then show structures and mechanisms of knowledge transfer diverging from those at the national level. The same would of course also hold true for the converse case of a traditional industrial sector as e.g. automotive within the institutional framework of a LME like the UK. As a consequence, in both cases a particularly strong piece of critique towards the VoC-framework would arise (Casper/Soskice 2004: 362).

For the pretended analysis of regional innovation systems in this study, therefore, a special focus will be put in all the six investigated regions on the two sectors of automotive and biotechnology representing one traditional industrial sector with a predominantly synthetic and highly cumulative knowledge base and an incremental mode of innovation, and one younger sector with a mainly analytical and lowly cumulative knowledge base and radical innovation processes, to "control" for sectoral specificities as explanatory factors for regional variation. On the other hand, this will allow for conducting a cross-regional and cross-national comparison of the predominant institutional structures and mechanisms of knowledge transfer in the two sectors with the aim to detect the respective variation, if it exists. The assumption is that the institutional structures and mechanisms in both sectors will considerably differ within one market economy model, according to their respective sectoral specialisms in terms of mode of innovation, type of knowledge base and degree of cumulativeness, but will show significant similarities across national and regional systems. Nevertheless, national institutional frameworks will still be assumed to exercise decisive influence on knowledge transfer processes in the different sectors, so that it will be the detailed analysis of the six regional case studies that is to show to which degree these national frameworks shape the structures and mechanisms of knowledge transfer in the two sectors within the six regions.

2.4.2.5.3 Metropolitan vs. Industrial Regions

When looking at distinct types of RIS, another differentiating dimension that will be addressed in this study is the one between metropolitan regions and industrial regions. It can be assumed that if there is a difference in the institutional structures and mechanisms of knowledge transfer in RIS within specific national institutional frameworks, it should be particularly strong and observable between these two quite distinct types of regions. Although all of the already mentioned characteristics of regions in general hold true for both types of regions, there might be specific features distinguishing them. These different characteristics of metropolitan and industrial regions will briefly be discussed in the following, pointing to possible consequences for deviating institutional structures and mechanisms of knowledge transfer.

The concept of *metropolitan regions* is a relatively new concept in regional planning that explicitly overcomes the view on urban agglomerations as pure metropolises representing the centers of a country's political, economic and cultural life. Rather, they should be seen as nodal points in an increasingly networked global knowledge economy and as important points of interest and coordination of internationally operating firms. Kujath (2005) defines in this connection metropolitan regions as strategic geographical places in which the command and control centers of transnationally acting firms with their respective supportive services structures are concentrated and from which their interlinked business networks are coordinated and controlled (Kujath 2005: 23ff.). It is particularly in metropolitan regions where national and global streams of information and knowledge agglomerate due to the presence of e.g. the headquarters of larger multinational companies, central financial places, umbrella organizations representing business and labor interests, or national or international governmental institutions and create "sticky places" for the concentration, recombination, application and transfer of knowledge (Allen 2000: 31; Kujath 2005: 34f.; Revilla Diez 2002: 64ff.; Brandt et al. 2008: 9ff.; Krätke 2009; Sassen 2001).

Metropolitan regions are characterized by a high density of population and represent prosperous service dominated economies (Krauss 2005: 110; Heidenreich 2003b: 323). This knowledge intensive service structures, complemented by a high concentration of higher education and R&D-infrastructures create and attract a pool of highly qualified workforce that can have positive externalities for the intensity and quality of knowledge transfer within these regions. Since the fluctuation of this highly qualified workforce within metropolitan regions, and between metropolitan regions and other national and international regions, can be assumed to be relatively high (Kujath 2005: 45; Audretsch 2003: 18), intra- and inter-regional knowledge transfer via this mechanism shall play an important role. This is supported by the *gateway*-function (NIW/ Nord/LB 2005: 100) that metropolitan regions exhibit within their respective national context, creating a privileged access to experts, knowledge and markets through an above average communicative and infrastructural connection and equipment e.g. with airports, media companies, trade fairs and exhibitions (Brandt et al. 2008: 11ff.; Schäfer et al. 2004: 15ff.; Fischer et al. 2001: 16f.).

In the present, these controlling-, learning- and gateway-functions of metropolitan regions can only in a very few cases be fully exercised within the political borders of one single city. Therefore, metropolitan regions not only consist of the respective national capital cities or other major urban agglomerations, but also of their surrounding regional environment. With that, the concept of metropolitan region explicitly highlights the attempt to connect major capital cities to their respective regional surroundings, creating favorable conditions for economic development in general and innovation and knowledge transfer processes in particular. This leads to an intra-regional division of labor in which the constituting subspaces might focus on specific sub-functions. This division of labor can be supposed to be particularly pronounced in polycentric metropolitan regions as are e.g. the German metropolitan regions Hannover-Braunschweig-Göttingen-Wolfsburg or the Rhine-Neckar-Triangle, in which the respective subspaces have distinct and complementary strengths with regard to the different metropolitan functions (NIW/Nord/LB 2005: 101f.; Blotevogel 2002: 345ff.; Brandt et al. 2008; Schäfer et al. 2004).

In contrast to that, *established industrial regions* are characterized by a considerably lower population density, a relatively low share of the service sector, and a high share of industrial production dominating the economic structure (Heidenreich 2003b: 323). These regions experienced a specific historical path of industrial development, including multiple periods of economic up- and downturns and technological lock-ins, adaptations and change. These historical industrial trajectories are to a certain degree path-dependent, and are accompanied with the creation of specific institutional frameworks specifically tailored for the needs and the support of the particular industrial focus (Whitford 2001). Nevertheless, also these regions experienced a process of tertiarization of industrial production in the last decades, leading to rising shares of industry-related services (Krauss 2005: 110ff.). Economic activities in general as well as innovation and knowledge transfer processes are strongly influenced among other things by the experience of these historical industrial trajectories, established practices, entrepreneurial mentalities of a "commercial community", and historically grown and formalized networks (Krauss 2005: 110ff.).

The business structure of traditional industrial regions is dominated by small and medium sized enterprises, although larger multinational enterprises might be present as well, often exercising the role of focal enterprises standing at the head of dense networks of customers and suppliers. A commonly shared socio-cultural identity creates an environment in which trust based relationships and cooperations between economic actors occur frequently and intensively, often stabilized by specific regional institutions and regional governments that play an active role in promoting cooperation and innovation activities. Even though also traditional industrial regions maintain intensive inter-regional relations, their focus is much more intra-regional in comparison to metropolitan regions. Examples for this type of region are the German regions of Baden-Württemberg and North Rhine-Westphalia, and the Italian region Emilia-Romagna (Krauss 2005: 110ff.; Heidenreich 2003b: 323; Koschatzky 2001a: 186ff.).

With regard to my research questions, one could then expect that if there is a match between national production system and regional innovation systems in the organization of knowledge transfer, it should especially appear in the established and historically grown industrial regions. One reason for this is the following: According to a neoinstitutionalist perspective and looking at regional innovation systems as specific organizational fields (see 2.2.3), one might have to expect that in the historical path of development of institutional structures in the field of innovation and knowledge transfer *processes of institutional and organizational isomorphism* create regional structures being more homogeneous to their national institutional environment. These regions might have paralleled the economic, structural and technological up- and downturns of their respective national production system and might, thus, have developed similar institutional structures and mechanisms in the fields of innovation and knowledge transfer. Consequently, if the case studies in these established regions would show knowledge transfer structures deviating from their national model, this would create a particularly strong piece of critique towards the VoC-approach and its claim of institutional coherence of national production models.

Conversely, the newly created institutional structures and mechanisms of supporting innovation and knowledge transfer in metropolitan regions can be supposed to be less influenced by these processes of institutional and organizational isomorphism. Since in most cases these structures are only partly historically grown and are, above all, the result of concrete political decisions and *planning*, it can be assumed that they are more likely to differ from their respective national institutional model. Moreover, and pointing to the debate on the convergence of market economies towards the liberal Anglo-Saxon model and its global spread (e.g. Hall/Thelen 2009; Streeck 2009), one should expect that it is especially the metropolitan regions in CMEs and MMEs that deviate in this sense from their respective national models, since the diffusion of globally dominant models should particularly and first be visible at the nodal points of global information and communication networks. Here, if the case studies in these metropolitan regions should show institutional structures and mechanisms of knowledge transfer matching to those on the national level, this would constitute a particularly strong layer of support for the VoC-approach and its assumption of institutional coherence.

2.4.2.6 Criticizing Innovation System Approaches

Substantial points of critique towards the concept of innovation systems in general have been formulated in the academic discussion on it, which will briefly be summarized in this section, deriving from these conceptual shortcomings evidence for the pretended analysis in this study. On the one hand, many scholars point out that the systems of innovation approach is "conceptually diffuse" (Edquist 1997: 26f.; Markusen 1999: 869ff.). One reason for this is that different definitions of the systems of innovation approach are used by different authors. Moreover, these definitions remain quite vague with regard to the single elements that should be included into a system of innovation, and they do not define the limits of innovation systems in an operational way. Consequently, an exact demarcation between a system of innovation and its environment is made more difficult, if not impossible. However, Edquist (1997) emphasizes that the conceptual weakness of this approach can in most cases be compensated by trying to identify the central elements of a *specific* innovation system and to focus on the relations between them (Edquist 1997: 26f.; Edquist 2005), what also appears to be useful with regard to the issues addressed in this study. The innovation system approach could then more precisely be described as a "heuristic model" (Iammarino 2005; Koschatzky/Kroll 2007: 1117), for which a certain degree of conceptual ambiguity or pluralism is natural and in which it cannot precisely be defined which elements are vital to the conceptual core of the approach, what their exact content should be and in which logical relation they stand to one another. Turned positively, the conceptual ambiguity of the systems of innovation approach could also be recognized as strength, since it allows for considerable openness and flexibility, giving room for competing perspectives and solutions (Lundvall 2007: 98f.; Edquist 1997: 27).

On the other hand, the innovation systems approach can certainly not be considered a formal theory (Nelson 2008: 1; Lundvall 2007: 98f.), since it does not provide convincing propositions on established and stable causal relations between different variables. Rather, it provides a conceptual basis for the formulation of conjectures on the functioning of different factors influencing innovation processes, like e.g. learning or institutions. As many other institutionalist approaches, also the systems of innovation approach is characterized by an almost uninhibited formulation of these conjectures. However, these hypotheses must, yet, be confronted with empirical evidence. Therefore, referring to innovation systems, Fischer and

Fröhlich (2001) rather speak of a *conceptual approach*, which, in spite of all the critique mentioned here, appears to have a high potential in deducing quite useful conclusions about innovation processes, their determinants and their consequences (Fischer/Fröhlich 2001: 1; Lundvall 2007: 98f.; Edquist 1997: 28f.).

Apart from its conceptual diffuseness, in most of the innovation systems literature the relation between national and regional innovation systems, as already mentioned, remains unclear. In particular, the important question to which degree the specific institutional order of RIS can differ from their respective national institutional framework is not sufficiently addressed so far (Le Galès/Voelzkow 2001; Crouch et al. 2001; Crouch et al. 2004). Although the implicit assumption in most contributions is that national systems of innovation and production are institutionally coherent and homogeneous units, this assumption appears to be at least highly questionable as could be shown from the discussion in the previous sections and needs further empirical investigation (Voelzkow 2007; Crouch/ Voelzkow 2009; Crouch et al. 2009), to what this study intends to contribute.

Moreover, and confirming once more the conceptual weakness of innovation systems approaches, processes of learning and knowledge transfer are "only" regarded as central explaining factors and important preconditions for successful innovation activities, that might emanate from certain national or regional advantages, might them be geographical, structural, institutional or cultural (Audretsch/Stephan 1996: 641ff.; Audretsch/ Feldman 1996: 630ff.; Jaffe 1989: 957ff.). What is still missing in the process of theory-building on innovation systems is a clear analytical framework for defining the specific roles single institutional actors or particular institutional configurations play for learning and knowledge transfer, and, therefore, to describe their concrete impact on innovation processes (Audretsch/Lehmann 2005; Acs et al. 1999; Ortiz 2008; Edquist 2005). As Audretsch and Lehmann (2005) formulate, authors like "[...] Jaffe (1989), Audretsch and Feldman (1996) and Audretsch and Stephan (1996) provided evidence concerning the spatial dimension of knowledge spillovers. In particular their findings suggested that knowledge spillovers are geographically bounded and localized within spatial proximity to the knowledge source. None of these studies, however, identified the actual mechanisms which actually transmit the knowledge spillover; rather, the spillovers were implicitly assumed to automatically exist (or fall like Manna from heaven), but only within a geographically bounded spatial area." (Audretsch/Lehmann

2005: 1192) Based on this conceptual shortcoming concerning the concrete structures and mechanisms of knowledge transfer in innovation systems, the study at hand derives the need to not only suggest a more systematic "heuristic model" of knowledge transfer systems (see section 2.5), but also to investigate these knowledge transfer systems in distinct national and regional contexts, intending to reconstruct specific national types of knowledge transfer systems and to empirically investigate the determinants of regional variation within these national systems from a *comparative* perspective.

Overall, the questions that arise from these shortcomings in the debate on innovation systems for this study are: *Which are the concrete structures and mechanisms that play a role for knowledge transfer and learning processes in concrete national or regional innovation systems? Are there differences in these structures and mechanisms between different regional innovation systems within particular national innovation systems? And how can these differences, if they exist, be explained?* It is the aim of this study to address these questions and to contribute to closing these gaps in the discussion on innovation systems so far.

2.4.2.7 Concluding Remarks

In sum, this chapter has shown that innovation, learning and knowledge transfer are social processes that can only be understood on the basis of their systemic character. In this, the region appears to be a central spatial entity to provide the actors involved in these processes with the social, cultural and institutional environment they need to learn, innovate and transfer knowledge successfully. Despite of its conceptual weaknesses, thus, the RIS-concept seems to be a useful approach to address the research questions outlined above. It not only allows for taking a differentiated look at the specific actors and mechanisms that shape processes of knowledge transfer in particular regions, but also at the specific regional governance models and institutional frameworks coordinating them. Moreover, the RIS-concept also makes it possible to investigate the relation of particular regions to their respective national innovation and production regimes, since it shares many of the same institutionalist assumptions that characterize the approaches focusing on these national frameworks.

In addition, from the discussion in this chapter several possible explanatory factors for regional variation within national systems of innovation and production have been identified. They suggest that in opposition to the assumption of institutional coherence and complementarity of national systems of innovation and production made by the VoC-approach, there are good arguments to assume that the institutional structures and mechanisms of knowledge transfer in RIS do vary within particular national institutional frameworks. For this purpose, in the next chapter a heuristic model of knowledge transfer systems will be introduced, which intends to capture the basic mechanisms of knowledge transfer in the context of innovation discussed in the literature in a systematic way by ascribing them to the six ideal types of social order discussed in section 2.4.2.5.1. This heuristic model will then, in a next step, be confronted with the VoC-approach to derive theoretical hypotheses on how knowledge transfer systems should predominantly be shaped in the different national market economy models. In a further step, these theoretical assumptions will then be confronted with the empirical evidence of the qualitative case studies in the investigated regional innovation systems.

2.5 A Heuristic Model of Regional Knowledge Transfer Systems

Building on this theoretical discussion on the nature of economically useful knowledge, its systemic character and its institutional embeddedness in national and regional systems of production and innovation, I suggest to apply a heuristic model of knowledge transfer systems in the context of innovation, capturing the basic structures and mechanisms of knowledge transfer in the different institutional dimensions of governance. I explicitly want to address with this model the theoretical gap in the literature in precisely describing the role of knowledge transfer processes in innovation systems, in naming and systemizing the concrete mechanisms clearly, and in reconstructing the respective structures through which knowledge is transferred in innovation processes. The aim is to present a "toolbox" of potentially observable mechanisms and institutional structures of knowledge transfer in innovation systems, which can serve as basis for the (comparartive) empirical analysis of regional case studies and through which the respective regional knowledge transfer systems can be reconstructed in their basic elements.

Fig. 1: A Heuristic Governance Model of Knowledge Transfer in Innovation Systems

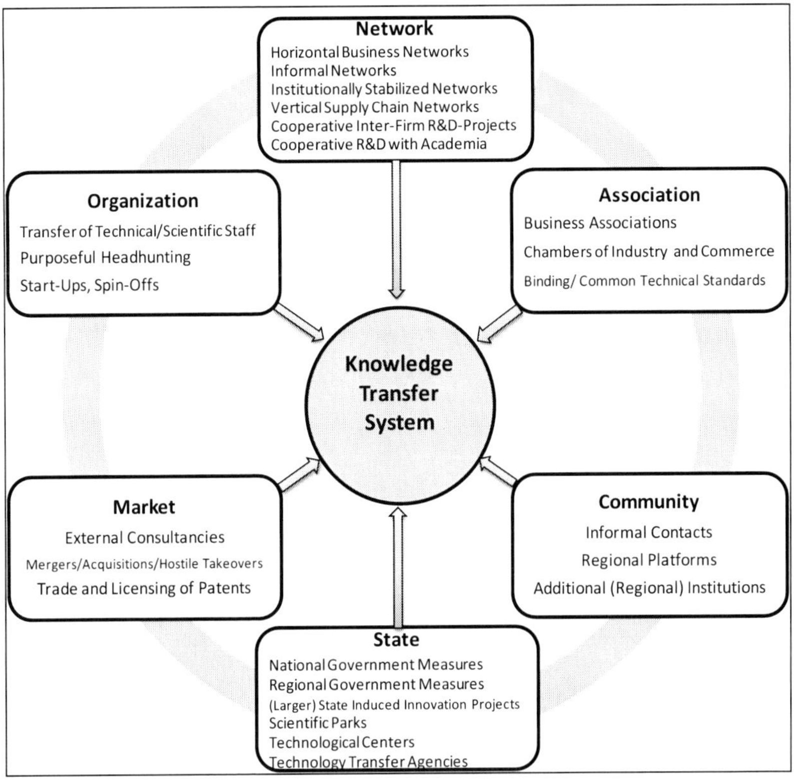

Source: Own Illustration

For this purpose, with Figure 2.1 I will introduce a heuristic model which uses the six dimensions of institutional governance presented above and assigns them concrete mechanisms of knowledge transfer which are predominantly shaped by one of the six ideal-typical forms of institutions (network, association, organization, state, market, community). Although this ascription of knowledge transfer mechanisms to institutional ideal types cannot provide absolutely clear divisions between the single dimensions, and even if the mechanisms included into the model might by far not be exhaustive, it still captures the majority and most important of knowledge transfer mechanisms usually discussed in the literature on innovation systems and national production systems (Figure 1).

Networks

In this, the *network dimension* has been defined as describing stable relations and interactions of individuals or institutional actors without losing their autonomy, and in which the coordination of economic transactions is neither realized through simple market based exchange processes, nor through instruction-based hierarchies, but rather trough trust in the fact that an own input into the network will be compensated at a later point with a more or less equivalent *quid pro quo*. It now appears plausible to ascribe this dimension *horizontal business networks* as one mechanism through which knowledge transfer in the context of innovation can take place. These networks can include competing as well as non-competing firms, might differ in their intensity and frequency of interaction, and might imply substantial flows of technical and scientific knowledge as well as low substance exchange of knowledge about the framework conditions of business activities. Furthermore, it should be distinguished between informal networks and institutionally stabilized networks in this connection, not least because the literature on innovation systems strongly emphasizes the informal component of knowledge transfers. *Informal networks* will then be defined as such networks without formalized membership and interaction, with a high contingency and fluctuation of participating actors, as well as with strong personal, cultural and trust based components as basis of interaction. *Institutionally stabilized networks*, on the other hand, will be those networks in which membership is formalized, interaction is regulated and stabilized according to established practices, and in which cooperation and knowledge transfer base on fixed contracts defining its goals and the use of the respective outcomes. Fourth, also *vertical networks* between firms and their customers and suppliers along specific value added chains fall under this dimension. In these vertical networks knowledge transfer takes place in various directions, from the suppliers to the customers or OEMs, e.g. through the improvement of components or the forwarding of practical experiences, and vice versa from the OEMs to the suppliers, e.g. through product guidelines, specifications, technical instructions and the exchange of research results, or among the suppliers themselves, e.g. through collaborative product development and improvement. The fifth network based channel of knowledge transfer is collaborative *inter-firm R&D-activities*, in which firms cooperate in concrete R&D-projects to concentrate their resources, use synergy effects and bundle their expertise to achieve innovation. Finally, the sixth mechanism is *cooperative R&D-projects between firms and scientific institutions* like research institutes and universities. Through this mechanism firms can access the expert knowledge and breaking point research results at the scientific institutions and, vice versa, the scientific sector can apply its research in a business context and access funding from the business sector. These cooperations would especially include the joint use of research facilities, laboratories, equipment and instruments as well as contract research.

Association

Concerning the *association based dimension* of governance, it has been stated that in this model corporatist actors like collective interest representation bodies coordinate their internal as well as their external interactions through the concertation of interests, assuming considerable regulatory potential, so that they can enforce taken decisions through effective organization structures, internally as well as in external collective agreements. Both *business- employers- and trade associations*, as well as *chambers of industry and commerce* will therefore be ascribed to the association based mechanisms of knowledge transfer, in the sense that they create enhanced levels of trust among their members and therefore foster intensified interaction and transfer activities. In addition, they also provide services like information, consulting, meetings, conferences and expert talks etc. to their members and other actors which function as channels of knowledge transfer. Third, associations and chambers are also active in elaborating and implementing *common and binding technical standards*. These technical standards are to be seen as mechanisms of knowledge transfer in the sense that they represent a common point of reference in the development of innovative technologies and provide a joint basis for interaction and cooperation easing knowledge transfer processes.

Organization

The third dimension of governance is the *organization*, in which the interactions between the actors are coordinated through hierarchies that are deduced from specific formalized as well as informal rights and duties. In this model of social order, knowledge transfer is realized through the creation and adaptation of organizational structures as well as the vertical integration of knowledge into the hierarchy. Therefore, the *transfer of highly qualified technical and scientific staff* from one organization to another and especially from academia to the business sector will be defined as one major organization based mechanism of knowledge transfer. Also *purposeful head-hunting* activities by firms to recruit highly qualified personnel from other firms or research institutes can be counted to this category. Finally, *business start-ups* will be ascribed to the organization based mechanisms of knowledge transfer, since they include transfers of knowledge either from the private sector or from the scientific world into a new organizational structure to bring new business ideas, technologies, research results or expertise to the markets. Especially academic spin-offs by scientific staff from universities and research institutes will receive special attention in this connection.

State

Fourth, the *state* has been defined to rely on hierarchical control to coordinate the interactions of the actors using its monopoly of power allowing him to implement decisions, if necessary by force. It will, therefore, be assumed that both *national* and *regional government policy measures* in the fields of innovation and knowledge transfer are central mechanisms of initiating, promoting, assessing, funding and organizing knowledge transfer processes. Also cooperations between private companies and public institutions, like national or regional administrative units, public initiatives and regional institutions for the promotion of economic development and trade, as well as the provision of economic actors with specific information, consulting and other services would be elements of these two mechanisms. The third mechanism would then be *(larger) state-induced innovation projects*, in which the state is the main initiator, funder, driver and sometimes also major customer of innovative activities, as it would be the case in sectors like aerospace and defense, for example. Through the cooperation of firms of different sectors and scientific institutions within the framework of these state-induced projects important knowledge transfer activities can be expected. Fourth and fifth, *scientific parks* and *technological centers* will be counted to the state induced mechanisms of knowledge transfer. Habitually, they represent publicly planned, organized and professionally managed physical areas and collections of facilities in which actors from the business sector and the scientific world can interact and collaborate intensively in close spatial proximity. In this, scientific parks mostly have a stronger focus on breaking point scientific research and advanced technology development, and are in most cases closely connected to universities or research institutes and are fully or at least partly funded by the public. Technological centers, on the other hand, are places in which predominantly technology oriented firms collaborate in joint R&D-activities, sometimes complemented by scientific actors and making use of the respective synergy effects. They are more business oriented, larger parts of funding stem from the business sector and their focus is stronger oriented towards advanced manufacturing, product development and applied research. Finally and sixth, the state also maintains *technology transfer agencies* at the public universities and research institutes which can be regarded as important channels of knowledge transfer. These agencies can be seen as central actors in commercializing the IPRs of the public scientific institutions, as well as in being an access point to the universities from outside and especially for firms.

Market

Furthermore, the *market* has been defined as a fifth type of social order in which interactions of rational and utility maximizing actors are coordinated through atomistic competition in the exchange of private goods and services by individuals and collectivities (particularly firms) being primarily concerned with profit. As market based mechanisms of knowledge transfer one can, therefore, define mechanisms through which knowledge is bought and sold by actors holding separate property rights in different stocks of knowledge on the basis of legally enforceable contractual exchanges. To this category of knowledge transfer mechanisms would, first, belong *external consultancies*, which can either sell their expertise in specific innovation related areas, or function as service providers in initiating, organizing, managing and assessing knowledge transfer activities. In addition, it is also possible that the stocks of knowledge incorporated in a whole firm are bought by another firm through *mergers, acquisitions or hostile takeovers*. In this, the knowledge incorporated in the absorbed firm's organizational structures, its staff, its technology, as well as in its IPRs is acquired. Finally, knowledge can also be transferred through *buying, selling and licensing of patents* and other forms of IPRs. These trading activities of IPRs must be differentiated clearly from patenting activities, which have the aim to protect one's own knowledge, while the trade or the licensing of such IPRs implies a transfer of knowledge.

Community

Lastly, in the *community* model of social order the central mechanism of coordination between the actors is informal, voluntary, and sometimes spontaneous solidarity among the members of a social unit, not being motivated by profit, but by the appreciation of the other members of the community or the desire to belong to a group. Therefore, to this model one should ascribe mechanisms of knowledge transfer which base on a strong and commonly shared historical, cultural and social identity as major driving force. *Informal contacts* of any kind would be one of these knowledge transfer mechanisms, through which the involved actors can exchange knowledge on a low level, flexible, unplanned and non-contractual way, being supported by commonly shared languages, experiences, practices and worldviews. Another would be *regional platforms*, on which actors from different fields can come together, interact and exchange knowledge and ideas, as well as establish more stable and formalized interaction. These platforms can be provided by a broad variety of players, as could be governments, administrative bodies, chambers, associations, clubs, trade fair organizers, regional and cultural initiatives, firms or individuals. Finally, and as a kind of residual category, I suggest to add to the community dimension the mechanism of *additional (regional) institutions*, which might be used as important knowledge transfer mechanisms and might not been captured so far. While the other presented mechanisms of knowledge transfer might be used to a certain degree in all empirically investigable regional innovation systems, this might give the opportunity to also include mechanisms into the analysis which are specific for single empirical regional cases and might only be revealed in the course of an intensive study of the respective regions.

In sum, this heuristic model offers a spectrum of six institutional dimensions of knowledge transfer in innovation systems, to which in total 24 mechanisms of knowledge transfer have been ascribed. This model will be the basis for the qualitative empirical analysis of six regional case studies in three different national market economy contexts, in which the qualitative relevance of these six dimensions and 24 mechanisms will be investigated to possibly detect regional variance of knowledge transfer systems within national institutional frameworks. Therefore, in a next step, this heuristic model of knowledge transfer systems will be brought together with the VoC-approach to formulate theoretical hypotheses on how this model should be coined in the three investigated ideal types of national market economies (liberal, coordinated and Mediterranean). These theoretical hypotheses on the national systems will then be confronted with the empirical results of the qualitative regional case studies searching for deviations from these assumed national models and for explanations for this possible regional variance.

3. Hypothesizing Structures and Mechanisms of Knowledge Transfer in Varieties of Capitalism

If the assumptions of institutional complementarity and coherence in national innovation and production systems made by the VoC- and innovation systems literature were correct, the structures and mechanisms of knowledge transfer in RIS should mostly correspond to their respective national institutional framework. To test these assumptions and show to which degree and why the coordination of knowledge transfer at the regional level deviates from the one at the national level, the approach followed in this study will be to confront the heuristic model of knowledge transfer developed in the previous section with the VoC-approach and, in a first step, to derive hypotheses on how knowledge transfer should be coordinated in the distinct types of national market economy models according to the VoC-literature. These hypotheses will then be used in the empirical case studies that will be discussed below, to show to which degree the structures and mechanisms of knowledge transfer at the regional level correspond to the respectively predicted national models of knowledge transfer.

The aim of this research strategy is twofold: On the one hand, an *abstract theoretical model* of specific types of "knowledge transfer subsystems" in liberal, coordinated and Mediterranean market economy models will be derived from the VoC-approach. This theoretical model will be the basis for investigating the regional variation of institutional structures and mechanisms of knowledge transfer in RIS, and, thus, addressing the overemphasis of institutional coherence in the VoC-literature as well as the presumed underestimation of regional variance within national innovation and production models discussed above. On the other hand, this theoretical model also addresses the need for further research in the literature on RIS in specifying which institutional structures and mechanisms exactly are important to determine processes of knowledge transfer in specific regional settings. It is one important contribution of this study's empirical analyses to identify at least the main institutional structures and mechanisms shaping regional systems of knowledge transfer and the central factors explaining their intra-regional variation.

3.1 Hypothesized Structures and Mechanisms in CMEs

For the CMEs, like for example Germany, the VoC-approach points out that, referring to the *financial system* or the *market for corporate governance*, the financing of companies not predominantly relies on balance-sheet criteria, but on access to private or inside information about the operation of companies. To secure this sharing of reliable information about the performance of firms, dense networks are established linking the managers and technical personnel within a company to their counterparts in other firms. This reliability is secured in a variety of ways. One might, for example, be sharing information with third parties, like business associations or chambers of industry and commerce, whose officials have an intimate knowledge of the industry. These institutions are in a position to exercise monitoring functions and to sanction firms for misleading them (Jackson/Deeg 2006: 15f.).

Another key factor in this is reputation, since in a system, where membership in a network is of continuing value, participants will avoid providing false information, given that they could lose their reputation and their access to the network. Moreover, investors in CMEs can get information about the operation of firms via the close relationships that these firms cultivate with major suppliers and clients, the knowledge secured from extensive cross-shareholding and, most important, from joint membership in active industry associations and chambers that collect information about firms in the course of coordinating technology transfer, vocational training and standard-setting. Furthermore, a big number of firms are engaged closely in joint research, product development etc. with other firms. These structures of corporate governance also discourage hostile takeovers, mergers and acquisitions as mechanisms of knowledge transfer, supported by tax provisions, securities regulations and networks of cross-shareholding (Hall/Soskice 2001b: 23f.; Hall/Gingerich 2009: 453). Thus, based on these considerations on the institutional sphere of the market for corporate governance, it is to be expected that knowledge transfer in CMEs is organized via the participation and cooperation in dense horizontal and vertical business networks and joint membership in business associations and chambers.

This structure of the financial market in many CMEs and in particular in Germany also impedes the intensive use of business start-ups and the creation of spin-off companies by academic staff as mechanisms of knowledge transfer. One important factor accounting responsible for this is a lack of companies providing seed- and venture capital, as well as business angels. The German venture capital market is dominated by business investment companies of banks, savings banks and insurance companies, which are mostly interested in capital gains and only invest little or no capital in start-up firms. Moreover, banks can lose their reputation in the event that a business they finance fails, which can have a negative impact on their core activities of saving, lending and issuing bonds and makes them restrictive against financing new businesses. However, in most cases they also not possess the necessary technological knowledge to adequately assess the risks of investing in start-up firms. In addition, legal restrictions concerning bankruptcy and liability also play a role: Bankruptcy is closely connected to the securities offered to and demanded by the banks, meaning that debtors are often made personally liable. Consequently, one failure in creating a business almost always means the end of all entrepreneurial activities. The frequently stated risk-averse mentality of German economic actors can be connected directly to these institutional structures (Abramson et al. 1997: 258ff.).

Second, also the *internal structure* of most firms in CMEs fosters this system of networking and, thus, knowledge transfer. In their major decisions, managers in CMEs depend on agreement from supervisory boards, and from other managers as well as major customers and suppliers (Hall/Gingerich 2009: 453; Hall/Soskice2001b: 24). This structure of consensual decision-making and network monitoring encourages the sharing of information and is expected to be an important feature in the organization of knowledge transfer in CMEs. In addition, it makes the use of external service providers, like consultancies, to manage and organize knowledge transfer processes, unlikely.

Third, because of an extensive use of long-term labor-contracts, firms in CMEs cannot rely as heavily on the movement of scientific and engineering personnel across companies to achieve knowledge and technology transfer. Instead, they cultivate intense *inter-company relations* facilitating the diffusion of technology across the economy. In the case of Germany, e.g. these relations are mediated by a number of institutions: "Business associations promote the diffusion of new technologies by working with public officials to determine where firm competencies can be improved and orchestrating publicly subsidized programs to do so. The access to private information about the sector that these associations enjoy helps them ensure that the design of the programs is effective for these purposes." (Hall/Soskice 2001b: 26) Associations exercising this function are e.g. employers and industry associations, but also chambers of industry and commerce and the chambers of crafts. They actively promote the exchange of knowledge among economic actors through innovation-oriented consultancy, publications, journals, meetings and informal networks (Abramson et al. 1997: 244).

Moreover, this system of strong industry associations is complemented by a *system of contract law* encouraging relational contracting among companies and promoting this type of knowledge and technology transfer. Since the contingencies always arising in such cooperations would make tightly written and formal contracts inadequate, in many cases open-ended clauses are formulated, but with the explicit condition that these have to be grounded on the prevailing standards of the relevant industry association. In this sense, strong business associations competent to promulgate standards and resolving conflicts are a decisive precondition for this system of contract law promoting relational contracting (Hall/Soskice 2001b: 26f.).

In addition, large part of industrial research is financed and conducted jointly by companies, often in collaboration with quasi-public research institutes, as e.g. institutes of the *Max-Planck-* and *Fraunhofer*-Societies, and the *Helmholtz Association*. In this, formal industrial R&D-consortia are one important form, in which the large number of industrial research associations and their umbrella organization, the *Federation of Industrial research Associations* (AiF), coordinate the realization of selected research projects being of interest for a larger group of their member companies (Koschatzky/Hemer 2009: 196ff.). Another form of cooperative R&D would be joint R&D-activities of firms with their major customers and suppliers along a specific value added chain (Abramson et al. 1997: 27ff.). In this, the common technical standards fostered by industry associations help to diffuse new technologies and knowledge, and they contribute to creating a common base of knowledge that facilitates collaboration among personnel from multiple firms, as do the industry-specific skills fostered by German training schemes (Hall/Soskice 2001b: 26; Jackson/Deeg 2006: 17f.). Finally, another form of collaboration would be contract research at universities and research institutes financed by firms, but, as in most other industrial countries as well, on a moderate level. Thus, third-party funds from the business sector represented 14% of the total German R&D-expenditures in 2007 (Polt et al. 2010: 116ff.).

Since the universities in Germany mostly do not see knowledge and technology transfer as one of their main missions, but focus on basic research and education, university-industry links are traditionally less developed. The most important channel of knowledge transfer from universities to other sectors of the economy is, thus, human capital, that is, well educated and learning-skilled science and engineering graduates. Other forms of transfer, if they occur, are mediated by intermediate technology transfer agencies at the universities, while direct contacts between scientists and firms are rare. Flows of staff between universities and the business sector are rare, especially because conflicting incentive structures do not encourage individuals to opt for career trajectories that include a substantial amount of movement among the two worlds (Wentland et al. 2011). In particular the civil-service status of German professors and the respective public employment regulations guaranteeing a higher level of job security as in industry and the fact that it is not possible to transfer supplementary public pension entitlements to private employment can be supposed to work as major barriers in this regard. At the same time, however, especially German technical universities have a long-standing tradition of also employing as professors high-level researchers from industry. Since labor markets are not very competitive and are mainly characterized by long-term labor-contracts, firms only exceptionally invest in purposeful head-hunting to hire qualified staff from other companies or scientific institutions (Polt et al. 2010: 91ff.; Abramson 1997: 16ff.; Wentland et al. 2011).

The role of buying and selling patents and licenses as more market oriented mechanisms can be assumed to be less important for knowledge transfer in CMEs in comparison to the association and network-oriented mechanisms presented above (Abramson 1997: 32ff.). In 2002, the German intellectual property rights regime has been reformed, taking away from the individual professor or inventor the privilege to commercialize his/her intellectual property him/herself and giving it to the respective employing institution, that can then decide if it wants to commercialize the intellectual property itself or leaves this to the professor or the inventor. As a consequence, most German universities created patent offices, mostly connected to the technology transfer agencies, to manage this commercialization of patents and licenses. However, since German universities and also other economic actors lack a tradition and experience of commercializing intellectual property, it is to be expected that these mechanisms only play a minor role for transferring knowledge in comparison to other mechanisms, which is also stated by actual studies on the German innovation system (Polt et al. 2010: 120f.).

Since in CMEs the state is defined as *enabling*, national and regional governments also exercise this enabling role in the fields of innovation and knowledge transfer. With creating specific national or regional public programs for promoting innovation and knowledge transfer together with the economic actors they initiate and facilitate cooperation in particular technological and industrial sectors and promote respective bottom-up activities. Although the state leaves the protagonism in these programs to the economic actors and intermediate institutions as, e.g., chambers of industry and commerce, fostering the creation of R&D-networks or cooperations between firms and R&D-institutions, in many cases the state contributes to their success with a certain amount of initial funding. Another important activity exercised by national and regional governments in this regard is the creation of national and, in particular, regional platforms, in which the actors can come together, interact, communicate and exchange knowledge, both in a formal and informal way (Schmidt 2006; Polt et al. 2010; Rammer/Weißenfeld 2008).

To summarize, for the CMEs and thus for the case of Germany, it has to be tested if knowledge transfer in the investigated regions will be organized according to the institutional features of the national level presented above. If this would be the case, consequently, cooperation in dense horizontal and vertical business networks including customers and suppliers, the use of intermediate institutions, like for example technology transfer agencies, chambers of industry and commerce, or business associations, joint R&D-activities, consensual decision making processes, common technical standards fostered by industry associations and a system of contract law promoting relational contracting are supposed to be the predominant characteristics of knowledge transfer in the regional innovation systems of CMEs. In Table 1 these mechanisms of knowledge transfer are summarized and categorized according to the heuristic model of knowledge transfer systems presented before, showing the specific occurrence of this heuristic model hypothesized for the German case.

Table 1: Hypothesized Mechanisms of Knowledge Transfer in Germany

MECHANISMS	GERMANY
NETWORK BASED MECHANISMS	
business networks	+
informal networks	+
institutionally stabilized networks	O
customers and suppliers	+
cooperative inter-firm R&D-projects	+
coop. R&D-projects with scientific institutions and universities	O
ASSOCIATION BASED MECHANISMS	
employers-, business-, and trade associations	+
chambers of industry and commerce	+
binding and common technical standards	+
ORGANIZATION BASED MECHANISMS	
transfer of technical and scientific staff	–
purposeful headhunting	–
business start-ups	–
STATE INDUCED MECHANISMS	
national government measures	O
regional government measures	O
(larger) state-induced innovation projects	–
scientific parks	O
technological centers	O
technology transfer agencies/companies	+
MARKET BASED MECHANISMS	
external consultancies	–
mergers and acquisitions, hostile takeovers	–
buying, selling and licensing of patents	–
COMMUNITY BASED MECHANISMS	
informal contacts	+
regional platforms	+
additional (regional) institutions	+

+ *Hypothesized as crucial mechanism of knowledge transfer*
O *Hypothesized as important but not decisive mechanism*
– *Hypothesized as not important at all*

3.2 Hypothesized Structures and Mechanisms in LMEs

Contrasting to that, in LMEs *markets for corporate governance* rely more on the current profitability and the actual price of their shares on equity markets. The financing of a firm heavily depends on its evaluation in the equity markets basing on publically available information, fostering short term entrepreneurial horizons and profitability oriented firm strategies. Closely linked business networks like in CMEs in which investors could access inside information on the performance of a firm do, by contrast, usually not play a central role, and networks in general should not be important mechanisms of knowledge transfer in LMEs. The regulation of financial markets is tolerant of mergers and acquisitions, including hostile takeovers. This allows firms to get access to new or radically different technologies and stocks of knowledge by acquiring other companies with relative ease. Moreover, the high presence of specialized seed- and venture- capital suppliers as well as business angel companies allows engineers and scientists to bring their own ideas and new technologies to the market by creating business start-ups and spin-offs. This is supported by a relatively risk-friendly system of company law, in particular in the field of bankruptcy law, which is part of a highly individualistic and entrepreneurial business culture that is fostered by private practices, public policies and a wide range of institutional mechanisms such as technology business incubators and venture capital firms encouraging risk taking (Hall/Soskice 2001b: 29ff.; Jackson/Deeg 2006: 13ff.; Abramson et al. 1997: 29f.).

Concerning the *industrial relations system*, the market relationship between individual worker and employer prevails, creating highly fluid labor markets. Representation of labor interest is less organized than in CMEs, and trade unions as well as business associations and chambers of industry and commerce are less cohesive and encompassing, giving them only a marginal role in structuring training schemes, inter-firm cooperation and technology transfer. These fluid labor markets give firms the chance to hire and fire labor to take advantage of new developments. This is also supported by the *education and training systems* in LMEs: Individuals are encouraged to invest in general and certifiable skills, transferable across firms and in career trajectorics including a substantial amount of movement among firms. Since labor markets are highly competitive, firms often invest in purposeful headhunting to hire qualified staff from other companies or scientific institutions (Hall/Soskice 2001b: 29f.).

Inter-company relations are mostly based on standard market relationships and enforceable formal contracts. The strict antitrust regulations in most LMEs make it much more difficult for firms to cooperate via relational contracts for example in R&D than in CMEs. Knowledge and technology transfer is then primarily realized through the movement of scientists and engineers from one company to another facilitated by the fluid labor markets. This also counts for movements from research institutions and universities to firms. This scientific staff brings its technical knowledge into the respective new company. Another prominent way to secure knowledge transfer in LMEs is licensing or sale of IPRs. These mechanisms are most feasible in sectors where effective patenting is possible. The importance of licensing is even reinforced by the character of standard-setting in many LMEs. Since most sectors lack of business associations that are strong enough to establish new standards collectively, standards are often set by market races, in which the winners profit by licensing their technology to many users (Etzkowitz et al. 2000; Hall/Soskice 2001b: 31).

Overall, these mechanisms of knowledge transfer in LMEs are supported by the model of the "entrepreneurial university", which does not see it as a university's only mission to engage in scientific discovery and education, but also to manage and promote technology and knowledge transfer, incubating new firms and networking with industry (Lerner et al. 2009: 269; Etzkowitz et al. 2000). This also enhances frequent *direct* interaction between firms and scientists, e.g. through telephone conversations, meetings, or faculty consulting, which are realized as a matter of course in these systems, and can expected to be at least as important as indirect contacts between science and industry mediated for example by technology transfer agencies/companies at the universities or research institutions. Moreover, as has been stated elsewhere (Jacobson et al. 2005; Werr/Stjönberg 2003; Mughan et al. 2004), professionalized commercial consultancy services play an increasing role for the management of knowledge transfer and innovation, particularly in LMEs. It is, thus, to expect that in the market-based LME-model these specialized external consultancy services are frequently used to manage knowledge transfer and innovation, both between firms and between firms and research institutions (Jacobson et al. 2005), while in CMEs like Germany this market based mechanism should be used less. Here, if anything, *internal* consultancies, above all in the larger corporations, should play a role.

Table 2: Hypothesized Mechanisms of Knowledge Transfer in the UK

MECHANISMS	UK
NETWORK BASED MECHANISMS	
business networks	–
informal networks	O
institutionally stabilized networks	O
customers and suppliers	O
cooperative inter-firm R&D-projects	–
coop. R&D-projects with scientific institutions and universities	O
ASSOCIATION BASED MECHANISMS	
employers-, business-, and trade associations	–
chambers of industry and commerce	–
binding and common technical standards	O
ORGANIZATION BASED MECHANISMS	
transfer of technical and scientific staff	+
purposeful headhunting	+
business start-ups	+
STATE INDUCED MECHANISMS	
national government measures	–
regional government measures	–
(larger) state-induced innovation projects	–
scientific parks	O
technological centers	O
technology transfer agencies/companies	O
MARKET BASED MECHANISMS	
external consultancies	+
mergers and acquisitions, hostile takeovers	+
buying, selling and licensing of patents	+
COMMUNITY BASED MECHANISMS	
informal contacts	O
regional platforms	O
additional (regional) institutions	O

+ *Hypothesized as crucial mechanism of knowledge transfer*
O *Hypothesized as important but not decisive mechanism*
– *Hypothesized as not important at all*

The *liberal* definition of the state in LMEs implies that national and regional governments should only exercise a peripheral role in the fields of innovation and knowledge transfer, leaving the coordination of these activities to the market. Neither (larger) state-induced innovation projects with the aim to bring different economic actors together on a project basis, nor public programs to foster innovation and knowledge transfer in specific technological fields or other national or regional government measures, for example with regard to the funding of innovation and knowledge transfer, should, thus, play a significant role. Rather, it is to be expected that activities in these fields are delegated to specific self-regulating bodies and regulatory agencies, as for example the *Regional Development Agencies* in the UK that might have certain but limited scope to create for example regional platforms to initiate interaction between economic actors in specific fields (Schmidt 2006; Polt et al. 2010).

In sum, also for LMEs and, thus, for the case of the UK, it is to be tested if knowledge transfer in the selected regions will match to the institutional features of the national level presented above. If this would be the case, thus, knowledge transfer in LMEs can be expected to be organized in particular through mergers and the acquisition of other companies, business start-ups by scientific personnel and from the private sector (supported by an extensive supply of seed- and venture-capital), strong interfirm fluctuation of scientists and engineers and fluctuation between research institutions and firms, frequent direct contacts between scientists and firms, the use of professionalized external consultancy services, as well as through licensing and sale of new knowledge and innovations in form of patents. Table 2 gives an overview on these hypothesized mechanisms of knowledge transfer in the LME of the UK according to the heuristic model of knowledge transfer systems presented above.

3.3 Hypothesized Structures and Mechanisms in MMEs

Finally, for the case of the Mediterranean MMEs like Spain, it is to be assumed that characteristics of both CMEs and LMEs, can be observed. This means that market- and non-market based forms of coordination exist simultaneously, whereby the state is a key actor in initiating and promoting coordination. The market for corporate governance in MMEs is characterized by the absence of a real *market* for corporate control, making hostile takeovers as well as mergers and acquisitions as mechanisms of knowledge transfer difficult. Instead, highly concentrated banks exercise a central role in the field of corporate governance. Family-owned small and medium sized companies dominate the business-structure, which are often closely connected to and dependent of regional and local "savings banks".

Lowly competitive product-markets allow for stable finance-industry relations. The financial market, thus, shows strong elements of strategic coordination, resulting from informal business networks and family financing. The provision of seed- and venture-capital is underdeveloped, such as the qualified assessment of entrepreneurs by business angel companies, making the state a central actor in financing larger innovation projects and business start-ups (Amable 2003: 113; Schmidt 2006: 4f.).

The *industrial relations* system is often characterized by a strong organizational fragmentation of employers' associations and trade unions, leading to high levels of social and industrial conflict. These confrontational industrial relations mostly impede bottom-up coordination and require intervention by the state (Molina/Rhodes 2007: 238; Ebbinghaus 1999: 19f.). Industrial relations are therefore shaped by centralized tripartite bargaining and social pacts between employers, trade unions and the state, what might also affect policies in the field of innovation and knowledge transfer (Royo 2007: 49ff.).

The workforce in MMEs is markedly divided, differentiating between a highly protected "core workforce" especially in large firms, and less protected flexible employment with temporary and part-time work. This divide of the workforce has also consequences for the *inter-company relations* in MMEs. In the case of large firms these are characterized by a higher internal than external mobility of the workforce. In the case of small and medium firms the external mobility is higher, due to the lower degree of protection. In this, the vocational training system is weak and also lifelong in-firm training is mostly absent, due to paternalistic firm structures, especially in small enterprises. The emphasis of the educational system is on general skills, but on a lower level than in LMEs, as demonstrate e.g. low enrollment rates in tertiary education. The mobility of researchers between the scientific and the business sector is low, due to the fact that a large share of scientific staff has a civil service status, making voluntary leaves or absences to industry difficult, because of the high barriers to return to the public sector and the risk to lose status or pension entitlements etc. (Amable 2003: 104ff.; Molina/Rhodes 2007: 238ff.; Royo 2007: 49ff.; Ebbinghaus 1999: 19f.; OECD 2010a: 11ff.; OECD 2007a: 59ff.).

Moreover, inter-company relations are competitive, but moderated by interventions of the state and business associations. Nevertheless, due to their fragmentation, corporate and intermediate actors as business associations, trade unions and chambers of industry and commerce can be supposed to play only a limited role in promoting inter-firm collaboration, innovation and knowledge transfer. Business collaboration in the field of innovation then often takes place in larger, state induced innovation projects. These short-term and project-based collaborations particularly between larger companies, to which the state frequently contributes with funding, can be assumed to be one important mechanism of knowledge transfer in MMEs. Cooperative business R&D-activities are traditionally low, and also cooperations between firms and universities or other scientific institutions are rare due to cultural and legal barriers impeding them. This is particularly true for the large share of low- and medium technology SMEs that dominate the business structure. Instead, the innovation landscape in many MMEs is characterized by large numbers of scientific parks and technological centers jointly funded by industry and national and regional governments that can be regarded as instruments trying to fill this gap and, thus, representing important mechanisms of knowledge transfer (OECD 2007a: 21ff.).

In the case of Spain, the main institutions performing research are universities and public research organizations, with the *Spanish National Research Council* (CSIC) as the main public research institution in the country, and several applied research institutes under the responsibility of different national ministries. Funding for teaching and operations at universities is provided by the respective autonomous regions, while funding for research is not or only little provided at the regional level. For this, university faculty has to apply for competitive funding from national, regional or European institutions. These different forms of funding find their expression in a wide range of programs promoting R&D, innovation and knowledge transfer, that are summarized in regularly renewed *Regional or National Plans for Scientific Research, Development and Technological Innovation.* These public programs and measures can be supposed to be important mechanisms of knowledge transfer in Spain (OECD 2007a: 12; OECD 2010a: 11ff.).

Concerning IPRs, patenting and licensing can be supposed not to play a central role for knowledge transfer in Spain. Here, again, cultural and legal barriers impede a more intensive use of these mechanisms. On the supply side, the results of public scientific research funded by the state are still considered by a large share of faculty as public goods that are not to be commercialized. On the demand side, the Spanish business structure mostly consisting of low- and medium technology as well as service oriented small and micro-firms, might affect only a low demand for patents or licenses from the scientific sector, and also only produce a low number of patents itself (Buesa 2006; OECD 2007a).

Table 3: Hypothesized Mechanisms of Knowledge Transfer in Spain

MECHANISMS	SPAIN
NETWORK BASED MECHANISMS	
business networks	O
informal networks	O
institutionally stabilized networks	O
customers and suppliers	O
cooperative inter-firm R&D-projects	–
coop. R&D-projects with scientific institutions and universities	–
ASSOCIATION BASED MECHANISMS	
employers-, business-, and trade associations	O
chambers of industry and commerce	O
binding and common technical standards	O
ORGANIZATION BASED MECHANISMS	
transfer of technical and scientific staff	O
purposeful headhunting	O
business start-ups	O
STATE INDUCED MECHANISMS	
national government measures	+
regional government measures	+
(larger) state-induced innovation projects	+
scientific parks	+
technological centers	+
technology transfer agencies/companies	+
MARKET BASED MECHANISMS	
external consultancies	O
mergers and acquisitions, hostile takeovers	O
buying, selling and licensing of patents	–
COMMUNITY BASED MECHANISMS	
informal contacts	O
regional platforms	O
additional (regional) institutions	O

+ *Hypothesized as crucial mechanism of knowledge transfer*
O *Hypothesized as important but not decisive mechanism*
– *Hypothesized as not important at all*

Summarizing this, for the Mediterranean MMEs and, therefore, for the case of Spain, it has to be tested if knowledge transfer in the investigated regions will be organized according to the institutional features of the national level presented above. If this would be the case, thus, knowledge transfer in MMEs is expected to be realized, on the one hand, via some fluctuation of technical personnel between firms and between firms and research institutions, although on a lower level than in LMEs. On the other hand, knowledge transfer should above all occur through cooperation of (especially larger) firms in often state induced larger innovation projects. In addition, it might also occur through cooperations in public technology centers and research parks funded by the state. Finally, also public programs defined by regional or national governments within the framework of the *Regional or National Plans for Scientific Research, Development and Technological Innovation* can be supposed to be central mechanisms of knowledge transfer in Spain. These mechanisms are illustrated and categorized in Table 3 according to the heuristic model of knowledge transfer systems detailed above, showing the specific occurrence of this heuristic model hypothesized for the Spanish case.

3.4 Summary

Summarizing the theoretical considerations discussed so far, it has been shown in a first step that processes of knowledge transfer in the context of innovation are characterized above all by their systemic nature and that they are strongly determined by the specific institutional environment they are embedded in. The point of view developed in this study is that this systemic and institutional environment of innovation and knowledge transfer is best described by the concepts of NIS and RIS, and that these concepts can also be read as theoretical approaches describing the institutional framework for knowledge transfer in the context of innovation. In this, an important gap in the literature on NIS and RIS has been identified, which uses the concept of knowledge transfer only as an explaining factor and central precondition for successful innovation activities, but still lacks a clear analytical framework for defining the specific roles single institutions or particular institutional configurations play for knowledge transfer, and, thus, to describe their concrete impact on innovation processes. In a second step, the national institutional framework of RIS has been discussed, using the concepts of NIS and VoC. It has been argued that these approaches overemphasize the notion of institutional coherence when describing national systems of innovation and production and thus underestimate the regional variance within national systems. The point of view taken in this study is that it is plausible to assume that the institutional structures and mechanisms of knowledge transfer in RIS might considerably differ from their respective national institutional framework.

To address these reciprocal gaps in the literature on innovation systems and VoC, in a next step a heuristic model of knowledge transfer systems has been introduced, which clearly identifies and systemizes the central structures and mechanisms of knowledge transfer in the context of innovation and ascribes them to the six ideal types of social order. With that, an analytic toolbox has been provided, allowing to reconstruct specific knowledge transfer systems through the analysis of empirical case studies. In a further step, hypotheses have been derived from the VoC-approach assuming which institutional structures and mechanisms of this heuristic model should determine processes of knowledge transfer in three different types of national market economy models, i.e. coordinated, liberal and Mediterranean. The aim of this proceeding is to test to which degree the regional innovation systems within these three national models show the same institutional structures and mechanisms of knowledge transfer hypothesized for the national level, to then detect and explain regional variation within national systems. Moreover, it is to be expected that this description and explanation of regional variation should be a useful contribution to a deeper understanding of processes of knowledge transfer in innovation systems per se.

For this purpose, an internationally comparative qualitative case study has been conducted, which will be presented and discussed in the next sections. In the following chapter, in a first step some methodological issues concerning this case study and the research design and will be outlined, before the case selection is discussed and the methods used as well as the data base are presented.

4. A Qualitative Approach to Innovation System Analysis

4.1 Conceptualizing an Internationally Comparative Qualitative Case Study: The Choice for a Qualitative Approach

The discussion of the theoretical concepts of NIS, RIS and VoC so far, as well as the elaboration of a heuristic model of knowledge transfer subsystems in innovation systems, and the formulation of theoretical assumptions for this model in different national market economy contexts raise the question of how to confront these theoretical and conceptual considerations with empirical evidence and how to approach them methodologically. This study will take the position that the guiding research questions are addressed best by an explicitly *qualitative research design*, using regional case studies and comparing them within and across national institutional models. In the following, some methodological considerations will be discussed by contrasting arguments for and against qualitative research methods in the fields of comparative political economy and innovation system research.

Broad parts of the literature on methodological issues in comparative political economy and innovation system research generally acknowledge that the empirical analysis of innovation and market economy systems requires both strategies: Quantitative and qualitative. Today, a consensus emerged that it is especially *qualitative within case analysis* which can create a particular added value in this field of study (e.g. Pierson 2007: 165f.; Mahoney 2007: 123; Peck/Theodore 2007: 758ff.; Boyer 2005: 17ff.). Already Freeman (1995) remarked in one of his early works that "qualitative factors affecting the national systems […] [have; MO] to be taken into account as well as the purely quantitative indicators." (Freeman 1995: 12) As Freeman argues, many of the institutional and technological aspects to analyze within national innovation systems would rather be subject to qualitative analysis and would be difficult to quantify (Freeman 1995: 20). Especially the over-simplification of purely quantitative R&D-comparisons, for example through patent-data (see also Nelson 2009: 995f.; Freeman/Soete 2009: 584ff.), would be an inadequate method, above all when pointing to "institutional differences in the mode of importing, improving, developing and diffusing new technologies, products and processes" (Freeman 1995: 20), as is also intended in the study at hand.

With a clearer focus on the regional level, also Markusen (1994) argues strongly in favor of qualitative approaches to investigate RIS. Especially because of the institutional embeddedness and the informal and network character of many innovation and knowledge transfer oriented interactions she highlights the need to apply qualitative methods when analyzing regional economies, while standard statistical data and quantitative surveys would be less adequate for capturing the *relational elements* which innovation and knowledge transfer are built upon (Markusen 1994: 478). Therefore, she regards qualitative interviews with regional actors such as firms, trade associations, business service suppliers, trade unions, and economic development officials as appropriate method to *intensively*, rather than extensively investigate regional economic interaction. As MacKinnon et al. (2002) accentuate, this would especially allow for the inclusion of distinct actors' perspectives into the analysis and to elaborate a differentiated view on innovation and knowledge transfer processes in RIS. They see "a need to employ a range of research methods—including corporate interviews, surveys and ethnographic approaches—that involve direct contact with the individuals, firms and organizations engaged in processes of learning and innovation within various regional and industry contexts. This can be seen as a necessary, if insufficient, condition for research to proceed beyond 'vague theory and thin empirics' [...] towards a firmer understanding of the key processes that underpin economic development at local and regional levels." (MacKinnon 2002: 305)

Especially for this study which intends to investigate the differences in the structures and mechanisms of knowledge transfer from an assumed theoretical model and intends to explain these differences comprehensively, this qualitative and interview-based approach appears to have decisive advantages when compared to quantitative approaches in innovation research. Following Siebenhüner (2007), quantitative approaches in innovation research can be seen as particularly weak since they are mostly based on comparatively superficial explanatory models. Moreover, statistical evaluations can only insufficiently detect causal relationships and are limited to fully quantifiable variables. Also individual interpretations and the specific knowledge of the actors participating in innovation and knowledge transfer processes can hardly be captured by the very rigid quantitative variables, so that unanticipated causal relations are basically excluded from the analysis (Siebenhüner 2007: 109). The same would be true for quantitative actor surveys as approach complementing official statistical data, which remain superficial and only focus on a limited range of factors influencing innovation and knowledge transfer processes, excluding the systemic interaction in these processes as well as the underlying perspectives and motivations, which is criticized by Beckenbach (2007) as the "iceberg-problem" in empirical innovation research (Beckenbach 2007: 58ff.).

Qualitative approaches, on the other hand, far more strongly allow for more substantial differentiations of analytical dimensions and intervening variables. According to Flick (2009), qualitative approaches are particularly well suited for the analysis of complex and insufficiently structured phenomena (Flick 2009: 26ff.), as is the case in complex national and regional systems of innovation and knowledge transfer, which require an explorative identification of relational structures and new explanatory factors (Lamnek 2010: 81). Through plausibility assumptions, ex-post analyses of the relation of distinct explanatory factors or the comprehension of conclusive narratives qualitative approaches allow for the explicit identification of the causal relationships in the underlying processes of innovation and knowledge transfer. Most qualitative studies in innovation research develop an *analytic model*, as has been done in this study with the heuristic model of knowledge transfer processes, which is then used as basis for the data collection process and from which items can be derived e.g. for guidelines for qualitative expert-interviews. This theoretical model not only allows for the formulation of hypotheses on the quality of specific factors, but also reduces the complexity of the investigated issues and cases considerably. Not least, qualitative research designs allow, through investigating more than one case, for comparative approaches and the testing of causal hypotheses (Siebenhüner 2007: 106).

Nevertheless, there are also some shortcomings of qualitative approaches: Research pragmatic factors limit the scope of most qualitative studies and the number of investigated cases (Gläser/Laudel 2009: 101; Flick 2009: 175f.), restricting a generalization of their results, even if the investigated cases are selected carefully (Lamnek 2010: 166f.; Flick 2009: 522ff.; Przyborski/Wohlrab-Sahr 2009: 311ff.). In this regard, the depth of the results obtained by qualitative designs at the same time impedes a testing of causal relationships across a higher number of cases. In addition, also the strength of qualitative research to include individual perspectives and perceptions into the analysis is accompanied with the weakness of being inadequate for addressing over-individual causal relationships in an objective way (Siebenhüner 2007: 106f.).

Despite these shortcomings, for this study a qualitative research design has been chosen, since it appears to be *adequate* (Flick 2009: 53f.; Lamnek 2010: 132; Strodtholz/Kühl 2002: 18) to address the research object, namely the heterogeneity of regional knowledge transfer systems within national innovation and production systems. Since these systems constitute complex social systems at the sociological macro level, in which innovation and knowledge transfer are conceptualized as highly relational processes depending on commonly shared languages, worldviews, cultures and mutual trust, the distinct perspectives of the investigated actors themselves are important to understand the differences among these systems. As has been shown before, the concrete institutional structures and mechanisms underlying these processes of knowledge transfer and innovation have neither on the micro- nor on the macro-level clearly and sufficiently been identified nor systemized, and also the causal relationships between distinct explanatory factors influencing these processes remain unaddressed and vague up to now. This makes an (at least partly) *explorative approach* necessary, allowing to identify unconsidered aspects, relational structures and explanatory factors, which can best be achieved through a *qualitative case-study design* (Lamnek 2010: 81ff., 277f.; Flick 2009: 177f.; Yin 2009).

These case studies allow for the systematic focusing on the most important structures and mechanisms of knowledge transfer in innovation systems as well as on the different actors involved in these processes and their cooperative relations to each other. This includes, e.g. national and regional cultures and identities, business structures and cultures, mentalities, and interpersonal aspects. In this, also a differentiated view on the impact of the distinct levels of governance and on the actors involved in innovation and knowledge transfer processes is possible, reflecting the perception of the interviewed actors and experts on which quality specific institutions, actors and mechanisms have on the knowledge transfer activities in their respective regional innovation systems. The *comparative analysis* (Ebbinghaus 2009: 485ff.; Ebbinghaus 2005a: 141f.; Flick 2009: 179f.) of distinct regional cases on the basis of a heuristic model of knowledge transfer systems can in this connection be used to reduce the complexity of the investigated issue, structure the analysis in a systematic way (Ebbinghaus 2009: 491f.), and guide and structure the data collection process, i.e. the interviews with regional experts and actors through constructing interview-guidelines (Liebold/Trinczek 2002: 41f.; Meuser/Nagel 2005: 77f.). With that, a semi-structured data-collection method like the *guideline-based expert-interview*, allowing for both structured hypotheses testing and explorative openness (Liebold/Trinczek 2009: 37ff.), appears to be adequate and most effective to address the research interest pursued in this study.

4.2 Sampling Strategies and Selection Criteria

To confront the theoretical hypotheses derived from the literature on NIS and VoC with empirical evidence, the study at hand applies an *internationally comparative qualitative case study design*, which will be detailed in this section. Since it is the aim of this study to analyze and explain regional variation of knowledge transfer systems in national institutional configurations, it appears appropriate to combine two basic designs of qualitative research: The case study and the comparative study design. While the *(within) case study design* can be used to precisely describe and reconstruct a single case (Flick 2009: 177; Lamnek 2010: 81ff., 277f.), as it is intended in this study for single regional knowledge transfer systems, the *comparative (cross-case) study* can be used to compare a higher number of cases with regard to specific research interests (Flick 2009: 179; Ebbinghaus 2009: 485ff.; Ebbinghaus 2005a: 141f.). For the research interest pursued in this study this allows for both comparing regional cases within national institutional models, and comparing them across national systems to analyze if distinct national institutional models show different degrees of regional variation.

As Ebbinghaus (2005a) points out, in these comparative case study designs with a low number of investigated cases a careful and deliberate *case selection* is central and intended: "In qualitative small-N comparison, the selection of cases assumes a particularly important role in the research design. Cases are not selected to represent observations (mere data points) for extensive analysis that allow for generalizing beyond the cases at hand, particularly in the case of a random sample drawn from a population of cases. Instead, qualitative comparative research selects cases for intensive within-case analyses, the results of which are then compared across cases." (Ebbinghaus 2005a: 134) In general, the main selection or "sampling"-criterion for the cases to investigate in qualitative research designs is, according to Flick (2009), the *appropriateness* for addressing the research interest and the research questions (Flick 2009: 169f.).

With regard to the research questions guiding the study at hand, regional innovation systems as defined in section 2.4.2.2 will be analyzed as cases, with a special focus on their respective knowledge transfer sub-systems. In this connection, as in most qualitative research projects, also this study faces the problem to decide for an adequate weighting of *intensity and extensity of the sampling* (Przyborski/Wohlrab-Sahr 2009: 182; Flick 2009: 167f.; Ebbinghaus 2005a: 142f.). Taking into account both the need to deeply understand and reconstruct the institutional structures and mechanisms of knowledge transfer in the single regional cases to detect and comprehensively explain possible variation within national institutional models, and the need to compare these regional cases intra-nationally with at least one further regional case, as well as to compare regional variation cross-nationally in distinct national market economies and their respective archetypical models, a sampling choosing two regional cases in each of the three types of market economies discussed in section 2.3.1 appears to be appropriate for addressing the research interest. Not least, this design also takes into account research pragmatic restrictions which lie, above all, in the high efforts connected to the method of qualitative expert-interviews, as well as in restricted time and financial resources (Gläser/Laudel 2009: 101; Flick 2009: 175f.).

With this form of qualitative research no statistical representativeness can be reached, but rather a certain degree of *analytic generalization* (Yin 2009: 15), which, however, strongly depends on the substantial representativeness of the investigated cases as well as their detailed functional analyses, making it possible to assign the obtained results also to other similar cases (Flick 2009: 169f.; Lamnek 2010: 161ff.). In this regard, the notion of analytic generalization does not point to a generalization to some predefined and sampled population, but rather to a theoretical model of the issue being studied, which may have wider applicability than the specific investigated cases per se (Yin 2009: 15). Therefore, adequate *case-sampling-strategies* are ascribed a central role in increasing representativeness and generalizability in qualitative case studies (Przyborski/Wohlrab-Sahr 2009:

173ff.). In the following, thus, some remarks on the sampling-strategies and the cases selected for this study will be discussed.

The starting point of the theoretical considerations guiding this study has been the assumption of institutional coherence in the approaches of VoC and NIS. The aim of this study is to test this assumption empirically and to demonstrate that there is more institutional heterogeneity and regional diversity than predicted by these approaches. Thus, in a first step, *national cases* have to be selected according to the market economy typology provided by the VoC-approach. In this, the intention has been to select one country from each of the three main types of market economy models discussed in section 2.3.1 (liberal, coordinated, mixed), allowing to investigate if regional variation within national models also varies with type of market economy model. In addition, the main *selection criteria* has been to choose one of the largest and strongest European economies within each of these three groups to include the most paradigmatic and representative cases into the analysis, what corresponds best to a sampling strategy of *sampling typical cases* (Ebbinghaus 2009: 492; Flick 2009: 165). Therefore, for the LMEs the UK has been selected, for the CMEs Germany and for the MMEs Spain, which are all discussed in the literature on VoC as especially paradigmatic and typical cases for their respective market economy model (Hall/Soskice 2001b: 19f.; Molina/Rhodes 2007; Schmidt 2002; Amable 2003; Hall/Gingerich 2009; Ebbinghaus 1999; Royo 2007). With that, the selection of these three national market economies corresponds to a "most-dissimilar-system-design-strategy" (Ebbinghaus 2009: 493f.; Przeworski/Teune 1970: 34ff.), allowing to exclude the type of market economy as explanatory factor for regional variation in national institutional frameworks, in case that all national cases result to show regional variation to the same degree.

Furthermore, within each of these three national economies *two regional cases* have been selected for in depth case studies to investigate to which degree they correspond to the assumed national institutional configurations in the field of knowledge transfer. Also these regional cases have been selected according to a "most-dissimilar-system-design-strategy" (Ebbinghaus 2009: 493f.; Przeworski/Teune 1970: 32ff.) by choosing in all three national cases one *established industrial region* and one *metropolitan region*, which have been described in section 2.4.2.5.3 as being types of RIS shaped by pointedly different structural conditions, what would correspond to the sampling-strategy of *maximal variation sampling* (Flick 2009: 165). However, from the group of available established industrial regions and metropolitan regions within each of the three national market economies, where possible, the most typical cases have been selected, respectively, applying again the typical cases sampling strategy (Flick 2009: 165), to include cases into the sample in which the respective industrial or metropolitan structural conditions are particularly well developed and observable.

Altogether, this will allow for the inclusion of specific sets of explanatory factors correlated to these two types of regions into the analysis, like e.g. the regions' economic structures, their business structures and their particular actors' constellations. The expectation is that if there is a match between national production system and regional innovation system in the structures of knowledge transfer, it should especially appear in the established and historically grown industrial regions, since they are assumed to have experienced a historical path of development of institutional structures in the fields of innovation and knowledge transfer (Heidenreich 2003; Whitford 2001; Krauss 2005; Koschatzky 2001a), in which *processes of institutional and organizational isomorphism* (DiMaggio/Powell 1983; Scott 2008; Meyer/Rowan 1991) have created regional structures being more homogeneous to their national institutional environment. One major criterion to select these established industrial regions is, therefore, that they should have paralleled the economic, structural and technological up- and downturns of their respective national production systems and might, thus, have developed similar institutional structures and mechanisms in the fields of innovation and knowledge transfer.

Conversely, the newly created institutional structures and mechanisms of supporting innovation and knowledge transfer in metropolitan regions can be supposed to be less influenced by these processes of institutional and organizational isomorphism. Since in most cases these structures are only partly historically grown and are, above all, the result of concrete political decisions and planning (Krätke 2009; Brandt et al. 2008; Kujath 2005; Krauss 2005; Schäfer et al. 2004; Revilla Diez 2002; Heidenreich 2003; Blotevogel 2002; Sassen 2001; Allen 2000), it can be assumed that they are more likely to differ from their respective national institutional model. Moreover, and pointing to the debate on the convergence of market economies towards the liberal Anglo-Saxon model and its global spread (see e.g. Hall/Thelen 2009; Streeck 2009), one should expect that it is especially the metropolitan regions in CMEs and MMEs that deviate in this sense from their respective national models, since the diffusion of globally dominant models should particularly be visible at the nodal points (Sassen 2001; Allen 2000; Kujath 2005) of global information and communication networks. Building on that, the main selection criteria for the metropolitan regions are that they must be characterized by the integration of one or several major urban agglomerations with their respective regional surroundings, by a high density of population, by predominantly service sector oriented economies (Krauss 2005: 110; Heidenreich 2003b: 323), as well as strongly developed controlling-, learning- and gateway-functions as defined in section 2.4.2.5.3 (see also Kujath 2005: 45; Audretsch 2003: 18; NIW/Nord/LB 2005: 100; Brandt et al. 2008: 11ff.; Schäfer et al. 2004: 15ff.; Fischer et al. 2001: 16f.).

Furthermore, a second criteria for the regions selected for this study is that they should have at least certain strengths in the two sectors of automotive and biotechnology. The aim is to put a special focus of analysis on these two sectors in all six investigated regions, to also "control" for *sectoral specificities* as explanatory factors for regional variation. This will make it possible to further differentiate the results obtained in the single regional case studies by sectors and to also investigate to which degree single industrial sectors adapt to their respective regional and national contexts and to which degree they also create their own systemic environments in the field of knowledge transfer. With the two sectors of automotive and biotechnology one traditional industrial sector with a synthetic and highly cumulative knowledge base and an incremental mode of innovation, and one younger sector with an analytical and lowly cumulative knowledge base and radical innovation processes are focused on, which are assumed to be shaped by quite different structural and functional requirements in the fields of innovation and knowledge transfer, and which are assumed in the theory of VoC to be found either in a coordinated or an liberal market economy context (Coriat/Weinstein 2004: 342ff.; Malerba 2004: 9ff.; Casper/Soskice 2004: 349ff.; Hall/Soskice 2001b). This would, again, correspond to the sampling strategy of maximal variation sampling (Flick 2009: 165). For this study, a special interest will be directed towards investigating to which degree sectors in an institutionally *inadequate* environment can adapt the structures and mechanisms of knowledge transfer available to their respective needs, or if they develop structures and mechanisms parallel to their institutional context, as it would be the case for the biotechnology sector in Germany or for automotive in the UK (Casper/Soskice 2004; Malerba 2004).

This will for conducting a cross-regional and cross-national comparison of the predominant institutional structures and mechanisms of knowledge transfer in the two sectors with the aim to detect the respective variation, if it exists. The assumption is that the institutional structures and mechanisms in both sectors will considerably differ within one market economy model, according to their respective sectoral specialisms in terms of mode of innovation, type of knowledge base and degree of cumulativeness, but will show significant similarities across national and regional systems. Nevertheless, national institutional frameworks will still be assumed to exercise decisive influence on knowledge transfer processes in the different sectors, so that it will be the detailed analysis of the six regional case studies that is to show to which degree these national frameworks shape the structures and mechanisms of knowledge transfer in the two sectors within the six regions.

Overall, in the comparative dimension, these sampling strategies on the regional and sectoral level correspond to a "most-similar-system-design", which, across national and regional cases, allows (in the sense of "Mill's Method of Difference") to "control" for these types of regions and sectors as explanatory factors for regional variation. However, also the limits of these sampling and design strategies must be highlighted: Other explanatory factors on the regional level remain powerful, and cannot be "controlled" through the chosen research design. Among others, these factors would include the role of multinational companies, supranational institutions (especially those related to the EU's R&D-, innovation- and technology-policy), as well as some national actors involved in innovation and knowledge transfer. Due to research pragmatic reasons, and especially to avoid an overcomplex research design, these factors shall not be "controlled" by this study's research design. Nevertheless, although not systematically "controlled", they are investigated in the course of the empirical case studies and are punctually discussed where necessary and useful.

Figure 2 summarizes this research design, which, altogether, can be described according to Yin (2009) as an "embedded multiple case design." (Yin 2009: 46ff.) In this design, the three selected national market economies constitute the investigated *cases*, in which the research questions are addressed (regional variation in national systems), while the six regional knowledge transfer systems within these national cases can be regarded as the *units of analysis*. However, the notion of units of analysis will be used synonymously to the notion of cases in this study, since they are investigated in depth as institutional contexts and configurations for the causal processes to be analyzed (Ebbinghaus 2009: 487; Yin 2009: 29ff.). Finally, within the regional cases (units of analysis) the *units of observation* (Przyborski/Wohlrab-Sahr 2009: 176; Ebbinghaus 2009: 487) within the framework of this research design are the interviewed experts, which have been addressed through the method of qualitative guideline-based expert interviews. This data collection method will be discussed in section 4.1.4. The following section, however, first discusses the selection of the investigated regions.

Fig. 2: Research Design: Embedded Multiple Cases

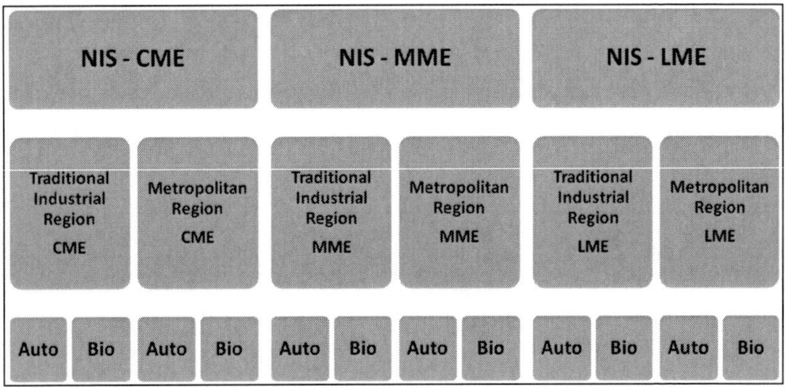

Source: Own Illustration

4.3 Selecting the National and Regional Cases

Overall, and basing on these sampling strategies and selection criteria, the following six regional cases have been selected: With regard to the established industrial regions, for the German case *Baden-Württemberg* has been selected, which fits particularly well into this category as will be shown in detail basing on quantitative structural economic data in section 5.2.1.1. As one of the country's strongest economic and industrial regions it is frequently used as a paradigmatic case to describe a RIS within a CME context (see e.g. Krauss 2009; Heidenreich/Krauss 2004) and is also used by many authors of VoC to describe the German national market economy as a whole, making it a well suited case to test its basic assumptions. In addition, Baden-Württemberg is not only shaped by a comparatively high share of industrial production in its economic model, but is also one of the country's major automotive regions and one of the strongest locations of German biotechnology (sections 5.2.1.2 and 5.2.1.3).

For the Spanish case, the autonomous region of *Catalonia* has been selected, since it is one of only two traditional and established industrial regions in Spain, together with the Ebro Valley (Aragon, Navarre and the Basque Country) (Bacaria et al. 2004: 64f.), and can be regarded as a typical case for the country's industrialization model. This is the reason why Catalonia is frequently analyzed as a classical example of a RIS in Spain. In addition, as the discussion in sections 6.1 and 6.2.1.1 will show, Catalonia is Spain's strongest pole of industrial activities and has more strengths in the two investigated sectors than the Ebro Valley, above all in biotechnology, so that this region has found its way into the sample (sections 6.2.1.2 and 6.2.1.3).

Third, for the case of the UK, the *East of England* region has been selected. This region experienced a specific path of industrialization with less tradition in early heavy industries, converting it in the present into a classical example of a high-tech industrial region within a LME context, as the discussion in section 7.2.1.1 will show. Therefore, it appears to be particularly adequate for being investigated as a case of an established industrial region in the study at hand, since it not only focuses strongly on New Economy sectors in high-technology fields, but since it also reflects the whole country's shift away for traditional and medium- and low tech to high-tech sectors and knowledge intensive services. However, while other UK industrial regions might still have comparatively stronger focuses on traditional industrial sectors, the East of England is the region with most prominent strengths in the two sectors of biotechnology and automotive, so that it has been selected into the sample (sections 7.2.1.2 and 7.2.1.3).

As metropolitan regions, for the German case the *Metropolitan Region Hannover Braunschweig Göttingen Wolfsburg (MR HBGW)* has been selected. This region is distinct from the other two investigated metropolitan regions in not containing the respective national capital city and in being a *multipolar* region composed of four urban agglomeration areas, which is, however, typical for the German case, where the historical development of the country's urban system created pointedly decentralized and multipolar metropolitan structures (see e.g. section 5.3.1.1; Metropolregion 2010; Brandt et al. 2008; Ortiz 2008; Stühring 2006; NIW/ Nord/LB 2005;

Priebs 2004). As the discussion on quantitative structural economic indicators will demonstrate, the region has, in addition, considerable strengths in both automotive and biotechnology (see sections 5.3.1.2 and 5.3.1.3). With regard to the research question, it furthermore allows for including a newly created regional entity into the analysis, which is based on explicit regional structure building efforts in the fields of innovation and knowledge transfer, making it possible to investigate to which degree these newly created and planned structures can deviate from their German national institutional framework. Finally, the author already conducted an intensive case study on the MR HBGW (Ortiz 2008), so that the selection of this particular region has been *convenient* (Flick 2009: 166; Lamnek 2010: 176f.), especially because of research pragmatic reasons and a partly already established access to the regional expertise.

For the Spanish case, as a metropolitan region the *Madrid Community* has been selected. As the autonomous region containing the Spanish national capital city, it connects the country's major urban agglomeration area with its regional surroundings, explicitly following a strategy of making the Madrid Community a metropolitan region and is, therefore, a good example of a metropolitan region in Spain as defined in section 2.4.2.5.3 (OECD 2007a; COTEC 2004; COTEC 2003; Buesa 2004; see also section 6.3.1.1 on Madrid's quantitative structural indicators). In this, it reflects quite well the highly centralized Spanish urban system. Apart from that, in the Spanish context it has considerable strengths in both investigated sectors, automotive and biotechnology (see sections 6.3.1.2 and 6.3.1.3).

Finally, for the case of the UK, the *Greater London* region has been selected as metropolitan region. Like Madrid, also Greater London contains the country's national capital city and is the UK's largest urban agglomeration area, including London's regional surroundings (GLA 2010). As a prime example of a "global city" it shows all major structural economic indicators characterizing a metropolitan region, and appears to be particularly well suited to being investigated as a case of a metropolitan region in the UK (see section 7.3.1.1). Within the UK, it reflects the country's typical centralized urban structure. In addition, Greater London has remarkable competences in biotechnology, and, even if only minor, activities in automotive (see sections 7.3.1.2 and 7.3.1.3). Figure 3 summarizes the outlined research design and gives an overview on the selected cases.

Fig. 3: Case Selection: Investigating Six Regional Knowledge Transfer Systems in Three National Settings

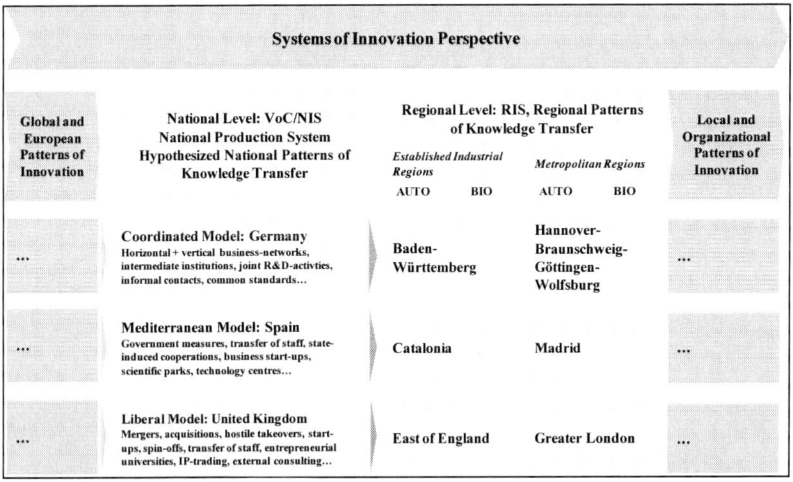

Source: Own Illustration

4.4 The Guideline-Based Expert Interview as Research Method

Since it is the aim of this study to investigate to which degree the structures and mechanisms of knowledge transfer within these selected regional cases vary from the assumed national models, the theoretical assumptions for the respective national market economies derived from the literature have to be confronted with empirical evidence from the regional cases. Many issues to be investigated in this regard are *subjective*, and are connected intensively to the perception and evaluation of the participating actors, as would be e.g. the relative importance of informal contacts, trust-based relationships and the use of network structures for knowledge transfer activities. Therefore, these aspects are difficult to quantify and are best addressed by qualitative research methods. In addition, the pretended use and empirical analysis of a theoretically constructed heuristic model of knowledge transfer systems requires an *explorative openness* towards the empirically observable regional knowledge transfer systems whose basic structures and mechanisms cannot be fully anticipated from theoretical assumptions, not least because of the high degree of *complexity* inherent to knowledge transfer processes. Thus, in all six selected regions *qualitative guideline-based expert-interviews* have been conducted, since they appear appropriate and most effective for addressing this study's research interest. In the following, some remarks on this research method will be discussed.

This research method is particularly well suited to reconstruct complex social structures, subjective meanings and perceptions, as well as to comprehend complex causal relationships (Strodtholz/Kühl 2002: 17; Gläser/Laudel 2009: 13ff.; Bogner/Menz 2005: 36ff.). Also the explorative research interest pursued in this study is well addressed by the method of guideline-based expert-interviews, since it allows for certain openness towards the units of observation (interviewed experts) and the respective research situations, allowing to also include unexpected information and patterns of perception into the analysis (Lamnek 2010: 19f.; Lamnek 1995: 22f.; Strodtholz/Kühl 2002: 16ff.). With the method of the qualitative guideline-based expert-interview this openness towards the complex and heterogeneous topic of regional systems of innovation and knowledge transfer can especially be reached through a flexible and context dependent use of the previously constructed *interview-guideline*, so that there is sufficient room for free narrative sequences and own priority setting by the interviewees. This can be further supported when questions motivating *narratives* are formulated and when the interviewees' narrative sequences are not or only limitedly influenced. Limiting this openness is, by contrast, a strict maintenance of the order given by the interview-guideline, or the formulation of closed questions with predefined answer categories (Liebold/Trinczek 2002: 39ff.; Gläser/Laudel 2009: 90ff.; Flick 2009: 203ff.; Lamnek 1995 22f.).

On the other hand, the use of an interview-guideline can also effect a certain *structuration*, which is essential to include the formulated theoretical hypotheses into the interviews and to confront them with empirical evidence (Schulze 2004: 105f.; Gläser/Laudel 2009: 90f.; Liebold/Trinczek 2009: 36ff.; Flick 2009: 203ff.). In addition, too much openness in the interview process would cause that in all interviews distinct variables would be addressed, so that the intended comparisons and reconstructions of patterns would be difficult. Therefore, the guideline based expert-interview combines structuration and openness, what has also been labeled "*closed openness*": "On the one hand, conceptual considerations structure the field, and, on the other hand, the structuration of meaning by the research subjects is maintained. Deduction and induction go hand in hand." (Liebold/Trinczek 2002: 42; translation: MO)

For the research interest of this study, in a first step, a "meta-interview-guideline" has been constructed, which builds on the heuristic model of knowledge transfer systems developed in section 2.5, asking for the qualitative relevance of the 24 mechanisms in their six analytical dimensions. This "meta-interview-guideline" has then been the basis to be slightly adapted to the respectively investigated region and also to the respective unit of observation, i.e. interview partner and the institution he/ she represents. The intention behind this proceeding was to, on the one hand, secure that all 24 mechanisms of knowledge transfer are addressed in all interviews in all six investigated regions, allowing to *compare* and to *aggregate* the results from the interviews across regions and also across the national models. On the other hand, the minimal adaptations allowed for putting special emphasis on previously known regional specificities, on issues concerning the institution represented by the respective interview partners, and on issues that particular experts were assumed to being able to judge especially well, not least because of their personal experience or their institutional background.

4.5 Selection of Experts as Units of Observation

This already points to the fact that in the qualitative research design presented so far, the selection of *units of observation*, i.e. the interviewed experts selected for the interviews in the six regional cases, has received special attention. In the following, some remarks on this selection process will be discussed. In general, *experts* can be defined as persons, which possess specific stocks of knowledge, and are regarded as specialists, connoisseurs or professionals. In the research process, therefore, those persons can be addressed as experts, which can be assumed to possess relevant knowledge with regard to the research interest, and which have a pointed advancement in knowledge in their particular fields of expertise compared to others. To this group mostly belong functional elites within organizational and institutional contexts, which are characterized by their responsibility for the development, implementation and the control of problem solving and decision making processes. In addition, these persons also possess privileged access to information on these decision making processes as well as to the groups of actors involved in them (Liebold/ Trinczek 2002: 35ff.; Meuser/Nagel 2009: 37ff.; Lamnek 2010: 655f.; Flick 2009: 214ff.; Bogner/Menz 2005: 39ff.).

For the study at hand, this implies that the experts selected as interview partners must either be *deep connoisseurs* for the respective regional knowledge transfer systems, or must be actively involved in knowledge transfer processes themselves, or be member of an organization which is so. This would include all relevant institutional actors in RIS as defined in section 2.4.2.2, as e.g. firms, academic institutions, political bodies, financial institutions and intermediary organizations. In addition, these experts must have a certain *decision making responsibility*, i.e. they must occupy posts at higher hierarchical levels of the institutions they represent, as it would be e.g. a departmental head of a chamber of commerce or a managing director of a SME. Apart from these aspects, the intention was to select a sample of interview partners for each of the six regions reflecting a broad spectrum of perspectives on the respective regional knowledge transfer system, so that interview partners from different types of institutions have been selected. However, because of research pragmatic reasons and limited resources, the total number of interviews conducted in each of the six investigated regions has been limited to between five and eight interviews. As a consequence, the interviewed experts only represent a small share of the institutions involved in the regional knowledge transfer activities. Yet, since the selected interview partners are also experts for the respective regional knowledge transfer system as a whole, they were also able to evaluate aspects beyond their own institutional context, allowing them to give a broad overview and deep insight on the functioning of the system as a whole. Finally, another selection criterion in this regard has been to choose for each region at least one explicit expert for automotive and one for biotechnology, to realize the intended differentiation between these two sectors. For these "sector-specific" interviews also the interview-guideline has been slightly adapted again, for addressing the respective sectoral specificities in depth.

Altogether, 35 interviews have been conducted in the period between July 2009 and March 2010, including 40 interview partners, since in five interviews the addressed expert invited another person to join the interview to give further or more detailed information on specific issues. All interviews have been conducted by the author in *personal face-to-face inter-*

views, except two, which have been conducted by phone. Due to confidentiality and data protection issues, the selected interview partners in the six investigated regions as well as the institutions they represent cannot be published at this point. In the vast majority of cases, the interviews had a duration of in between 60 and 90 minutes. These interviews have been recorded on digital audio files (Przyborski/Wohlrab-Sahr 2009: 79ff.), and have then fully been transcribed by the author (Flick 2009: 379ff.; Przyborski/Wohlrab-Sahr 2009: 160ff.; Brüsemeister 2008: 131ff.; Liebold/Trinczek 2002: 47ff.) in the respective original language (German, Spanish, English), using the transcription software "f4" (Dresing/Pehl 2011: 30ff.; Kruse/Wenzler-Cremer 2008), obtaining more than 700 pages of interview transcripts as qualitative data base, which have then systematically been evaluated. This systematic evaluation of the interview transcripts will briefly be discussed in the following.

4.6 Evaluation Methods: Aggregation and Triangulation

The aim of this evaluation, whose results will be presented in chapters 5, 6, and 7, is to confront the theoretical assumptions on the national knowledge transfer systems with the empirical evidence of the interview transcripts. This methodological proceeding is in line with common practice especially in the fields of regional economics and innovation research, where hypotheses-testing research strategies building on qualitative interviews are frequently used, as for example Markusen (1994) already indicates in an early work: "Although many advocates of the interview technique stress its inductive and exploratory nature, I would like to suggest that interview data can, when properly generated and collated, be used to test hypotheses. At a minimum, interview findings can suggest either consensus (for or against) or lack of it on hypothesized causal connections." (Markusen 1994: 478) However, this "hypotheses-testing" character of the evaluation process needs, in a first step, some further attention. Although, in general, there is a strong tradition in qualitative research to avoid an ex-ante formulation of hypotheses derived from the theory to not limit the openness of the applied methods and towards the research objects (Meinefeld 2008: 266f.; Flick 2008: 258; Gläser/Laudel 2009: 77ff.), newer debates on qualitative methodology emphasize less orthodox approaches and acknowledge the role of formulating hypotheses in structuring qualitative analyses (for an overview on these debates see e.g. Meinefeld 2008: 271ff.). According to Meinefeld (2008), this structuration not necessarily means a restriction of openness of a qualitative method, since it only provides a conceptual framework of analytical dimensions, but the concrete occurrences of their contents remain open. In addition, data collection methods like semi-structured interviews leave broad space for unexpected results per se. Even further, some qualitative research interests explicitly require structuration through an ex-ante formulation of hypotheses, not least because they address highly complex issues, as is the case with national and regional knowledge transfer systems, and because they address a critical investigation of already existing or self-developed theoretical concepts with regard to specific aspects and allow, therefore, for a certain analytical focusing (Meinefeld 2008: 272ff.; also: Gläser/Laudel 2009: 77ff.).

It is this need for *structuration* of the highly complex topic of knowledge transfer and innovation systems as well as the aim to qualitatively test theoretical assumptions made in the approaches of VoC and NIS, which stand behind the decision to opt for a formulation of *ex-ante hypotheses* and their testing in the study at hand. For the evaluation of the interview transcripts, this structuring effect arises from the use of the systematically developed and adapted interview-guidelines in all conducted interviews (Liebold/Trinczek 2002: 39ff.; Gläser/Laudel 2009: 90ff.; Flick 2009: 203ff.; Lamnek 1995: 22f.; Schulze 2004: 105f.), basing on the heuristic model of knowledge transfer systems and the respectively formulated theoretical assumptions for the three national cases. This structure of the interview-guidelines, consequently, also coins all interview-transcripts, not only allowing to evaluate the single interviews issue by issue, i.e. mechanism by mechanism, but also to *compare* the statements on the single issues across interviews. Where the statements given by the interviewed experts have been consensual and consistent, this has been used to condense the statements to *aggregated evaluations* (Meuser/Nagel 2005 86ff.), which have then been presented in the discussion of the results. This aggregation is in line with the "qualitative cluster analysis", as has been suggested among others by Liebold and Trinczek (2002): "The material is investigated along the case specific combinations of occurrences in the single relevant dimensions, analyzing if clearly differentiable combinations can be detected, which are characterized by a certain internal logic and which can then be used to reconstruct the typology as a whole." (Liebold/Trinczek 2002: 55; translation: MO)

This reconstruction of the respective regional knowledge transfer systems builds on the heuristic model, summarizing for all six investigated regions the results for the six institutional dimensions and the respective 24 mechanisms of knowledge transfer, making, in a next step, also cross-regional (and cross-national and cross-sectoral) comparisons of the results possible. In this, the *aggregation* of the interview-statements is to be seen as an interpretative process by the author in the sense of a "sociological conceptualization" (Meuser/Nagel 2005: 88f.), which eases these *reconstruction* efforts by coming to one evaluation for each investigated issue (Liebold/Trinczek 2002: 54; Gläser/Laudel 2009: 270ff.), whereby, however, affirmative and negative arguments are discussed in a balancing way. Taking into account the total number of interviews and the amount of transcript material, this evaluation strategy also displays considerable research pragmatic advantages, especially in terms of a conclusive and comprehensible presentation of the results (Liebold/Trinczek 2002: 55f.; Mayer 2008: 48).

This presentation of the obtained results takes the form of a text, in which the condensed qualitative evaluations given by the interviewed experts are summarized, i.e. as how important they consider the different mechanisms asked for in the interviews in the respective regions, as well as the explanations they provide for their estimations. However, to achieve a maximal reference to the transcript material, but still taking into account research pragmatic considerations, certain, but limited, parts of the argumentation are underlined, reinforced and illustrated by citing interview passages, what has been described elsewhere (admittedly with a certain skepticism) as "selective plausibilization" (Mayer 2008: 56; Flick 2009: 488f.). Altogether, the chosen evaluation strategy broadly corresponds to the *pragmatic evaluation* method suggested by Mayer (2008), in which the main goal is to extract the over-individual similarities through comparing the transcribed interview material (Mayer 2008: 47ff.; Mühlfeld/Windolf 1981: 325ff.; Meuser/Nagel 2005: 80ff.).

Transcending this pragmatic evaluation method, and for the purpose of a further condensing, summarizing and visualization of the obtained results, an evaluation scheme has been developed, in which the qualitative assessments of the interviewed experts concerning the single mechanisms of knowledge transfer are classified according to an ordinal scale (Vogel 2000: 5; Lamnek 2010: 240f.; Ragin 2007), differentiating between crucial

(+), important but not decisive (O), and those mechanisms which have been evaluated as not important at all (–). This proceeding corresponds to the evaluation strategy of detecting "case-specific occurrences in dimensional typologies", as suggested by Liebold and Trinczek (2002), in which typical occurrences in the different analytical dimensions (here: qualitative evaluations of knowledge transfer mechanisms) are condensed and a consistent typology is reconstructed in a contrasting way (Liebold/Trinczek 2002: 54). Since this proceeding can be interpreted as a certain quantification of qualitative data, it has to be treated very carefully (for a further discussion on quantifying qualitative data see Flick 2009: 47ff.). This quantification is not only highly interpretative and depends on the aggregating judgment of the author, but it also only displays, if anything, a quantitative aggregate of interview statements for the respective regions, what should by no means be confounded with the kind of statistical representativeness pursued in explicitly quantitative studies. This form of statistical representativeness is neither intended nor possible on the basis of the interview data obtained within the framework of this study.

Rather, this proceeding relates to the form of "*logical formalization*" suggested by *fuzzy-set qualitative comparative analysis (fsQCA)*, ascribing qualitative score values between 0 and 1 to single cases (here: mechanisms) belonging to a specific set of cases (Ragin 2000; 2007; 1987; Schneider/ Wagemann 2007: 173ff.). Consequently, in this study crucial (+) mechanisms would be ascribed a value of 1, those being important but not decisive (O) a value of .5 and the ones being irrelevant (–) a value of 0. From the point of view of the author, both the logical formalization and a quantifying condensing of qualitative evaluations appear to be particularly well suited for efficiently summarizing and illustrating the results of the qualitative case studies and for addressing the research questions, not least through allowing an efficient hypotheses testing concerning 24 mechanisms of knowledge transfer in six regional cases with regard to their respective national institutional context, as well as a comparison of the results in a cross-regional, cross-national and cross-sectoral perspective.

One critical aspect in this regard is the question to which degree the six analytical dimensions and the 24 selected mechanisms of the heuristic model can be divided from each other clearly in both analytical and empirical terms. As already stated in section 2.5, the ascription of the 24 knowledge transfer mechanisms to the six dimensions is sometimes ambivalent, and other studies might come to distinct ascriptions. The decisions taken in this study, however, are solidly rooted in theoretical and analytical considerations related to the research question which are extensively discussed above. What has to be taken into account, yet, is that in the empirical analysis of the investigated cases single mechanisms and analytical dimensions cannot fully be divided from each other and might sometimes mutually intervene in their effects as explanatory factors for regional variation. This is, of course, one immanent limitation of the chosen research design and methodological approach of this study and should be kept in mind when valuating the empirical results and interpretations of the case studies. Partly, this can be compensated in the course of the analysis and interpretation of the empirical data by recurring on the theory, as well as on available secondary sources. And still, as the results presented below demonstrate clearly, the chosen research design, and the heuristic model of knowledge transfer systems it bases on, are well suited to approach the research question and, in sum, do deliver a clear, detailed and differentiated picture of the respective regional knowledge transfer systems, their systemic particularities, their regional variation and the major explanatory factors behind it.

Furthermore, to support and complement the evaluation and interpretation of the interview-material, as introduction into each of the six regional cases a brief *quantitative socio-economic profile* is discussed, providing descriptive statistics on major structural economic indicators for the regions and characterizing their knowledge generating capacity. The motivation of presenting these socio-economic profiles is that substantial parts of the argumentation extracted from the interviews refer back to and explicitly emphasize structural conditions as explanatory factors for specific causal relations, which are, of course, not further specified in the interviews in the sense of providing actual, reliable, comparable and detailed statistical data. The purpose is, therefore, to address this shortcoming in the chosen qualitative approach (Flick 2009: 48f.) and to give a brief statistical overview on the major structural indicators mentioned in the interviews as explanatory factors, as it would be e.g. economic structure, industrial structure, business structure, overall and sectoral R&D-intensities, as well as some more specific indicators of the two investigated sectors of automotive and biotechnology. The majority of these data stems from official data bases like, above all, national and regional statistics offices and have then, where necessary, been further processed and calculated by the author. In addition, also other (secondary) data sources have been used, as further studies on the respective regions, information material, or information provided on the websites of regional institutions (Schulze 2004: 4).

Where possible, these data refer to the year 2008, to secure a high degree of comparativeness across the investigated cases. This has two reasons: First, newer data for the years 2009 and 2010 would strongly be biased by the effects of the financial and economic crisis and would display an imprecise picture on the respective structural conditions, and, second, many specific evaluations and data-sets from the official statistics offices were only available for 2008 but not yet for the subsequent years. Overall, this strategy of complementing qualitative interview material with quantitative data corresponds to the "*triangulation* of qualitative and quantitative research" as it is described by Flick (2009: 44ff.; see also Flick 2010; Lamnek 2010: 245ff.), and appears to be well suited to provide a deeper, broader and more comprehensive understanding with regard to the research interest pursued by this study.

4.7 Quality Criteria and Analytical Generalization of the Results

The methodological discussion so far raises the question to which degree the results obtained through the application of the presented research design and methods fulfill the quality criteria of empirical research and to which degree they are generalizable with regard to the research interest. Pointing to the quality criteria first, the qualitative design chosen in this study explicitly follows a strategy of achieving *validity* of the results through carefully sampling the investigated cases, the units of analysis, the units of observation, as well as the data collection and evaluation methods, with the aim to obtain empirical evidence which is as close as possible to the research objects (Lamnek 2010: 134ff.; Flick 2009: 429ff.; Mayer 2008: 55f.). The rationales behind these decisions are sufficiently discussed in the previous sections, so that they shall not be repeated here. Second, and following Flick (2009), *reliability* in qualitative research is basically secured through making the research design and the applied methods transparent and comprehensible (Flick 2009: 489ff.; also: Lamnek 2010: 149ff.; Mayer 2008: 55f.), what has been realized to the largest possible extent in the previous sections. Where this has not been the case, the reasons were restrictions rooting in confidentiality commitments towards the interview partners (Gläser/Laudel 2009: 279ff.), as is e.g. the case with the anonymization of direct and indirect citations from the interview transcripts in the presentation and discussion of the results. However, this restriction concerning the criterion of reliability has to be accepted when applying the data collection method of expert-interviews, since it has to be weighted up against the possibility to include the most adequate and best qualified experts into the sample (Gläser/Laudel 2009: 280f.; Flick 2009: 59ff.). Finally, *objectivity* in qualitative research can be defined as *emergentistic* (Lamnek 2010: 158), in the sense that qualitative objectivity arises from the analysis of subjectivity of the interaction partners (interviewees, interviewer/researcher). This represents, therefore, an "inter-subjective" notion of objectivity, through which the subjectivity of qualitative research is overcome through generalization in the research process and the inter-subjective traceability of the results (Lamnek 2010: 154ff.; Flick 2009: 499f.; Przyborski/Wohlrab-Sahr 2009: 40ff.). According to this definition, the discussion of the methodology and the methods applied has shown that the research design chosen for this study fulfills to the largest possible extent also the criterion of objectivity in qualitative research.

However, even if this study's research design and the obtained results fulfill the criteria of validity, reliability and objectivity of qualitative research to the described degree, the question remains open to what extent the results for the single case studies are *generalizable* to other cases? While in quantitative research generalization is secured through the concept of statistical representativeness, in which the typical is assigned to the general, basing on random sampling of a high number of investigated units of analysis (Lamnek 2010: 166f.), in qualitative research generalization emerges out of extracting the general out of the typical, especially through investigating fewer, but typical cases being selected by theoretically-systematically grounded sampling strategies and through creating types, as e.g. ideal types, extreme types or important types (Lamnek 2010: 166f.; Flick 2009: 522ff.; Przyborski/Wohlrab-Sahr 2009: 311ff.). This leads to a form of type-building in the sense of *representation*, rather than representativeness in a statistical sense (Lamnek 2010: 167). Taking these qualitative notions of representation and generalization into account, the study at hand intends to achieve representation through purposefully sampling typical cases (national market economies, established industrial regions) and atypical cases (metropolitan regions; biotechnology or automotive sector), as well as appropriate units of observation (interviewed experts) according to the case selection criteria discussed in the previous sections, which seem to be adequate for addressing the research questions.

Although according to the previously outlined standards of qualitative research a high degree of representation is achieved, the extent to which the obtained results can be generalized requires some further specification. Starting with the level of the units of observation, i.e. the interview partners first, a total number of interviews per region varying between five and eight might be considered as too low for capturing a broad overview on the existing qualitative evaluations concerning the research interest, given that it is directed towards highly complex social systems on the macro level, as are regional knowledge transfer systems with their multiple and variegated actor structures. Even if a sample of about 20 interviews per region would have allowed to at least include one representative of all central institutions involved in the respective regional knowledge transfer systems, research pragmatic considerations as well as limited resources impeded a broadening up of the sample. However, the results demonstrate that even with a small scaled, but broadly diversified and purposefully sampled spectrum of interviews highly consistent views and qualitative evaluations given by the interviewed experts have been detected. This could be interpreted in a sense that regarding the research interest a certain empirical as well as theoretical *saturation* (Przyborski/Wohlrab-Sahr 2009: 182; Strauss 1991: 21) is achieved even with this limited scale of the sample, so that the relatively small total number of interviews not necessarily limits the extent of representation of the investigated units of analysis (here: investigated regional knowledge transfer systems). Therefore, a generalization of the results obtained from the interviews to the respective regional knowledge transfer systems as a whole appears to be appropriate and possible, and the research object well captured by the chosen research method.

Concerning the second level of analysis, the investigated regional knowledge transfer systems, the question arises to which degree the results of two investigated regions per national market economy model can be generalized to the respective national economies as a whole? A higher level of representation could of course be reached if more regions per national case would be included into the sample, what has, again, on the one hand, not been practicable because of research pragmatic reasons. On the other hand, and taking into account these restrictions, the limitation of the number of investigated regions per national case had to be weighted up against also investigating regions within three distinct national market economies, as they are discussed in the VoC-approach, to analyze if regional variation also varies across distinct national institutional frameworks. As a consequence, of course, it must be pointed out clearly that the obtained results for each national case only reflect a limited part of the possible regional variation in these cases and the respective explanatory factors behind it. Even if other regions within the same national cases might also vary to a similar extent from their national institutional frameworks, the particular explanatory factors as, e.g. structural conditions or actors' constellations might considerably differ from those in the regions investigated by this study, as is already indicated by the variation detected among the investigated regions themselves. However, through the sampling of most different types of regions (established industrial vs. metropolitan regions), and through the selection of the most typical representatives of these two types of regions within each national case, it has been intended to increase the level of representation, by making the results (to some extent) generalizable (at least) to other established industrial and metropolitan regions within the same national cases.

Therefore, the kind of generalization concerning the national cases allowed by these results is limited to the detecting and stating of regional variance, as well as to the specific explanatory factors behind this variation in the investigated regions, which might also explain variation in other, not investigated regions. Nevertheless, further forms of regional variation and further explanatory factors for this variation might exist in each of the investigated national cases, which have not been captured by this analysis, and for which to detect an alternative research design would have been necessary, including a larger number or even all of the regional innovation systems within the respective national market economies. However, since it is not the aim of this study to give a full overview on the full range of regional variance in national systems and their explanatory factors, but rather, to challenge the assumption of institutional coherence of national market economy models and to explain the existing regional variance in a comprehensive way, the degree of generalization made possible by the obtained results appears to be absolutely sufficient for addressing the research questions guiding this study.

Finally, for the third level of analysis, the investigated national market economies, it must be asked to which degree the results of the investigated national cases can be generalized to the respective national market economy models (coordinated, Mediterranean, liberal) as a whole? Again, also for this level of analysis, research pragmatic reasons impeded an enlargement of the sample, and again it could be asked if an alternative research design focusing on only one market economy model and investigating three national cases within it would have increased the possible level of representation. However, as already discussed, this study intends to also compare regional variation across national models, so that the chosen research design appears to be well substantiated altogether. In addition, the already discussed sampling strategy of selecting the most paradigmatic cases of European economies discussed in the literature on VoC allows for drawing also generalized conclusions on the respective models as a whole, provided that the typology of market economy models suggested by VoC correctly classifies the single cases to the respective types. Yet, one might have to differentiate at this point between the three investigated cases: While Spain and especially Germany are also in international comparison discussed as the most paradigmatic cases for their respective model (Mediterranean and coordinated) and have even been exemplary cases, upon which the respective models have been developed, the case of the UK is different. Although the UK is the most important representative of the liberal model in Europe, the most paradigmatic case in international comparison in this category is the USA, which has mostly been used as exemplary case to develop the category of LMEs as a whole. Therefore, it is to be assumed that the case of the UK might not only show certain deviation from the ideal type of a LME as defined by the VoC-approach, but that it might also be less congruent with the ideal model it represents than its Spanish and German counterparts. Thus, from a theoretical-conceptual view of representation, it can be assumed that the possible degree of generalization might be lower in the case of the UK than in the two other national cases.

5. The German Case—Knowledge Transfer Systems in a Coordinated Economy

5.1 The German Economy—Conserving Industrial Structures

In the previous chapters, Germany has been described as a CME according to the VoC-approach (Hall/Soskice 2001b). Building on this, Germany has been selected as an example of a CME in Europe in the case study at hand. This section will give a brief overview on some fundamental empirical indicators to describe Germany's current economic structure. Germany is the largest country in the EU with 81.9 million inhabitants in 2009 (IW 2010: 7) and Europe's strongest economy with a GDP of 2.46 trillion Euros in 2008 (Destatis 2010: 5; Eurostat 2010), representing 19.7% of the EU's GDP. Much stronger than in other EU-countries, Germany's economic structure is still pointedly shaped by the industrial sector, which in 2008 covered 25.6% (without construction) of the country's total gross value added (GVA), a value ranking considerably above those of Spain (17.3%) and the UK (17.6%). By contrast, Germany's primary sector is comparatively small with a share of 0.9% in total GVA, being less than one third of the value for Spain (2.8%), but almost equaling the value for the UK (0.8%). However, also in Germany the most important economic sector is services, with a share of 69.2% in the total GVA, a value almost equaling the one of Spain (68.4%), but ranking markedly below the UK (75.4%) (Figure 6; Eurostat 2010: 100).

The German business structure is shaped by both small and medium sized firms, as well as large corporations. In 2008, 89.9% of all firms in Germany were micro-firms with less than ten or no employees. Small firms with ten to 49 employees represented 7.9% of all German firms, medium sized firms from 50 to 249 employees 1.8%, and large firms with more than 250 employees only 0.3%. In this, the micro-firms only employed 17.2% of the country's total workforce in the private sector, while small, medium and large firms employed 23.2%, 27.6%, and 31.9%, respectively, pointing to a comparatively strong SMEs business segment and a well balanced business structure of SMEs and large firms[1] (Figures 4 and 5).

Fig. 4: Business Structure Germany I

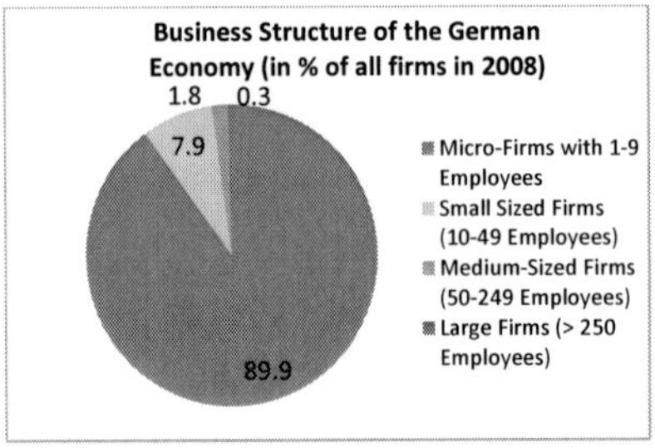

Source: Own Calculations and Illustrations based on Bundesagentur für Arbeit (2011)

Fig. 5: Business Structure Germany II

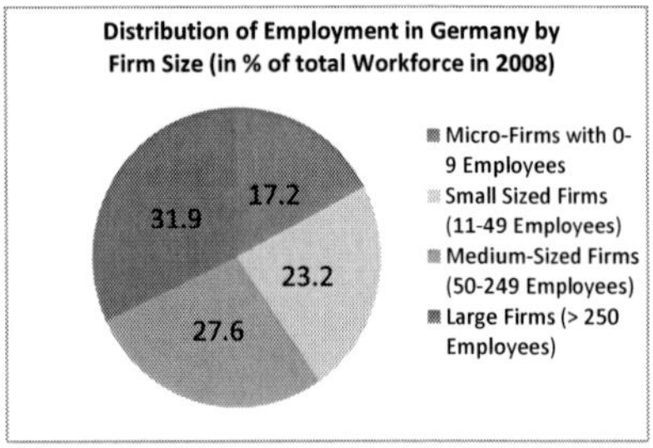

Source: Own Calculations and Illustrations based on Bundesagentur für Arbeit (2011)

A closer look at the sectoral structure of the German industry mainly discloses medium and medium-high technology sectors. In this, the industrial sector is characterized by a high degree of diversity, but with a clearly dominating automotive industry as the country's core manufacturing sector (Figure 7): In 2008, automotive had a share of 21.6% in the total sales volume of German industry, followed by mechanical engineering (13.2%), the metalworking industry (12.8%), food processing and luxury foods (10.0%), electrical engineering, (9.5%) and chemicals (9.2%)[2].

Looking at the R&D-activities of the German economy as indicator of its knowledge generating capacity, in the year 2007 506,450 persons were employed full time in the field of R&D in Germany, representing 1.27% of the total workforce, of which the business sector had a share of 63.6%, while 36.4% were employed at universities and other non-commercial public and private research institutions[3]. In terms of this R&D-personnel-intensity, Germany ranks clearly above the EU-average of 1.04% and the values for Spain (0.94%) and the UK (1.1%) (Eurostat 2011a; Figure 7). The total expenditures for R&D-activities in Germany were about 61.5 billion Euros in the same year, representing 2.5% of the German GDP, and of which 69.9% were covered by the business sector and 30.1% by the

1 Own calculations based on data from the *Beschäftigungsstatistik der Bundesagentur für Arbeit* (2011); data for 2008

2 Own calculations based on data from Destatis (2011), data for 2008

3 Data from Destatis (2011), data for 2007

public sector. The business R&D-intensity (share of R&D-expenditures in total sales volume) of all German firms was of 3.9% (IW 2010: 95ff.), the R&D-intensity of the manufacturing sector of 3.2%.

Fig. 6: Sectoral Structure of Germany, Spain and the UK

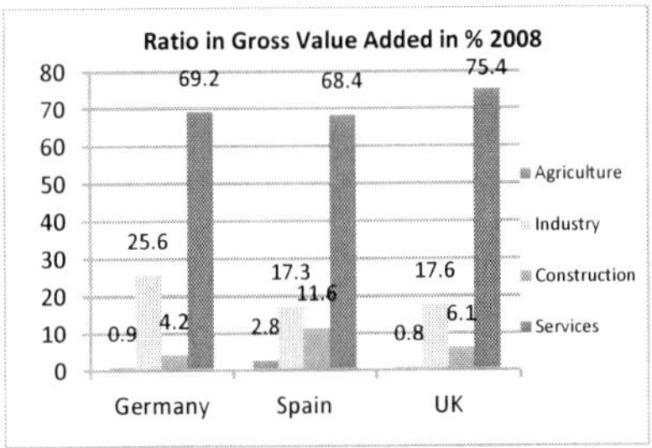

Source: Own calculations and illustration based Eurostat (2010): 100

Fig. 7: Germany's Industrial Structure

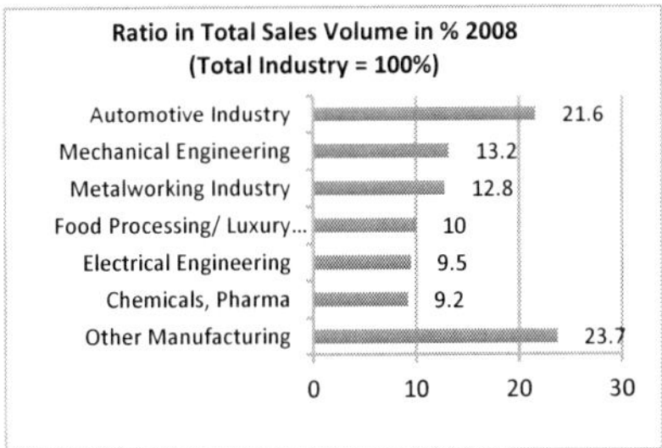

Source: Own calculations and illustration based on on data from Destatis (2011)

Fig. 8: Sectoral R&D-Performance in the German Industry

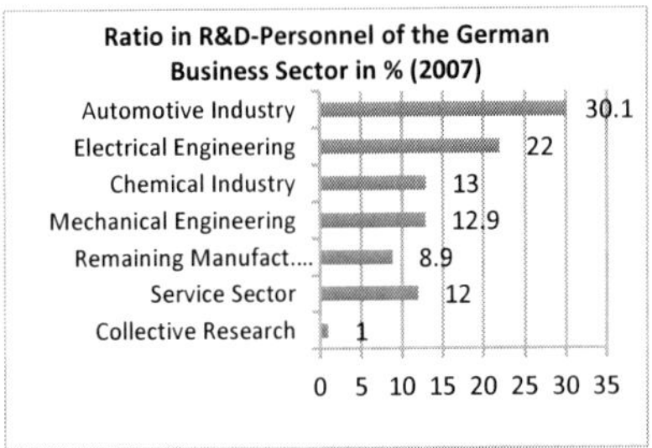

Source: Own Illustration based on Legler/ Schasse (2010): 20

Fig. 9: R&D-Personnel-Intensity in Selected Countries and Regions

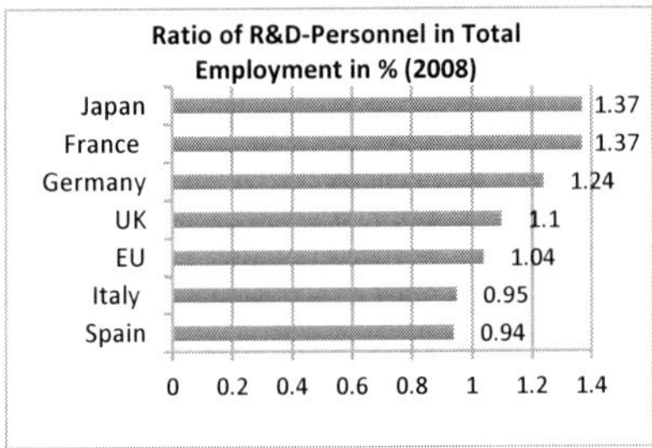

Source: Own Illustration based on Data from Eurostat (2011)

When it is also differentiated in this context between SMEs (< 500 employees) and large firms (> 500 employees), the dominance of the latter in the country's R&D-activities becomes obvious: While the SMEs had a share of 14% in the total R&D-expenditures of the business sector, and employed 22% of its R&D-personnel, the large firms covered 86% of the expenditures and 78% of the R&D-personnel, respectively. Differentiated by sectors, Germany's key industries are also leading in terms of their R&D-activities: Measured by their share in the R&D-personnel of the business sector, automotive (30.1%), electrical engineering (22%), chemicals (13%) and mechanical engineering (12.9%) together covered about 78% of the employees in the field of R&D (Legler/Schasse 2010: 17ff.; IW 2010: 95ff.).

The first sector this study focuses on, automotive, is Germany's key industrial sector. Automotive contributed 12.2% to Germany's total GDP in 2009, representing 22.9% of the added value of the country's manufacturing sector. These are more than the double of the average values for the other western industrial countries, demonstrating not only the extraordinary significance automotive has for the German economy, but also its high international competitiveness. In 2007, the German automotive sector employed 848,200 persons, representing 14.2% of the total employment in manufacturing (Legler et al. 2009: 19; IW 2010: 35). In this, the German automotive sector consisted of 1,319 companies (Destatis 2009: 371), of which the most important in terms of employment and sales volume are the large OEMs VW, Daimler, BMW, Porsche and MAN, as well as the large suppliers Bosch, Continental, and ZF (Legler et al. 2009: 95f.).

The sector's business structure is markedly shaped by a strong medium sized business segment (50–249 employees), which represents 43.6% of the total number of firms in the sector. Small firms with less than 50 employees represent 26.9%, and the large firms with more than 250 employees 29.5% of all firms in the sector. However, the distribution of employment in the sector reveals the high average firm size of the large German automotive firms: While the small firms employ only 1.7%, and the medium sized firms 9.5% of the total workforce in the sector, the large firms by far have the major share of 88.8% in the sector's total employment[4]. In this, German automotive is a highly networked industrial sector, in which large and small firms, OEMs, customers and suppliers are interconnected intensively with each other as customers and suppliers. In 2004, more than 78% of the sector's total production value stemmed from supplied advanced performances (components, materials, services), reflecting the strong intra- and inter-sectoral division of labor. This implies not only supplies of suppliers among each other, but also supplies between suppliers and OEMs and supplies within the OEMs themselves (Legler et al. 2009: 51ff.).

When looking at some indicators of the R&D-performance characterizing German automotive, it becomes clear that it is a highly R&D-intensive sector: In 2007, around 30% of the total internal R&D-expenditures

4 Own calculations based on data from Destatis (2011), data for 2008

of the German economy have been realized by automotive, contributing decisively to the overall innovativeness of the country. Its comparatively high intra- and international innovativeness is also illustrated when looking at its R&D-intensity (share of R&D-expenditures in total sales volume): With an R&D-intensity of 5.9% German automotive is not only high above the value for German manufacturing as a whole (3.2%), but also considerably above the mean for the automotive sectors of the largest OECD-countries (3.3%) and their respective manufacturing sectors (2.5%) (Legler et al. 2009: 90ff.; Figure 21).

In this, the R&D-activities of the German automotive industry are mostly concentrated in the large companies with more than 1000 employees, up to 98.1% in 2007. Despite the increasingly higher capital linkages with companies abroad, investments in R&D by foreign companies or supranational institutions only cover 3.2% of the total R&D-activities in the sector. Even less significant is the role of the state, which only has a contribution of 0.3% to the sectoral R&D-activities. Differentiating between internal (within a company) and external R&D-activities, more than one fourth of the total R&D-activities are conducted externally by cooperating partners or specialized service providers, and, above all (more than two thirds) by affiliated companies within the corporation. 77% of these external R&D-activities are conducted by domestic companies, around 10% by universities and other research institutes in Germany, and about 13% by partners abroad. All R&D-activities in the automotive sector are strongly dependent on researchers and engineers, who conduct about 60% of these activities, reflecting the increasing academization of the R&D-personnel in the last years, creating a strong demand for qualified staff on the labor market (Legler et al. 2009: 104ff.).

German biotechnology, on the other hand, is a comparatively young sector. Its emergence and growth in Germany are closely connected to the two public programs of *BioRegio* and *BioProfile* funded by the Federal Ministry of Education and Research. In the second half of the 1990s and the beginning of the 2000s these programs had the aim to initiate regional competition between networks of actors engaged in biotechnology to promote the growth of the sector in Germany, which until then had been underdeveloped in international comparison, but was expected to be a key industry of the future. Altogether, the BioRegio-Regions Munich, Rhineland, Rhine-Neckar-Triangle and Jena, as well as the BioProfile-Regions Braunschweig-Göttingen-Hannover, Potsdam/Berlin and Stuttgart

(STERN) were funded with about 200 million Euros through these programs between 1997 and 2006 (Stähler et al. 2007: 4ff.; Krauss/Stahlecker 2001; Kaiser/Prange 2004).

The importance of these public programs for the development of the German biotechnology becomes clear, when looking at some structural data: For example, the dynamics of business start-ups is significantly higher in the funded regions than it is in the other German biotech regions, which can be explained in particular with the provision of seed- and venture capital through these programs. Moreover, also the number of biotech companies has grown significantly stronger in the funded regions than in the other biotech regions, in terms of the average annual growth rate for the period between 1997 and 2005 by 9.6% in the former and by 7.4% in the latter (Stähler et al. 2007: 4ff.). Altogether, around 60% of the biotech companies that were active in the year 2009 started their activities before or around 2001 in the context of the two competitions (biotechnologie.de 2010: 6). Looking at the regional distribution of biotech activities in Germany, 57% of the biotech companies in Germany are located in one of the seven funded regions, while 43% are located in the remaining 14 biotech regions. This regional distribution can also be observed when looking at the long term (period between 1998 and 2006) growth rates of employment in biotech companies: While the employment rose by 11.8% in the BioRegio-regions and 3.3% in the BioProfile-regions, it only grew by 2.4% in the not-funded regions (Stähler et al. 2007: 38ff.).

However, and despite this strong public support, today German biotechnology still only has a marginal share in the overall performance of the German economy, especially when compared to its leading sectors like automotive. In the year 2009, 531 firms were active in the biotech sector, employing 14,500 persons (Ernst & Young 2010: 6), representing only 0.036% of the total German workforce (Bundesagentur für Arbeit 2010: 18). These firms created a total sales volume of 2.2 billion Euros (Ernst & Young 2010: 6), representing a share of 0.09% in the German GDP[5]. In this, most of the German biotech companies remain small in size. Almost half of them (45%) have less than ten employees, while a comparatively large percentage (42%) has between ten and fifty employees. Only 28 of the German biotech companies had more than one hundred employees in 2009, and the largest German biotech company with more than 3,500 em-

5 Own calculation on the basis of Destatis (2010): 9

ployees worldwide is the purification and diagnostics specialist *Qiagen*, located in North Rhine-Westphalia (biotechnologie.de 2010: 9).

Looking at indicators describing the sector's knowledge generating capacity, German biotechnology can be characterized as extraordinarily R&D-intensive. In 2009, the German biotech companies invested around one billion Euros in their R&D-activities (Ernst & Young 2010: 6; Biotechnologie.de 2010: 15), representing almost half of their total sales volume. In the year 2005, around 53% of the employees in German biotech companies were R&D-personnel (biotechnologie.de 2007: 2). Apart from basic research, in particular in the field medical biotechnology the development of new products, and especially new pharmaceuticals, requires long clinical test phases (phase I–III) of up to ten years or more, before a new pharmaceutical is approved for commercial distribution. In total, German biotech firms had 141 drug candidates in phases of clinical trials in 2009, clearly less than the UK (235), but considerably more than Spain (35). In addition, the German biotech firms had 197 drug candidates in pre-clinical phases, but only two in the phase of approval. Looking at the distribution of drug candidates in the three clinical phases, 30% of the drug candidates were in phase I, 60% in phase II and only 10% in phase III. In comparison to the mean of all other European countries active in biotechnology, thus, the German biotech firms have comparative strengths in drug candidates in phase II, but are behind in drug candidates in phase III, pointing to certain weaknesses of the sector in bringing its efforts in R&D to the markets, but also to its still young business demography (Ernst & Young 2010: 42).

Finally, also in Germany biotechnology, cooperation with other actors plays a role. In the year 2009, the most important partners in these cooperations were research institutes with a share of 46% in all cooperations of the sector, followed by other biotech firms (22%) and industrial partners (in particular pharmaceutical firms) with a share of 30%.[6]The main fields of cooperation were research (37%), development (28%), validation (13%) and sales (13%), whereby about 29% of these cooperations took part with foreign partners[7] (biotechnologie.de 2010: 14).

6 Own calculation on the basis of biotechnologie.de (2010): 14

7 Own calculation on the basis of biotechnologie.de (2010): 14

5.2 The Case of Baden-Württemberg

5.2.1 Locating Baden-Württemberg in the German Context

5.2.1.1 Selected Facts on Baden-Württemberg's Economic Structure

Baden-Württemberg is one of the sixteen German *Länder*, being located in the south-west of the country and bordering on France in the west, Swiss in the south, on Bavaria in the east and on Hesse and Rhineland-Palatinate in the North, and has a population of 10.7 million. Coming from a mainly agricultural tradition, the early development of an intensified textiles sector in the 19th century paved the way for the emergence and growth of a mechanical engineering industry, which was an important precondition for Baden-Württemberg's extraordinarily successful course of development in the early post-war decades. After the destructions of World War II and the constitution of the Land in 1951, Baden-Württemberg could experience a dynamic and catching up strategy of industrialization, which was responsible for extraordinary economic growth rates in the 1950s and 1960s. Especially in the 1950s the Baden-Württemberg economy expanded much faster than its West German counterpart, not least because it was able to export a comparatively large share of its products (Krauss 2009: 232). The agricultural sector, which still covered one quarter of the workforce at the beginning of the 1950s, declined very quickly, covering in the year 2007 only 1.8% of the workforce (Krauss 2009: 232; SLA BW 2009: 20), while the number of employees in the manufacturing sectors rose from 1.4 million in 1950 to 2.3 million in 1970 as a peak value, representing 55.9% of the workforce, compared with 48.9% for Germany as a whole.

Baden-Württemberg's successful strategy of industrial development converting it into one of Europe's strongest industrial regions was especially based on the investment goods industry, with mechanical engineering, automotive and electrical engineering in its center. In these three branches the number of employees more than tripled since 1950, from 225,000 to 761,000 in 2007 (Heidenreich/Krauss 2004: 218f.; Krauss 2009: 232; SLA BW 2009; BW MoE 2008: 7). These branches stand in the center of two relatively autonomous industrial clusters which are mainly located in the Stuttgart region: Automotive with its suppliers and mechanical engineering, which covers, apart from some larger companies, many medium sized firms producing a broad spectrum of production technologies. In addition, and complementing these traditional clusters, a new cluster of business related software and service firms developed, including firms like SAP, IBM or Hewlett-Packart (Krauss 2009: 232f.).

Fig. 10: Sectoral Structures: Germany and Baden-Württemb. in Comparison

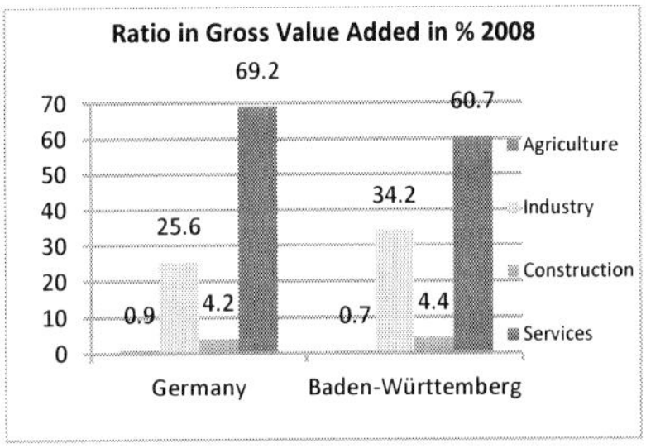

Source: Own Calculations and Illustration based on Eurostat (2010): 100 and SLA BW (2011)

Fig. 11: Baden-Württemberg's Industrial Structure

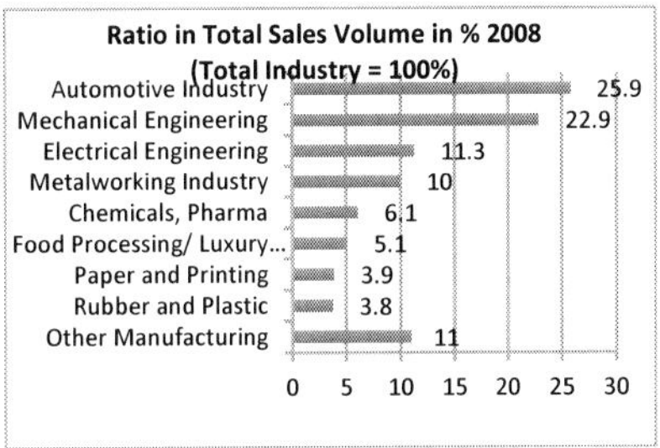

Source: Own Calculations and Illustration based on data from SLA BW (2011)

Today (2008), Baden-Württemberg's sectoral structure is still strongly shaped by industrial production, which has a share of 34.2% (without construction) in the region's total GVA, a value which is highly above the value for Germany as a whole of 25.6% and identifies it as one of the country's major industrial poles. Services, on the other hand, covers 60.7% of the region's GVA and ranks clearly behind the value for Germany of 69.2%, pointing to a comparatively less tertiarized regional economy. Finally, the primary sector covers 0.7% of the region's total GVA (Figure 10; SLA BW 2011; BW MoE 2008: 8).

Looking at Baden-Württemberg's industrial structure, it becomes obvious that the region's core branches of mechanical engineering, automotive and electrical engineering constitute the heart of the regions industry model, and concentrate around 60% of the total sales volume of the region's industry. In detail, automotive is the region's strongest industrial sector, representing a share of 25.9% in the total sales volume of the region's industry, followed by mechanical engineering (22.9%), electrical engineering (11.3%), metalworking (10%), chemicals (6.1%), food processing and luxury foods (5.1%), paper and printing (3.9%), and rubber and plastic (3.8%) (Figure 11; SLA BW 2011). Especially the three regionally concentrated investment goods industries are characterized by extraordinarily high export rates, making the region much more export-oriented and -dependent than the country as a whole. In the year 2008, 48.6% of the regional industrial production was exported, with automotive exporting

62.5%, mechanical engineering 59.2%, and electrical engineering 49.5% of their production[8] (SLA BW 2011; SLA BW 2009: 10).

Fig. 12: Business Structure BW I

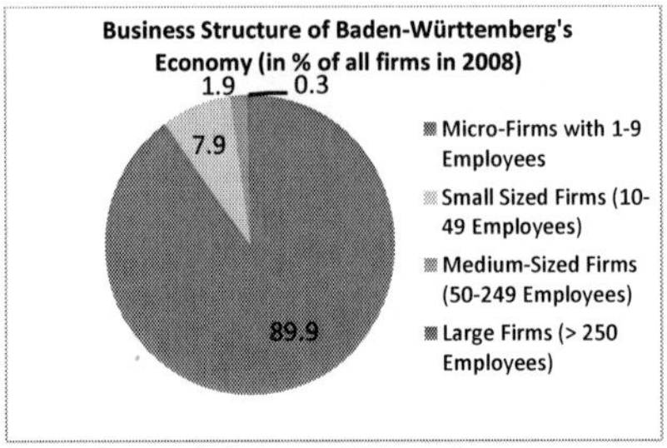

Source: Own Calculations and Illustration SLA BW (2011)

Fig. 13: Business Structure BW II

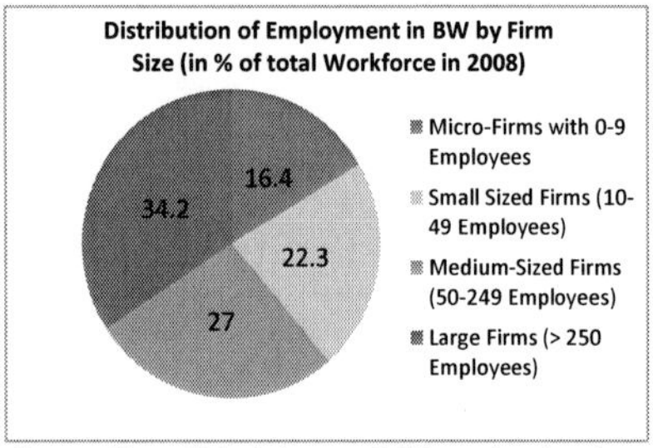

Source: Own Calculations and Illustration based on based on SLA BW (2011)

Baden-Württemberg's business structure very much resembles its national German counterpart in being shaped by both small and medium sized firms, as well as large corporations. In 2008, 89.9% of all firms in Baden-Württemberg were micro-firms with less than ten or no employees. Small firms with ten to 49 employees represented 7.9% of all firms in Baden Württemberg, medium sized firms from 50 to 249 employees 1.9% and large firms with more than 250 employees only 0.3%. In this, the micro-firms only employed 16.4% of the region's total workforce in the private sector, while small, medium and large firms employed 22.3%, 27.0%, and 34.2%, respectively, pointing to a comparatively strong and diversified SMEs business segment and a well balanced business structure of SMEs and large firms in the region[9] (Figures 12 and 13; SLA BW 2011).

One decisive factor of this economic strength is Baden-Württemberg's R&D-activities. In 2007, the R&D-intensity measured as gross domestic expenditures in R&D (GERD) was of 4.2%, being one of the highest values among all European NUTS-1-regions. Within Germany, Baden-Württemberg has by far the highest R&D-intensity of all *Länder*, ranking clearly above Berlin (3.4%), Bavaria (2.8%), Saxony (2.6%) and Lower Saxony (2.5%), and almost doubles the German average R&D-intensity of 2.5%. Also Baden-Württemberg's total expenditure for R&D activities of

8 Own calculations based on data from Statistisches Landesamt Baden-Württemberg SLA BW (2011); data for 2008

9 Own calculations based on data from the *Beschäftigungsstatistik der Bundesagentur für Arbeit* (2011); data for 2008

15.7 billion Euros is the highest in Germany, representing more than 25% of Germany's total expenditure in R&D (Figure 14, SLA BW 2011; Egetemeyr/Werner 2008: 17ff.).

Fig. 14: R&D-Intensities in Selected German Regions

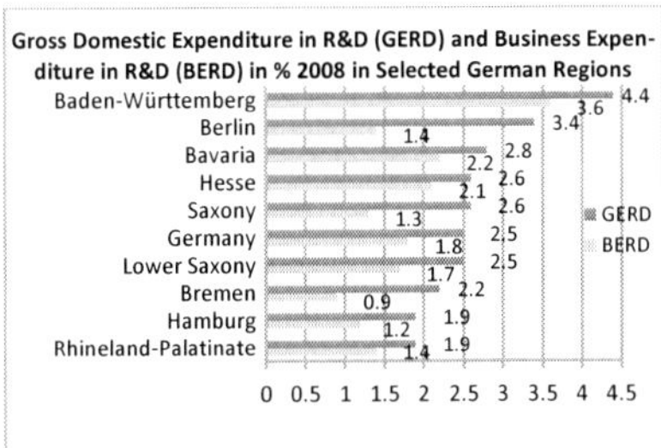

Source: Own Calculations and Illustration SLA BW (2011)

Fig. 15: Sectoral R&D-Performance in Baden-Württemberg's Economy

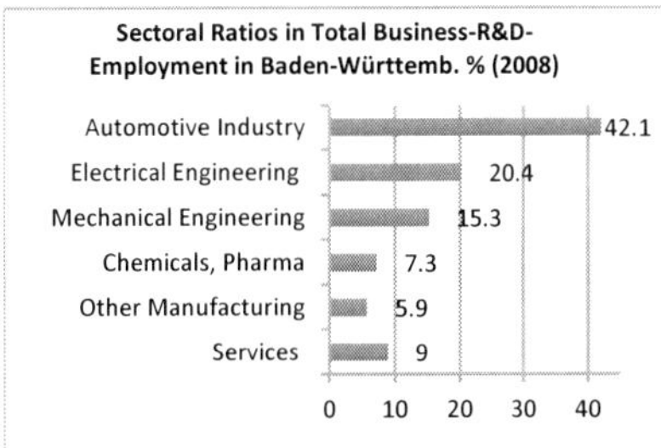

Source: Own Calculations and Illustration based on based on SLA BW (2011)

Herein, the private sector accounts for 81.4% of R&D-expenditures, a value which ranks clearly above the national German average of 70%. This represents a business R&D-intensity (BERD) of 3.6%, more than doubling the national value of 1.6%, and points to an extraordinarily R&D-intensive business sector in the region, which is further underlined by the fact that almost 30% of the German private sector's total R&D-expenditures are realized by Baden-Württemberg firms alone, clearly leaving behind Bavaria (22.3%) and Northrhine-Westphalia (13.9%) on the subsequent ranks. In this, Baden-Württemberg's business R&D-activities are concentrated clearly in the three core manufacturing branches, with automotive accounting for 42.1% of the region's total R&D-employment, followed by electrical engineering (20.4%) and mechanical engineering (15.2%) (Figure 15; SLA BW 2011; Egetemeyr/Werner 2008: 17ff.).

5.2.1.2 Baden-Württemberg's Automotive Sector—Structural Facts

Baden-Württemberg constitutes the center of the German automotive industry. With a total sales volume of about 106 billion Euros in 2009 it created almost one third of the sectors total sales volume in Germany, more than any other of the German *Länder*. Baden-Württemberg's automotive sector consists of around 740 firms, which represent about one third of the firms in the sector in Germany. The most important of these firms are the two large OEMs *Daimler AG* and *Porsche AG*, but also the *Audi AG* maintains an important production and development location in Baden-Württemberg. Moreover, also in the field of utility vehicles some of the most important OEMs are located in the region, as e.g. the *Mercedes-Benz LKW* (as a part of the *Daimler AG*), *Volvo Busse Deutschland GmbH* and the *Kässbohrer Geländefahrzeuge AG*. In addition, a large number of important suppliers to these OEMs are located in Baden-Württemberg, of which the most renowned are the *Robert Bosch GmbH*, *ZF Friedrichshafen AG* as well as the *Mann & Hummel GmbH*. These automotive OEMs and suppliers account altogether for almost one quarter of the employment in the regions manufacturing sector and for about 41% of the total workforce in German automotive (SLA BW 2011; BW-Invest 2011a).

The automotive industry in Baden-Württemberg appears to be highly networked and cooperative and is characterized by a large variety of business networks, cooperation projects (including automotive firms and other (institutional) actors) and cluster initiatives. In the region, 16 sectoral networks and cluster initiatives do exist, which are complemented by two interregional cluster initiatives (*Commercial Vehicle Cluster Southwest* and *Automotive Cluster Rhine-Main-Neckar*) which also include other regions, although the majority of the participating firms and actors in these two clusters are located in Baden-Württemberg. The nodal point of these networks is situated in the Stuttgart region, where six cluster and network initiatives are located with more than 300 participating firms and research institutes, making Stuttgart the most important automotive cluster in Europe. Other important agglomerations are Karlsruhe and Heilbronn with two cluster initiatives, respectively. The most important of these clusters are the *CARS-Initiative*, the *Fuel Cell Alliance Baden-Württemberg* and the *Cluster Initiative Automotive Southwest*, which are all located in the Stuttgart region. In addition, two region-wide cluster initiatives do exist, which try to coordinate the sectoral activities at the level of the *Land*. *Autoland-bw* as a cluster initiative by the Baden-Württemberg Ministry of Economics focuses on supplier firms, while the cluster initiative of *automotive-bw* tries to establish a centralized, region-wide and independent cluster management (BW-Invest 2011b; BW MoE 2010a).

Looking at some R&D-related indicators, automotive in Baden-Württemberg is characterized by the largest expenditures in R&D of all the German *Länder*. In 2009, the region's automotive firms had a share of about 40% of all R&D-expenditures in the sector in Germany, and they employed about 37% of its R&D-personnel. Apart from the sector's internal R&D-activities, research in automotive-relevant fields is conducted in 67 research institutes within the region, of which 46 are connected to universities or other higher education facilities (*Fachhochschulen*). In addition, also the Max Planck Society (1), the Fraunhofer-Society (8) and the Helmholtz Association (3) maintain important research institutes in Baden-Württemberg. The most renowned of these institutes are the *KIT* in Karlsruhe, the research centers of the *DLR* and the *Max Planck Institute for Metals Research* in Stuttgart. This R&D-landscape is complemented by two science parks in which sector-relevant research is conducted, namely Ulm (hosting the *Daimler Research Center*) and the *Stuttgart Engineering Park* (BW-Invest 2011c; BW MoE 2010b; BW MoE 2007).

5.2.1.3 Baden-Württemberg's Biotech-Sector—Emerging Structures

The biotech sector in Baden-Württemberg is concentrated in five bioregions, which are the BioRegion Rhine-Neckar-Triangle, the BioRegio STERN (Stuttgart, Tübingen, Esslingen, Reutlingen und Neckar-Alb), the BioRegionUlm, BioRegio Freiburg and the BioLago Region around the Lake Constance. In four of these regions five biotech-parks have been established with the support of the Baden-Württemberg government, which are located in Freiburg, Heidelberg, Ulm, Tübingen/ Reutlingen and Esslingen and in which research institutes, biotech firms, supporting institutions and service providers have the chance to closely interact with each other. All of these five bioregions have their own cluster-management organizations, whose work is complemented by the *BIOPRO Baden-Württemberg GmbH*, a region-wide public service agency founded by the regional government and promoting business development, location development and public relations in the biotech sector (biotechnologie.de 2010; Ernst & Young 2010; bw-international 2011; BIOPRO 2010; BIOPRO 2011).

In the year 2009, the biotech sector in Baden-Württemberg consisted of 84 companies, representing 15.8% of all biotech companies in Germany (531). In the comparison of the German *Länder*, only Bavaria had a larger number of biotech companies (105). These 84 companies had 2,303 employees, representing 15.9% of all employees in the German biotech sector, and being the third largest contingent of employees in the sector after Bavaria (3,279) and North Rhine-Westphalia (3,210). As in the rest of

Germany, also the biotech sector in Baden-Württemberg is characterized by a majority of small businesses: Around 45% of the firms have between one and ten employees, another 45% have between eleven and 50, and only 10% have more than 50 employees. The biotech sector in Baden-Württemberg generated a total sales volume of about 240 million Euros, representing 11% of the total sales volume of the sector in Germany, and contributing 0.07% to Baden-Württemberg's GDP[10] (Figures 16 and 17; biotechnologie.de 2010; Ernst & Young 2010; SÄdL 2011; bw-international 2011; BIOPRO 2010; BIOPRO 2011).

Fig. 16: Regional Distribution of Biotechnology Firms in Germany

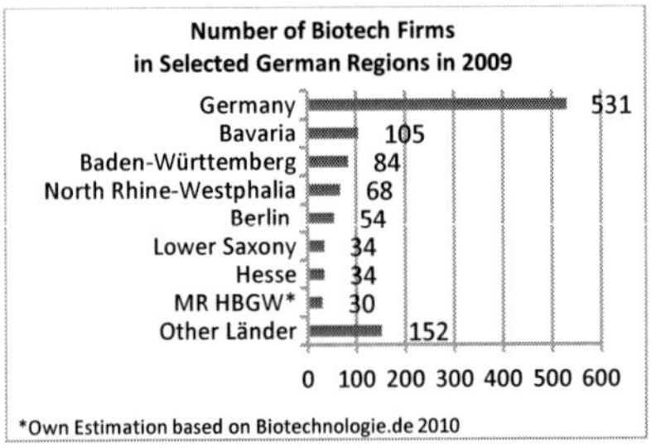

Source: Own Illustration based on Biotechnologie.de 2010: 8

Fig. 17: Regional Distribution of Biotechnology Employment in Germany

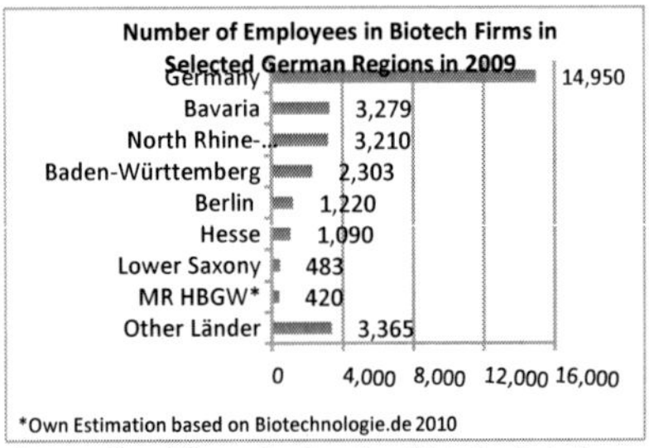

Source: Own Illustration based on Biotechnologie.de 2010: 8

Apart from their internal R&D-activities, research in the field of biotechnology is conducted in a large variety of research institutions such as universities, other higher education facilities, or research institutes. In 2009, 27 of these research institutions conducting research in the field of biotechnology were located in Baden-Württemberg, representing 8.4% of these institutions in Germany. These institutions in the region employed 4,499 persons involved in respective research activities, representing a share of 16.8% of the total number of employees in this field in Germany (Figures 18 and 19). The most renowned of these institutes are the *German Cancer Research Centre* (DKFZ) and the *European Molecular Biology Laboratory* (EMBL) in Heidelberg as well as the *Max Planck Institute for Developmental Biology* in Tübingen (biotechnologie.de 2010; BIOPRO 2010; BIOPRO 2011).

10 Own calculations based on biotechnologie.de (2010), bw-invest.de (2011) and Statistische Ämter der Länder und des Bundes (SÄdL) (2011)

Fig. 18: Scientific Institutions in German Biotechnology: Regional Distribution

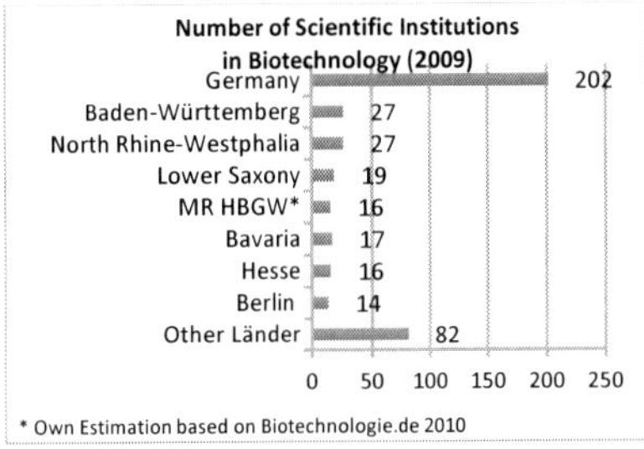

Source: Own Illustration based on 2010: 20

Fig. 19: R&D-Employment in German Biotechnology: Regional Distribution

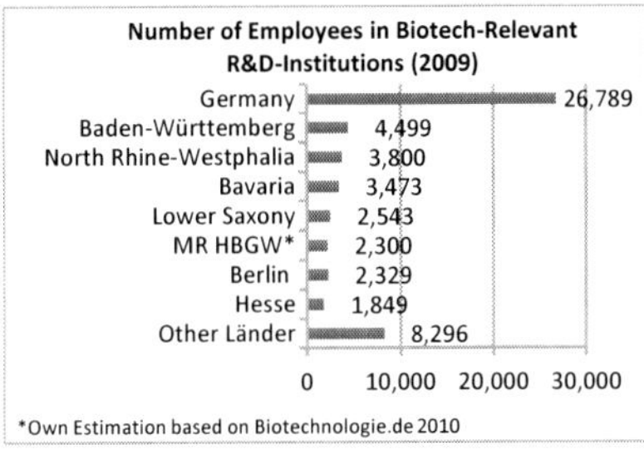

Source: Own Illustration based on Biotechnologie.de Biotechnologie.de 2010: 20

5.2.2 Structures and Mechanisms of Knowledge Transfer in Baden-Württemberg: Six Governance Dimensions

5.2.2.1 Networks as Mechanisms of Knowledge Transfer

In the following section, the results of the empirical qualitative case study of Baden-Württemberg's knowledge transfer system will be discussed. Starting with the network-based mechanisms of knowledge transfer first, the theoretical assumption derived from the literature on VoC and NIS was that business networks, and in particular informal networks and vertical networks between firms and their customers and suppliers should be central mechanisms of knowledge transfer in regional innovation systems of CMEs like Germany. Moreover, also cooperative R&D-activities should be important network based mechanisms of knowledge transfer in this context, with cooperative inter-firm R&D-projects as central channel of knowledge transfer, and cooperative R&D-projects with scientific institutions and universities as less significant mechanism.

Horizantal Networks

These theoretical assumptions have mostly been confirmed by the results of the case study. All interview partners reported that networks of actors would have at least some significance for knowledge transfer activities in the region, although they attributed the different dimensions of knowledge transfer via networks to have different impacts. With regard to *horizontal cooperations between competitors*, the interview partners pointed to the fact that these cooperations would indeed be important, and that their significance as mechanisms of knowledge transfer might even rise in the future. This would particularly be the case in the region's core industries, like electrical engineering, automotive and mechanical engineering, where current and future technological developments might particularly take place in cross-sectional technologies, as it would be the case e.g. with battery concepts in the field of electro mobility (Interviews BW 1: 12f.; BW 2: 7ff.; BW 6: 5f.). As one interview partner from the *Steinbeis Beratungszentren GmbH* points out, the participation in networks and the intensive exchange of knowledge might also be particularly important for firms not to miss new technological developments and being left behind by their competitors (Interview BW 1: 3ff.). But also factors hindering knowledge transfer through these horizontal cooperations were mentioned by some experts: In particular, the fear of losing technological advance was reported to be the main reason for not cooperating with competitors. As one interview partner states:

> "We have to differentiate clearly at this point: There are many projects, in particular when new products, processes and services are concerned, in which also a negative attitude towards networks can exist, simply because one wants to keep the know-how for oneself. [...] This is a decisive point for the firms, because many of them say that they do not want to talk to others, and that they do not want to apply for a patent, because then everything would be disclosed." (Interview BW 2: 8, translation: MO)

The forms these horizontal cooperations between competitors take would be, therefore, mostly formalized with fixed contracts and would be project specific, meaning that only in a few cases they are lasting over a period of many years and are institutionally stabilized (Interviews BW 2: 9ff.; BW 3: 8).

Vertical Networks

As a particularly crucial mechanism of knowledge transfer and as a specific feature of Baden-Württemberg's knowledge transfer system almost all interview partners mentioned the high importance of *vertical cooperations of firms with their customers and suppliers*. These networks were reported as frequently being institutionally stabilized, long lasting (sometimes for many years or even decades) and basing on fixed contracts. The interview partners highlighted that although these dense networks of customers and suppliers would indeed characterize Baden-Württemberg, they might be especially relevant in the traditional sectors as electrical engineering, chemicals, automotive and mechanical engineering, and less in other parts of the economy (a.o. Interviews BW 1: 4ff.; BW 2: 15f.; BW 3: 11f.). This is also confirmed e.g. by studies of Krauss (2009) and Fuchs and Wassermann (2004), who claim that in a comparison of the mechanical engineering sectors in all western German *Länder* Baden-Württemberg has a particularly high share of firms cooperating in vertical cooperations, but that horizontal cooperations do not happen more frequently than in the other *Länder* (Krauss 2009: 237ff.; Fuchs/Wassermann 2004: 10).

However, both horizontal and, above all, vertical cooperations between firms and other firms and actors were valued as central mechanisms of knowledge transfer in Baden-Württemberg by the interview partners, and this is also reflected by the broad variety and high number of business networks, cluster initiatives and cooperation projects in the region, which constitute one of its most distinctive characteristics. Overall, more than 100 local and sectoral clusters and cluster initiatives can be identified, as well as a considerable number of region wide clusters and cluster initiatives promoted by the regional government (BW MoE 2010b: 5ff.). Many of these networks would have emerged bottom-up, out of the production process and sectoral dynamics, as e.g. the *IT-Forum Rhine-Neckar* or most of the automotive clusters, others would have been initiated top-down by regional government, as are e.g. networks in the fields of fibre-based materials, satellite navigation and mobile-IT, environmental technology, logistics, automotive, mechatronics and creative industries, that emerged out of the competitions to strengthen region wide networks (*Wettbewerbe zur Stärkung landesweiter Netzwerke*) (BW MoE 2010b; Prognos 2009).

Moreover, also cluster initiatives promoted by the federal government were mentioned by the interview partners to be important for knowledge transfer in this context, and in particular the *Leading-Edge Cluster Competition* (*Spitzencluster-Wettberweb*) (a.o. Interviews BW 1: 12f.; BW 2: 10f.). Especially these publicly promoted cluster initiatives would mostly be institutionally stabilized, in the sense that they would have a professional cluster management organizing the network activities and coordination among the actors. Even more important, as was highlighted by some interview partners, is the fact that this institutional stabilization creates trust and enhances knowledge transfer activities. In addition, as one interview partner points out, these cluster projects would also intensify knowledge transfer between firms and actors from other parts of society:

"I have to say, that very positive effects emerged [from these clusters; MO], since many firms [...] often do not know anything from each other, because the communication is not optimal. [...] The advantage is crystal clear, that the firm's communication is intensified. Moreover, clusters also have a political aspect, the administration, governments and local authorities are participating as well, leading to stronger communication. The third party is the universities and the scientific world. If you have an active person in charge, then it can be very successful. [...] The more complex a project becomes, the longer its realization will take, because the actors have to start trusting each other. It is a question of trust. [...] Clusters must have a certain critical mass of participants and actors from politics, science and business and a person in charge, keeping the whole thing alive. Clusters do not live by themselves. There has to be someone initiating them and keeping them alive." (Interview BW 1: 12f.; translation: MO)

When differentiating between the two investigated sectors, business networks were reported to being clearly more important for knowledge transfer processes in the automotive industry than in the biotechnology sector. For the automotive sector, in particular vertical cooperations of firms in densely interacting customer-supplier networks have been valuated as an essential channel of knowledge transfer (Interview BW 6: 3ff.). As one expert for the automotive industry from the Baden-Württemberg Ministry of Economics explicates:

"Inter-firm networks play a very central role. That is certainly one of the most important points. In the field of automotive the OEMs are essential, and fortunately we have some of them here in Baden-Württemberg. These OEMs stand in the center of networking activities in the automotive sector. But also the suppliers are networked among each other and cooperate. Hereby, there are different cluster organizations that coordinate these cooperations at the local level. Recently, these local cluster initiatives are networked as well, new management structures are created, which shall embrace the different clusters." (Interview BW 6: 3; translation: MO)

These networks would mostly be institutionally stabilized, above all by the already mentioned cluster organizations and would base on fixed contracts. According to the interview partner, the endurance of these cooperations would result from a logic of highly specialized supply chains, in which presently only a few individual pieces are supplied, but instead, above all, whole modules. Developing an efficient cooperation in this context would then only be possible in the long-term. This already refers to horizontal cooperations between the supplier firms, which have also been valuated as important, although the interview partner also accentuates the importance of competition among the suppliers:

"In the pre-competition phase there is certainly an intensive exchange of knowledge. But of course there is also an enormous competition when it is about concrete products that are to be brought to the markets. In some cases, the OEMs even play the suppliers off against each other, naturally with the aim to beat down the prices." (Interview BW 6: 4; translation: MO)

Cooperations among the OEMs, on the other hand, were reported to only occur on a predominantly topic- and project-specific basis, since everything that would be connected to brands would be treated with great caution by these firms. An example would be the cooperation of AUDI and Porsche (although the recent acquisition of Porsche by VW could intensify the cooperation in this case), or cooperations of the OEMs in the development of new battery-technologies (Interview BW 6: 5).

In the biotechnology sector, horizontal cooperations among biotech firms were reported to only be of medium range importance for knowledge transfer processes. Since the vast majority of biotech firms in the region would still be young and would find itself in the phase of conducting basic research and product development activities, the protection of IPRs would be a major obstacle to cooperations with competitors. If firms cooperate, this would only occur on a project-specific basis, keeping the necessary distance to protect their business interests, as the executive director of the *BioRegio Rhine-Neckar-Triangle e. V.* illustrates:

"They are first of all project-related, but that is not so easy to judge, since, naturally, everybody has its own needs. In this, intellectual properties also play a role, which cannot simply be passed. That implies that it has to be reconsidered very exactly with whom one would cooperate. But they work together in large complexes, as e.g. cancer stem cells, whereby everybody works on different fields and they know about each other and also discuss their progress. However, only to a degree that is in the interest of their own firm." (Interview BW 5: 8; translation: MO)

Vertical cooperations in the biotech sector would have different dimensions: On the one hand, the majority of biotech firms would be suppliers to the pharmaceutical or chemical industry, depending on their field of specialization. The entrepreneurial goal of a large part of the "red" biotech firms in Baden-Württemberg, for example, is, therefore, to develop an agent for a new drug to a point when its effectiveness is approved in first clinical test phases and then to sell a patent or license (or even the whole firm) to one of the large pharmaceutical firms that would subsequently be responsible for product development and commercialization. Although this specific division of labor would be vital for knowledge transfer in the biotech sector in general, the regional dimension of this would be strongly restricted, because of the lack of large pharmaceutical companies in Baden-Württemberg (Interview BW 5: 9).

On the other hand, and again depending on their field of specialization, many biotech companies would also maintain relations to own suppliers, which were reported to be an important, although not essential part of their knowledge transfer activities. As a representative of a larger biotech company in the region explains:

> "These firms often have knowledge in specific fields, as for example a firm that specializes in optical design. Although we also have internal expertise in optics, we do not simulate in the field of optics frequently, so that we fall back on them. And they again know our products and requirements very well. Also in the realization of clinical studies we are supported for example by a small regional firm. [...] I think that this is a very important factor, simply because it allows us to be fast and flexible. [...] These are often regional firms, but there is also a strong local component, since cooperation is fostered by spatial proximity. The mutual exchange is faster, you can just visit each other, you can just get instruments or work results across." (Interview BW 4: 3f.; translation: MO)

In any case, these cooperations would be formalized with fixed contracts, above all to secure confidentiality, as the interview partner continues:

> "The simplest contract we would always need is confidentiality. Even in the case of a doctoral candidate we have confidentiality agreements. Especially in the field of research there are always things that could be patented. And if we would work together with somebody who did not sign confidentiality, it is as if it would be published. Therefore, we could not afford to work without contracts." (Interview BW 4: 5f.; translation: MO)

Inter-Firm R&D-Cooperations

With regard to the importance of *inter-firm cooperations in the field of R&D* as mechanisms of knowledge transfer, the interview partners draw a differentiated picture for the region as a whole. First, they emphasize that not all firms conduct own R&D, and that therefore many of them, if anything, only apply the results of other actor's R&D-activities. Second, they point out that firm size matters in this context. While larger firms would often have the resources to conduct their own R&D-activities, and might be less likely to cooperate with other firms, for smaller firms cooperative R&D-projects might be an efficient strategy to compensate their limited resources (a.o. Interviews BW 1: 14ff.; BW 2: 24f.; BW 3: 16ff.). These results are also confirmed by a quantitative study of the IAW (2008), which finds out that the propensity of firms in Baden-Württemberg to cooperate with other firms in their R&D-activities rises with firm size (IAW 2008: 32f.).

Referring to the forms these cooperative inter-firm R&D-activities take, all interviewed experts agreed that they would always be formalized and fixed with contracts, to set up the mutual obligations and access rights on the respective outcomes (a.o. Interviews BW 1: 16; BW 2: 25). One example would be the industrial collective research (*Industrielle Gemeinschaftsforschung*), which are joint and sector-wide research activities of especially SMEs at a pre-competitive stage (AiF 2010). Overall, these results make clear, that inter-firm cooperations in R&D can indeed be considered as important mechanisms of knowledge transfer in Baden-Württemberg, but that their contribution to the knowledge transfer activities in the region is limited by the factors discussed above.

Also when focusing on the automotive sector, these results were fully confirmed (Interview BW 6), while in the biotechnology sector, again, firms would be more restrictive to conduct joint research activities. Apart from the fact that most biotech companies would conduct intensive own research activities as their core entrepreneurial mission, and would, therefore, not need to cooperate with other firms, the fear of "losing" knowledge would impede these cooperations in many cases (Interviews BW 4: 2ff.; BW 5: 8ff.).

Cooperations between Firms and Universities or Research Institutes

As an even more important mechanism of knowledge transfer in Baden-Württemberg the interview partners regarded *cooperations between firms and universities or research institutes*. With that, the significance of this mechanism is evaluated clearly stronger by the experts as it would be to expect from the theoretical assumptions derived from the literature. The main reasons given can be summarized as both demand and supply dependent: On the one hand, the region would be strongly shaped by research intensive industries requiring a dense cooperation with the scientific world, and, on the other hand, the region would offer to the business sector a singular scientific infrastructure with highly renowned universities and research institutes. Nevertheless, also in this context, the motivation to transfer knowledge via this channel would be quite different, as one of the interview partners highlights:

"There surely are two or three classes of forms of cooperation in R&D. *Fraunhofer* is certainly an important research partner, partly the university institutes are research partners, partly *Steinbeis*, and the whole thing is driven by the respective support programs of the EU, the federal government and the respective regional governments. And here it holds true again: Many firms are not able to conduct their own basic research, so that they cooperate with research institutes. […] This will even become more important in the future, since the technologies are more and more knowledge driven. […] This is a structural change penetrating all economic sectors. Therefore, you do not have another chance as to cooperate in R&D, because even large firms cannot maintain the knowledge that would be necessary to realize such developments." (Interview BW 1: 14f.; translation: MO)

Moreover, also in this context firm size would matter again. As one interview partner from the Baden-Württemberg Chamber of Industry and Commerce points out, particularly SMEs would profit from these cooperations, since they would often lack the necessary resources to conduct sufficient R&D-activities themselves:

"On the other hand, these small firms need these cooperations to access know-how. At the universities a lot of know-how is produced […]. It is their specific advantage that they can try a lot of things that might perhaps not work out at the end, but at least they tried. Firms cannot operate like that. That means that in particular these small firms need the universities and they need them to work as they do and that they are willing to cooperate and to externalize their knowledge. This is essential for Baden-Württemberg as a location as well as for the majority of the firms, because most of them are SMEs." (Interview BW 2: 24f.; translation: MO)

On the other hand, as factors impeding cooperation of SMEs with scientific institutions the interview partners reported the lack of knowledge how to approach the adequate experts and certain organizational barriers that might be rooted in the fact that many SMEs would not have an explicit R&D-department and that therefore cooperation is complicated. In this regard, cooperations with scientific institutions would be easier for larger firms, as is also confirmed by the already mentioned study of the IAW (2008), which finds out that the propensity of firms in Baden-Württemberg to cooperate with scientific institutions rises with firm size (IAW 2008: 32f.). In addition, another factor that might play a role is the higher share of academic personnel in larger firms that could contribute to lower mental barriers in cooperation with the scientific world (a.o. Interviews BW 1; BW 2; BW 6).

Concerning the forms that these cooperations between firms and scientific institutions take, the experts particularly emphasized the classical forms of cooperative research projects and contract research, which would usually be strictly formalized with fixed contracts, limited in time and sometimes also be promoted through public funding, as it would be for example the case with the "innovation voucher" of the regional government:

> "Another possibility would be the classical contract research, in which a firm tells a professor or an institute please develop me this or please conduct a study on that. Then they pay for that and perfect. This would also be a typical case for the innovation voucher." (Interview BW 2: 25; translation: MO)

Other forms mentioned by the experts are endowed chairs, which would particularly be used by larger firms, and diploma or doctoral theses, which would be used frequently and intensively by SMEs as well. Moreover, and pointing to knowledge transfer in the direction from the firms to the scientific world, these cooperations would also include guest lectures by visiting lecturers from the firms. In addition, practices of (doctoral) students in the firms have been mentioned, whereby both partners would mutually profit from each other, the students would contribute with some knowledge to specific projects in the firms and the universities might get some know-how and a publication back, and might have a chance to place one of their graduates in these firms (a.o. Interviews BW 1: 16; BW 2: 25).

The informal dimension of these cooperations between the business sector and the scientific world (estimated as less important by half of the interview partners) would be accompanying events, conferences, workshops, talks and congresses, or contacts to former personnel or students:

> "If you wrote your diploma thesis at an institute and then you are in a firm and you get a problem, of course, you call your former colleagues. […] Then you have got an informal space there, which works much easier in a bilateral way. […] In this case you do not have fixed processes." (Interview BW 1: 16; translation: MO)

A detailed look into the automotive sector reveals that although cooperative R&D-projects with universities and research institutes play a role for knowledge transfer, they are clearly less relevant than the internal sectoral R&D-activities within the firms or through inter-firm collaborations. One interviewed expert for the automotive industry in Baden-Württemberg illustrates this and also points to project-based cooperations and contract research as the usual forms these cooperations would take:

"[C]ooperations in R&D are very important. But you also have to view this in a very differentiated way. A large part of R&D-activities is conducted internally. External cooperations then mostly emerge on a topic-specific basis. In the field of developing new engines and drives, e.g., it is natural that the firms cooperate with research institutions. In Baden-Württemberg the focus of external research is clearly on applied research, with the strong *Fraunhofer-Institutes*, the institutes of the *Helmholtz Society* with the *DLR*, or the *Hahn-Schickard Society*.

In this regard, the Ministry of Economics supports the *Innovationsallianz Baden-Württemberg*, which is a roof organization for the contract research institutions in the region. Mostly, these are non-university research institutes conducting contract research for the firms in the region. They work as bridges between basic research at the higher education facilities and the technical development in the firms. Overall, my estimation is that cooperations in R&D emerge more likely on a project base and are intended for a short to medium term duration. But, naturally, also follow-up projects might emerge from that." (Interview BW 6: 9; translation: MO)

For the highly research intensive biotechnology sector, on the other hand, cooperations with universities and research institutes are vital. Its strong dependence on basic research activities and results requires intense interaction with the scientific world by almost all types of biotech firms (biotechnologie.de 2010: 14). In this connection, the interviewed experts for the sector mentioned a broad variety of forms that these cooperations would take. With regard to cooperations with universities, a representative of a large biotech firm in the region specifies e.g. the exchange of diploma and doctoral students, cooperations with single professors, including the funding of scientific positions at the respective chairs, endowed chairs, or the exchange of scientific instruments. Cooperations with research institutes, like the *Max Planck Institutes* or the institutes of the *Fraunhofer Society*, would mostly take the form of contract research (Interview BW 4: 10). Moreover, also public-private partnerships would be relevant in this connection. Two examples for this model of cooperation would be the *Hi-Stem GmbH* (Heidelberg Institute for Stem Cell Technology) as a partnership of the *German Cancer Research Center*, the *Dietmar Hopp Foundation* and the *BioRegio RND*, or the *Catalysis Research Laboratory* as a cooperation of the *BASF AG* and the *University of Heidelberg* (Interview BW 5: 12).

5.2.2.2 Association Based Mechanisms of Knowledge Transfer

With regard to the association based mechanisms of knowledge transfer, the theoretical assumption derived from the literature on VoC and NIS was that all the three of them (employers-, business-, and trade associations; chambers of industry and commerce; binding and common technical standards) should be of high importance in the German regions. These assumptions have basically been confirmed by the interviewed experts when looking at the region as a whole.

Employers-, Business- and Trade Associations

As expected, the services provided by *employers-, business- and trade associations* were reported to be frequently used as mechanisms of knowledge transfer (a.o. Interviews BW 2: 18f.; BW 3: 13ff.). As one interview partner remarks, their effect in enhancing knowledge transfer would be particularly strong if these associations would be highly specialized and subject-oriented, as e.g. in the case of the business associations in the fields of microelectronics and mechatronics (Interview BW 1: 9).

Chambers of Industry and Commerce

The role of the *chambers of industry and commerce* has also been described as central for knowledge transfer processes in the region, although with some qualifications. First, in contrast to the majority of business associations, the membership in these chambers would be obligatory, so that some firms would be members, but would not actively use the provided services. Second, these services would be of particular interest for SMEs and less for larger firms. Third, the majority of the interview partners reported as the most important mechanism through which the chambers would contribute to knowledge transfer their role in defining and securing the quality of further education and training. Other important and frequently used mechanisms of knowledge transfer provided by the chambers (apart from congresses, conferences, expert talks, internet platforms, newsletters, and magazines) would be consultancy and information services in a large variety of business relevant fields (innovation, finance, funding etc.). In this, the firms would get access to specific information through personal attendance or by phone, what would be, again, particularly interesting for SMEs, and less for larger firms (Interviews BW 1: 9ff.; BW 2; BW 3: 13ff.). One representative of the Baden-Württemberg Chambers of Industry and Commerce remarks in this connection:

"The typical questions that we are asked concern e.g. the CE-labeling, patents, intellectual property rights protection in general and support programs. [S]o if at the end of the year a new CE-guideline for machines is approved, one can be sure that there will be many requests on this topic. [...] Only in the last year alone there have been 8,000 telephone consultations and 1,200 personal consultations in Baden-Württemberg, next to all the other things, as e.g. events etc." (Interview BW 2: 7; translation: MO)

As a decisive strength of the chambers of industry and commerce in Baden-Württemberg (and also of many of the employers-, business-, and trade associations) their strong focus on the region as whole has been highlighted. Through the principle of lead management (*Federführungsprinzip*) the distinct local chambers would coordinate their activities in specific fields on the regional level, taking into account region-wide developments and perspectives, also allowing them to articulate their interests in a concerted way towards the regional government. Moreover, this coordination of the local chambers is further supported by the Federation of the Baden-Württemberg Chambers of Industry and Commerce (*Baden-Württembergischer Industrie- und Handelskammertag*) (Interview BW 2: 2ff.).

For the automotive industry, especially the sectors' national business association, the *VDA*, was attributed a decisive role for knowledge transfer processes in Baden-Württemberg. Although this association is organized on the national level, it would exercise important functions for the regional firms, above all in securing approvals and funding from the national government and the EU. Other sector relevant associations and the chambers would particularly contribute to enhancing knowledge transfer through their participation in the various local and regional cluster initiatives, where they would offer their services to the members, like internet platforms, newsletters, publications, conferences and, most important, cluster and networking forums, which would be used intensively by the participating actors (Interview BW 6: 7). In addition, one interviewed expert for the automotive industry in Baden-Württemberg highlights the role of chambers and associations in initiating cooperation, but also refers to the differences between the large OEMs and the SMEs in their attitudes towards these activities:

"They play a very important role, these associations often initialize cooperations. Hereby, [...] also the trust transmitted by these associations plays a role. They are seen as a neutral actor, so that they can initiate and moderate cooperations. This especially applies for the SMEs. For the OEMs this is less the case, they have a critical position towards the associations. Naturally, that is a question of power, the really large companies have considerable powers themselves and with regard to certain topics they want to go their own way, independently of the association." (Interview BW 6: 8; translation: MO)

In the biotech industry, the role of business associations and chambers in knowledge transfer processes has been estimated as only limited by the experts (Interviews BW 4: 8f.; BW 5: 11). One reason would be that many firms would indeed use business associations to exchange knowledge and information, except that this would mostly not include R&D and technology relevant topics, but, above all, general topics as interest representation, business processes, marketing or market movements, to not endanger own know-how advances. The chambers, on the other hand, would often be perceived as not specialized enough, but would contribute to the cluster initiatives, as the executive director of the *BioRegio RND e. V.* explains:

"I would say that they play a role, but not an outstanding one, simply because they do not have the specific know-how you would need for the biotech sector. The chambers are [...] contact partners when it is about the economic aspects of business start-ups. We have a very close relationship to the chamber, and they are also a strategic partner in the management of the *Leading Edge Cluster*. And we compare notes on events. But they are surely not the first and most important partner for the biotech sector in this regard. That would also hold true for the *Metropolitan Region* or the *German Chemical Industry Association*. The latter would be more focused on the larger and more mature industry. There are connections, but they are not very intense. [...] Since some of these organizations are also members of our cluster, we do specific tasks or events together. [...] there is also an exchange of information. E.g., we act as a broker for the programs which the chambers offer through their start-up center or their start-up assessment." (Interview BW 5: 11; translation: MO)

Binding and Common Technical Standards and Norms

Finally, the importance of *binding and common technical standards and norms* fostered by these associations has been estimated as less significant by the interviewed experts than assumed from the theory. Although some experts indeed attributed binding and common technical standards a positive and reinforcing effect on cooperation and knowledge transfer, its significance for the success of knowledge transfer processes in the region as a whole has been described as only average. The main reason mentioned is that in particular in highly innovative fields of the region's economy firms would always find themselves in a hard competition to be the first to set a technical standard themselves, as e.g. in the case of the *Blu-Ray* standard, and would not wait for an association to set a commonly accepted standard. These standards would, if anything, help firms to access and cooperate on the basis of the technological state of the art. Nevertheless, the experts highlighted that there would be strong sectoral differences in the importance of technical standards for knowledge transfer (Interviews BW 1: 13f.; BW 4: 8; BW 6: 8).

For the automotive industry, technical standards and norms were reported to absolutely play a role. As an expert for this sector in Baden-Württemberg comments, in many cases they would be a central precondition allowing cooperation in the first place. In this context, technical standards would represent predefined interfaces for the actors and create the framework conditions under which cooperation could take place. Moreover, they would reduce the administrative affords (Interview BW 6: 8). Although not explicitly reported be the interview partners, it is to be expected that also in Baden-Württemberg's highly regulated biotechnology sector, large parts of interaction and knowledge transfer processes would be influenced by common technical (as well as medical and chemical) standards, norms, regulations and clinical test requirements, as it would be e.g. *ISO-Norms* or regulations on blood sugar measurement, among many other things (Interview BW 4: 8).

5.2.2.3 Organization Based Mechanisms of Knowledge Transfer

With regard to the organization based mechanisms of knowledge transfer the interviewed experts have been asked for their estimation on the role of the transfer of technical and scientific staff, purposeful headhunting and business start-ups. Building on the literature on VoC and NIS it has been assumed that all of these mechanisms should not play an important role for knowledge transfer processes in the regional innovation systems of Germany's CME. These assumptions have basically not been confirmed by the interview partners. Although all three organization based mechanisms were not attributed a decisive importance, they were all evaluated by the experts to have a medium range significance for knowledge transfer processes in Baden-Württemberg.

Transfer of Technical and Scientific Staff

Concerning the *transfer of technical and scientific staff*, the interviewed experts particularly emphasized the direction of transfers from scientific institutions to the business sector as important, while the opposite direction from firms to the academic sector was considered as almost irrelevant. However, the importance of these transfers has been reported to be rising in the last years. Due to a reduction of supply for qualified staff on the labor market and the beginning effects of the process of demographic change firms would find themselves more and more in a competition on these highly qualified persons. This would not only open up more opportunities for scientific and technical staff from the scientific world to transfer to the business sector, but also enhance the firms' recruiting activities in the scientific sector (Interviews BW 1: 17f.; BW 2: 27ff.; BW 3: 18ff.). One important source for these transfers would be joint R&D-projects, or specific programs by regional institutions, as a representative of the *Steinbeis Beratungszentren GmbH* explains:

> "When the research project is over, it is about the implementation and then it could be that I hire the whole crew or parts of it for my firm. In this case, I would also already know the persons. Moreover, there are instruments by the regional government, which they call "*Innovationsassistent*": If somebody hires an innovation-oriented person, then he partly gets the personnel costs subsidized. At our *Steinbeis-University* we developed another concept, a so called "*Projekt-Kompetenz-Studium*." In this, we look for potential executives for the firms, graduates or people with a university degree and some professional experience that will then do a bachelor or a master degree parallel to their jobs. In this, they are in the firms and work on concrete projects. At the end, this is nothing else than knowledge transfer." (Interview BW 1: 17; translation: MO)

As one representative of the Baden-Württemberg Chamber of Industry and Commerce points out, the intensity of personnel transfer activities might differ between technology based firms and other firms, as shows the strong demand for engineers in the region's core industries of. Moreover, in this context it must also be differentiated between the capabilities of large and smaller firms to hire qualified staff. Larger firms would have much more resources to actively recruit staff from scientific institutions, as the interview partner continues:

> "For the small firms the question is how to access qualified staff. They might less often get the chance to do so and they will then rather try it via cooperations. The large firms can simply offer much more to the younger persons, particularly in fi-

nancial terms. That is why many of them go to the large firms, not least because they can get ahead there easily and quickly. On the other hand, the small firms got the advantage that you are more flexible and that you get more responsibility, the hierarchies are flatter. So there are also advantages, even if I would earn less there at the beginning […] than […] in large firms. Thus, the small firms always have to fight to be attractive for qualified staff." (Interview BW 2: 28f.; translation: MO)

Purposeful Headhunting

Purposeful headhunting of other firms' staff would be mostly restricted to the larger firms. In particular for the SMEs the experts stated that there still would be a mentality considering active headhunting from other firms as inappropriate. Overall, this mechanism would indeed be used by regional firms, but it would not play a dominant role in their knowledge transfer activities (Interviews BW 2: 27ff.; BW 3: 18ff.; BW 6: 9f.).

A sectoral perspective reveals that these results also apply to the automotive industry. Firms in the sector were reported to undertake the strongest recruiting efforts to get the best qualified graduates from the region's higher education institutes, but also country- and worldwide. Another important channel of personnel transfer would be the recruitment of already renowned researchers from the sector relevant research institutes, like the *Fraunhofer Institutes* or the *DLR*. However, purposeful headhunting of qualified staff from other firms would be less important, not least because of a "code of honor" among the firms in the sector, as an expert for the regional automotive industry emphasizes (Interview BW 6: 9f.).

Also in the biotechnology sector the transfer of qualified staff has been described as being strongest through the mechanisms of graduate recruitment and the hiring of scientists from universities and research institutes. Purposeful headhunting from other firms, on the contrary, would also in this sector play a less significant role. As a representative of a large biotech firm in the region accentuates, firms would, in case they needed highly qualified staff for a short term, prefer to use the services of external firms, and, thus, conduct contract research (Interviews BW 4: 12f.; BW 5: 12f.).

Business Start-Ups

A quite similar picture is disclosed when focusing on *business start-ups* as mechanisms of knowledge transfer: Looking at the region as a whole, business start-ups are realized, but only to a limited extend, and, as almost all interview partners add, much less as would be desirable and necessary from their point of view. However, they admit that in the last years the propensity to create start-ups has risen, not least because of public funding for entrepreneurs, high-tech incubators at some universities or institutional support, as e.g. by the regional government or the *Steinbeis* network (Interviews BW 1: 18f.; BW 2: 29; BW 6: 10). As one representative of the *Steinbeis Beratungszentren GmbH* explicates:

"In the past, it was almost impossible that a private firm could be established on the campus. Recently, also in Germany private firms and start-ups begin to be realized on the campuses in the respective centers. In Baden-Württemberg we actually found a good solution for that, since the *Steinbeis* network intervened exactly in this point […], namely if the framework conditions are favorable or not. We accomplished that we as *Steinbeis* can offer the professors and also the scientific personnel from the university environment a legal framework in which they can carry out projects, services and orders with industry. And they can do that in cooperation with the university, but independently of it." (Interview BW 1: 18f.; translation: MO)

One could summarize that business start-ups in Baden-Württemberg are still strongly dependent on external funding and support and that only to a limited extent they emerge out of market dynamics. As a reason the experts do not see responsible legal barriers, also when comparing Baden-Württemberg with LMEs as the USA or the UK (Interviews BW 4: 14; BW 1: 18). One interview partner states in this connection:

"I assume that for small companies with a size up to eight employees the difference will not be particularly big. There you do not need a works council. The requirements are relatively low. You can register a business quickly and easily. So that should not be the reason." (Interview BW 4: 14; translation: MO)

Rather, they see cultural differences at work and emphasize distinct mentalities with regard to entrepreneurship and risk aversion. As a representative of a large pharmaceutical firm points out:

"[W]e do not have a culture in this direction, while in the USA it is normal to start-up a new business. Everybody stands on one's own feet and takes care of one's self. In Germany, one feels more comfortable in the role of a doctoral candidate or an employee; that is much safer." (Interview BW 4: 14; translation: MO)

And another interview partner adds:

"[T]he reason is […] the entrepreneurial culture. There is an actual inquiry: On average, professors in the USA created at least one or two business start-ups together with their students. In Germany, this share is minuscule. In the USA 50% of the students think about creating a business themselves, in Germany only 8%. You see, it is a cultural question." (Interview BW 1: 18; translation: MO)

In particular, and confirming in this point the assumptions derived from the literature on VoC, all experts mentioned the insufficient *provision of seed- and venture capital* as an explanation for the limited use of business start-ups as mechanisms of knowledge transfer. As one representative of the Baden-Württemberg Chamber of Industry and Commerce highlights:

"What is certainly underdeveloped from the point of view of the *DIHK*, is the venture-capital market. That is something that is common practice in the USA and the UK and absorbs almost everything that happens in this direction there. In Germany, this is underdeveloped, and almost all experts see it this way. [...] Naturally, in this phase a lot depends on funding. [...] Responsible are the general structures and the history of society. In the USA, the scientific world is structured in a completely different way. There, a lot is carried out from outside, by firms, donations and foundations. This is completely different in Germany. Also the market for venture-capital has very different roots and a distinct mentality as in Germany. Here in Germany the nature of people is to be focused on security, classically, they prefer to go to their bank." (Interview BW 2: 29ff.; translation: MO)

Therefore, what is missing in the region according to the interviewed experts are seed- and venture-capital firms and institutions, as well as business angels, but also a functioning alumni-culture at the universities. When asked if in a bank-driven corporate governance system like Germany the banks could not exercise this role, they reported that indeed first steps in this direction are undertaken, but that in general the banks still follow different strategies. Banks in Baden-Württemberg still rely strongly on the "*Hausbankprinzip*," and are experts for debt financing, in which project funding is mostly based on the provision of securities by the borrower (Interviews BW 1: 19ff.; BW 2: 30ff.; BW 3: 22ff.; BW 6: 15). On the other hand, as one of the interview partners explains, venture-capital provision requires very different know-how and valuation methods:

"In Germany we do not have enough qualified people that are able to evaluate a project according to its assumed success. In other words: If the projects will generate the expected returns and interest payments they have to bear. And it is a difference if you have a large body of bankers that learned to do debt financing, or if you need people, that are specialists for equity financing. Both have to be able to assess if a project will be rentable. The one tells you to bring securities, and then you get the money. But in the other case it could be that you do not have the staff to assess the project, and then many projects are rejected." (Interview BW 1: 19f.; translation: MO)

Nevertheless, for the case of Baden-Württemberg some private investors and patrons were mentioned that try to fill this gap. Patrons as *Dietmar Hopp*, their respective foundations and some private firms as SAP and BASF, e.g., provide seed- and venture capital for young entrepreneurs and university spin-offs. While the considerable number of these players in the region brings Baden-Württemberg into a privileged position within Germany (although still on a low level), and also the *Landesbank Baden-Württemberg* (LBBW) offers some venture-capital via its *LBBW Venture Capital GmbH*, some interview partners reported that the access to this capital would often not be easy enough (Interviews BW 3: 25; BW 5: 14; BW 6: 15).

Differentiated by sectors, also in automotive business start-ups would only play a limited role for knowledge transfer, particularly since this sector would be too mature to easily establish oneself with a new business on the market. This would hold true for start-ups out of the sector itself, as well as for sector-relevant academic spin-offs from scientific institutions (Interview BW 6: 10). In addition, the insufficient provision of seed- and venture-capital can be supposed to cause effects in this connection as well. In biotechnology as a strongly analytic knowledge driven sector, business start-ups and, above all, academic spin-offs from scientific institutions should play a decisive role for knowledge transfer from academia to the market. Nevertheless, as already discussed, the foundation of new biotech firms in Baden-Württemberg is predominantly connected to public programs like *BioRegio*, *BioProfile* and the *Leading-Edge Cluster Competition* and occurs only in a limited number of cases out of market dynamics. As an extraordinarily research intensive sector, in which firms often spend up to ten years or more only on conducting R&D-activities without producing any marketable product, funding would play a crucial role in this connection. In this, Baden-Württemberg's biotech sector would still be strongly dependent on public funding by national and regional government and lack a sufficient provision of seed- and venture-capital via the market, a fact that would strongly effect entrepreneurial activities in the sector (Ernst & Young 2009; Ernst & Young 2010; Interviews BW 4: 13ff.; BW 2: 30; BW 5: 13ff.). As the executive director of the *BioRegion RND e. V.* comments:

"What we need, is, above all, a financial basis. We particularly need seed-capital, what you basically only get via venture-capital. There are some public programs, like *ProBio*, that can help a bit at the beginning, but that is not much. In the early phases there simply is not enough money available. [...] At the moment, venture-capital is almost not available, apart from private financiers as for example *Hopp* etc." (Interview BW 5: 14; translation: MO)

Also in the case of Baden-Württemberg's biotechnology sector, therefore, it is private investors and their respective foundations compensating to some (limited) extent the gap in funding through the capital market.

5.2.2.4 State Induced Mechanisms of Knowledge Transfer

The state intervenes in knowledge transfer processes in regional innovation systems above all through mechanisms such as public programs, differentiated by national or regional measures, through (larger) state-induced innovation projects, the creation of scientific parks and technological centers, as well as the maintenance of technology transfer offices or companies at the universities and public research institutes. Basing on the discussion of the literature on national production and innovation systems, it has been assumed that in the German regional innovation systems national and regional government measures, scientific parks and technological centers should only play an average role for knowledge transfer activities, according to the role as a bystander that the state exercises in the German national market economy, only creating the framework conditions in which the economic actors shall act. (Larger) State-induced innovation projects should therefore not be important in this context, while technology transfer offices/companies as intermediary institutions, on the contrary, were assumed to be central mechanisms of knowledge transfer between academia and business.

National and Regional Government Measures

The results of the qualitative interviews only confirmed some of these assumptions, while others were not approved. Looking at the single state induced mechanisms in detail, *national and regional government measures* were confirmed by the experts to play only an average role for knowledge transfer in Baden-Württemberg. Although the state was acknowledged to exercise an active function in the fields of innovation and knowledge transfer, most of the interview partners pointed out that the typical Baden-Württemberg way of being successful in these fields is not top-down action by the state, but bottom-up activity by the economic actors (Interviews BW 1: 5; BW 2: 10f.). As one interview partner emphasizes:

"From my point of view it is the completely wrong way to initiate and institutionalize this top-down. [...] Of course there are other examples, like e.g. the regional cluster competitions here in Baden-Württemberg or the *Leading-Edge Cluster Competition* by the federal government. Hereby, something is installed top-down and one will see how long this will work at the end. [...] in these cases the temporal experience is not that advanced to evaluate this in a definite way. But this surely contains an artificial component. A network must exist in this, and the firms must confess that they want that, that they need that and that they want to get involved in it. But it is just different, it did not emerge out of itself, but was initiated because of this competition. One will see how sustainable something like this can be. Our opinion [...] is that it is not the right way to initiate this top-down. [...] The state has no place in the field of innovation, he is not the better entrepreneur, and he is not the better innovator, and definitely he does not know which products might perhaps be demanded in ten years and which topics will dominate the market in the future." (Interview BW 2: 10f.; translation: MO)

Nevertheless, publicly funded measures by national government as e.g. the *Leading-Edge Cluster Competition*, *KMU-innovativ* to support SMEs, *SIGNO* to support IP-protection, or *EXIST* to promote business start-up activities would absolutely show helpful effects. The same would hold true for measures supporting innovation and knowledge transfer by regional government, as it would be e.g. the regional cluster competitions, regional cluster and network initiatives, promotion of regional economic development (*regionale Wirtschaftsförderung*), the *IFEX* initiative to support business start-up activities, the support of collaborative research, or the "*Innovation Assistant Program*" to promote the employment of innovation-oriented personnel and the "*Innovation Voucher Program*" to support innovative and R&D-activities, both with a special focus on SMEs (Interviews BW 6: 12f.; BW 2: 10ff.; BMBF 2009; BW MoE 2010a).

From a sectoral perspective, as expected, automotive as a traditional and mature industrial sector does not deviate much in this regard from the conditions in the region as a whole. Also for this sector, the state would predominantly exercise an enabling role in the fields of innovation and knowledge transfer. As an expert for the automotive industry in Baden-Württemberg emphasizes, the main task of regional government in this context would be the provision and maintenance of highly efficient and specialized research institutes to support sector relevant R&D-activities. Furthermore, the *Land* would actively promote local and regional cluster and network initiatives, where it would also forward funding from EU-programs like the *European Regional Development Fund*, as e.g. in the case of the *Automotive Engineering Network* (AEN) in Karlsruhe. In addition, also the creation of region-wide cluster initiatives in the field of automotive would be supported by regional government, as e.g. *autoland-bw*, which especially focuses on automotive supplier firms, and *automotive-bw*, which has the aim to establish a professionalized, centralized, region-wide and independent cluster management for Baden-Württemberg's automotive sector (Interview BW 6: 3ff.; BW-Invest 2011c; BW MoE 2010b).

The biotech sector in Baden-Württemberg, on the other hand, would depend much more on public measures and programs in the field of knowledge transfer than automotive. As the executive director of the *Bio-Regio RND e. V.* points out, already the emergence and also the growth of the sector in Baden-Württemberg was tightly connected to public measures, like *BioRegio*, *BioProfile* and the *Leading Edge Cluster Competition* by the federal government, as well as the cluster initiatives by the *Land*:

> "[M]ost of the biotech firms are still very young. Apart from only a few exceptions most of them did not start until 1997. The majority of these firms emerged in the course of public support. And that did not happen through supporting the firms, but scientific projects that have a clear focus on application." (Interview BW 5: 4; translation: MO)

As the two interviewed experts for the biotech sector emphasize, apart from providing funding and seed- and venture capital (although only to a limited extent, as discussed before), these public programs and cluster initiatives would, above all, aim to foster knowledge transfer and innovation through the establishment of professional cluster management organizations being able to coordinate and evaluate the single projects centrally and being responsible to manage the interaction with the funding bodies, might they be at the EU-, national or regional level.

Scientific Parks and Technological Centers

As two interview partners commented, from their point of view *scientific parks* would be important mechanisms of knowledge transfer where they exist, but there would be too few of them operating in the region to ascribe them a decisive role for Baden-Württemberg's knowledge transfer system as a whole. *Technology centers*, on the other hand, would be more numerous, with about 50 of them operating in the region (ADT 2011). They would be important, but not central mechanisms of knowledge transfer, in particular in providing a physical space for collaborative research (Interviews BW 6: 2ff.; BW 3).

Differentiated by sectors, the picture is slightly different. In automotive, the two already mentioned sector-relevant science parks in Ulm (hosting the *Daimler Research Center*) and the *Stuttgart Engineering Park* (BW-Invest 2011c; BW MoE 2010b) can be supposed to be at least one element among others in the overall knowledge transfer activities in the sector. Technology centers were not reported to play a role by the interviewed expert for the automotive industry in Baden-Württemberg (Interview BW 6).

For biotechnology, the relevance of science parks and technological centers as knowledge transfer mechanisms could not be evaluated by the interviewed experts. Nevertheless, it can be assumed that in particular science parks should be important instruments of knowledge transfer in Baden-Württemberg's biotech sector, since its strong dependence on basic research and product development activities requires a close connection of the firms to the regions renowned universities and research institutes, which can particularly well be achieved within the framework of these installations. The fact that in four (Freiburg, Heidelberg, Ulm, Tübingen/ Reutlingen and Esslingen) of the five bioregions in Baden-Württemberg these science parks have been established with the support of the regional government to foster interaction between universities, research institutes, biotech firms, supporting institutions and service providers, underlines their significance (biotechnologie.de 2010; Ernst & Young 2010; bw-international 2011; BIOPRO 2011).

(Larger) State-Induced Innovation Projects

The *state as an initiator and executor of (larger) innovation projects* in which the participating actors could transfer knowledge was not ascribed an important role by the interviewed experts. As one representative of a large biotech-firm accentuates, the role of the state is not the one of an active innovator, but of an enabler supporting innovation and knowledge transfer processes carried out by the economic actors:

> "As I said, the way it would work is vice versa. If we wanted something, then we would have to suggest to the political actors that things should develop into this or that direction. And if then they think that it is good and that it fits into the agenda, something develops." (Interview BW 4: 17; translation: MO)

When looking at the two investigated sectors, this would particularly hold true for biotechnology. In automotive, on the other hand, the role of the state is slightly more active in promoting concrete innovation projects and pushing technological development into specific directions. As one interview partner labeled it, the state would be active in the "management of change" (Interview BW 6: 6). One example would be the field of electric mobility, which is promoted intensively by the federal government (NPE 2010) and by the *Land*, among other things through the *Landesinitiative Elektromobilität* (BW MoE et al. 2010b; Interview BW 6: 6).

Technology Transfer Offices

Finally, the state is also active in promoting knowledge transfer through the maintenance of *technology transfer offices* or companies at the universities and public research institutes. In this point, the interviewed experts contradict clearly the high importance that has been assumed for these institutions. All experts consistently stated that they would neither consider technology transfer offices as central channels of knowledge transfer, nor that they would frequently be used by the regional actors, and especially not by the firms (a.o. Interviews BW 1: 21ff.; BW 2: 33f.). The reasons given are manifold: First, the majority of these offices would lack sufficient human, organizational and financial resources to play an active and efficient role in transferring knowledge between the scientific sector and the business world. Second, their focus would be more on commercializing marketable knowledge from the universities and research institutes, and less on bringing the firms and researchers together to resolve specific requests, initiate concrete projects, and, therefore, foster knowledge transfer through these channels, which is criticized by one interview partner as follows:

> "This is an exasperating topic. Originally, the technology transfer offices that have been established by the *Land* were created to work in another direction. They were given the task to externalize the patentable knowledge of the university. That means that as a firm you actually cannot call there and say that you are looking for somebody. That is not their task. But nobody is satisfied with that, and it does not work the way that it should. The University of Karlsruhe, e.g. has such a position which is occupied by only one person, and, moreover, this person also has to fight with bureaucracy. Therefore, the majority of actors has a very critical opinion towards this and currently the whole thing appears to be pretty underdeveloped. Many actors demand that the *Land* should take measures to organize and equip these

positions in a different way. Up to now, these offices did not play a role for the firms, mostly they do not even know them." (Interview BW 2: 33; translation: MO)

Concordantly, the experts stated that from a firms' perspective there is a great need to grade up the technology transfer facilities by a *switchboard function*, so that they could provide a competent overview on the current research activities in their respective institutes and could ease the access to the universities, research institutes and single researchers. This would especially be useful for smaller firms, which would often lack the knowledge how to access experts at universities and research institutes and might also have general reservations to interact with the scientific world. In addition, the interviewed experts identified a general preference of firms to directly interact with individual researchers, without having to choose the way through intermediate institutions like the technology transfer offices, which would be frequently perceived as additional bureaucratic barrier. Nevertheless, these shortcomings would partly be compensated by the numerous knowledge and technology transfer centers of the *Steinbeis* network, which, in terms of number and density, can be seen to constitute a distinctive element in Baden-Württemberg's knowledge and technology transfer system (a.o. Interviews BW 1: 21ff.; BW 2: 33f.).

This picture also resulted for the automotive sector. As one interview concisely summarizes:

"The technology transfer agencies at the universities are used to a lesser extent, particularly because they are less known by the SMEs. You must take into account that in these smaller firms, if anything, only the proprietor or the executive director did universities studies, and probably nobody else. There you simply got little knowledge about the universities and the possibilities of transfer. And maybe there are also fears of getting into contact in certain points. The firms that belong to this group would then use more likely job exchange forums (*Jobbörsen*) to get into contact with the universities, something that is used intensively. On the other hand, these transfer offices could also work as door openers to the universities, especially in those cases in which inhibitions do exit. Many transfer offices recognized that and are pretty successful in it." (Interview BW 6: 10; translation: MO)

For the large OEMs in automotive, the interview partner continues, these transfer offices would be less relevant, since they would prefer to have direct contacts to the researchers and institutes. In these cases, the socialization of the individual actors would be different and also the organizational capabilities to find adequate experts would be higher (Interview BW 6: 11). Also in biotechnology, the technology transfer offices would only be marginally significant for knowledge transfer. From the point of view of the biotech firms, one of their representatives emphasizes that these offices would in most cases only offer isolated patents to industry. But as he explains, an isolated patent would never have such a broad scope of protection that it could not be circumvented. What the transfer offices would not offer, but what would be of interest for the firms, would be consistent concepts pillowed by experts. Only this way, successful knowledge transfer would be to expect from the firms' perspective. Therefore, also the majority of the biotech firms would prefer to maintain direct personal contacts to the chairs, which would sometimes last for many years and result into stable partnerships (Interview BW 4: 11).

Moreover, as another interviewed expert for the biotech sector remarks, the quality of the services provided by these offices would strongly differ from case to case: While the transfer offices at the universities would not work in a satisficing way, those of the renowned research institutes, or the one's of the more specialized institutes, would work much better. According to the expert, the main reason for that would be their better equipment with organizational, personal and financial resources (Interview BW 5: 15f.).

5.2.2.5 Market Based Mechanisms of Knowledge Transfer

Mechanisms, through which knowledge is "bought" and "sold" via the market have been defined as market based mechanisms of knowledge transfer. For capturing this market dimension of knowledge transfer, the interviewed experts have been asked for the role of external consultancies, mergers, acquisitions and hostile takeovers as well as the buying, selling and licensing of patents in knowledge transfer processes. The assumption derived from the literature on VoC and NIS is that most of these market based mechanisms should only play a marginal role in the regional innovation systems of a predominantly strategically coordinated market economy like Germany.

External Consultancies

Looking at the region as a whole, these results were fully confirmed by the interviewed experts. Starting with *external consultancies*, the experts concordantly pointed out that the services these firms provide only play a marginal role as mechanisms of knowledge transfer in Baden-Württemberg (a.o. Interviews BW 1: 25f.: BW 2: 37; BW 3: 26). Although also some of the large commercial consultancy firms like *McKinsey*, *Ernst&Young*, *Roland Berger* and, above all, the *Steinbeis Beratungszentren GmbH* would offer services in the fields of innovation and knowledge transfer in the region, they would only be used to a very limited extent by the regional actors and firms. Thereby, the interview partners emphasized that firm size would matter in this regard: If consultancy services would be used, then they would be used more intensively by the large firms, and much less by the SMEs. And even in this case, many important large and multinational firms created their own *internal* consultancies in the last years, as one interview partner explains:

"These external consultancies are also used here, but only by the large OEMs. On the one hand, these firms have their own internal consultancy departments, or even own consultancy firms, on the other hand, they just use the services of the external consultancies. But from my point of view this does not play a role for the SMEs. The reason for this surely is a certain critical distance towards consultancies in general, but then again also reservations. And surely cost-benefit considerations also play a role." (Interview BW 6: 14; translation: MO)

Furthermore, a representative of the Baden-Württemberg Chambers also identifies a supply problem in this connection, since there are only very few smaller consultancies specialized in innovation management and knowledge transfer in the region which could offer attractive services in these fields, especially to the important group of SMEs (Interview BW 2: 37).

These differences between larger firms and SMEs in terms of using external consultancy services as mechanisms of knowledge transfer also hold true when focusing on the automotive sector. As an interviewed expert for consultancy services in this sector formulates:

"In the smaller companies they mostly do not want to make the afford and in the larger companies an internal consultancy developed in the meanwhile, which partly also offers its services externally. *Porsche* has its own *PorscheConsulting*, *Daimler* has its *MBTech*. These are the big ones, and the smaller ones are resistant to consulting." (Interview BW 1: 25; translation: MO)

Despite this critical distance and reservations towards consultancy services in the SMEs of the automotive sector, the barriers to use these services would be lower when they are offered by the *internal consultancy* firms of the large OEMs, which would also provide their services to external clients, especially within the densely integrated customer-supplier networks. Moreover, as an expert for the automotive industry in the Baden-Württemberg Ministry of Economics adds, also some public consultancy services would be used in this connection:

> "However, what is used intensively and works pretty well are the consultancy services provided by the various cluster initiatives in the region. These are accepted by the firms and above all by the SMEs as actors at the same level, which maybe do not act so much top-down." (Interview BW 6: 14; translation: MO)

Also in the biotech sector, the interview partners pointed to the difference between large and smaller firms in the use of external consultancies. While the few larger biotech firms in the region would use external consultancies in the field of knowledge transfer, although only to a very limited extend, for the majority of the smaller firms these services would not be relevant. If anything, they would use the consultancy services provided by the publicly founded cluster organizations, as e.g. the *BioPro Baden-Württemberg GmbH*. Internal consultancies in (particularly larger) firms were reported to not play a role at all (Interviews BW 2: 37; BW 4: 18; BW 5: 17).

Mergers, Acquisitions and Hostile Takeovers

With regard to *mergers, acquisitions and hostile takeovers* as mechanisms of knowledge transfer, the interviewed experts consistently indicated that these mechanisms would not play an important role in Baden-Württemberg (Interviews BW 1: 6ff.; BW 2: 38; BW 3: 27). In this regard, they named a lack of business culture and tradition as one major reason, while they supposed legal restrictions to not be a decisive barrier in this connection. This would also hold true when looking at the automotive sector, as one interview partner explains:

> "Mergers and acquisitions surely are less developed than in the Anglo-Saxon countries, at least the ones that have the explicit aim to broaden one's own knowledge base. In my judgment, this is not a dominant goal in the few cases in which a merger takes place. Legal regulation might not impede that more than it does in other countries […]. What is really opposed to mergers and acquisitions here in Baden-Württemberg is our business culture, particularly in automotive. Here you have got a strong sense of belonging together with which particularly hostile takeovers are not compatible." (Interview BW 6: 14; translation: MO)

In biotechnology the picture would only be slightly different, although the reasons would be dissimilar: In general, mergers and acquisitions would be crucial mechanisms of knowledge transfer in the sector, but only in the advanced phases of product development and clinical trials, when e.g. a smaller biotech firm might be bought by a large pharmaceutical firm to bring a new drug to the market. Since in Baden-Württemberg there are only a few larger pharmaceutical firms that could exercise this function, and since a great majority of the region's biotech firms is still very young, with a low share of firms in advanced phases of product development (Ernst & Young 2009), mergers and acquisitions would (still) not play an important role for knowledge transfer. Moreover, a vast majority of biotech firms in the region is not listed on the stock exchange, a fact that would strongly impede strategies of hostile takeovers (Interviews BW 3: 18; BW 5: 18). However, in this regard, one interviewed expert also points to a specificity of biotechnology in Baden-Württemberg, which would be related to the funding of some firms by private financiers:

"It might be interesting for you to know that some firms which were financed, or rather saved, by *Dietmar Hopp*, enforcedly come into closer contact. These are definitely different firms, but since they have the same financier, and since also the same persons sit in their supervisory boards, it is absolutely possible that the result could be the merger of two firms. Although this did not happen yet, I absolutely consider this as possible." (Interview BW 5: 18.)

Buying, Selling and Licensing of Patents

Finally, the *buying, selling and licensing of patents* have been attributed a medium range importance as mechanisms of knowledge transfer (a.o. Interviews BW 2: 38ff.; BW 3: 27f.). Most of the interview partners accentuated that only the larger firms might have the organizational, financial and personnel resources and capabilities to acquire and sell knowledge through the market for IPRs. For the smaller firms these channels might, thus, be less important and practicable (Interview BW 1: 25). Furthermore, as one representative of the Baden-Württemberg Chambers adds, many firms already use their searching-for-patents-activities as strategic instruments in their knowledge transfer efforts:

"Normally, in the larger firms from 100–200 employees upwards there is somebody who is responsible for this topic, who does the searching and looks for the new registrations in the patent data banks. Patents are absolutely used by many firms as strategic instruments. Indeed, one can read many things out of these researches for protected properties. One can look what the competitors are doing, even if I do not have patents myself. If, e.g., my three most important competitors suddenly apply for patents in other fields than the one's they were active before, then one can recognize in which direction a technology develops [...]. One can also recognize, who applies for what, which are the best experts appearing there again and again, applying for patents concerning the same topic again and again, and one can try to headhunt them. This is particularly true for the technology oriented firms [...]. The vast majority of firms will, on the other hand, not be aware of that, partly because of ignorance that this exists, and partly because of not knowing how to do it. Thus, this will not be a topic for the smaller firms with 10–20 employees." (Interview BW 2: 39f.; translation: MO)

In some cases, transfer via patents and licenses would be supported and mediated by intermediate institutions, like e.g. the "*Technologiebörse*" provided by the chambers, the *Technology Licensing Office of the Higher Education Institutions of Baden-Württemberg GmbH* or the *Steinbeis Transferzentren.* Partly, these intermediate institutions would actively try to help in particular the SMEs to access and commercialize IPRs, and partly they would try to improve the commercialization of research results from the region's universities and higher education facilities (Interview BW 2: 40; TLB 2007).

When differentiating between the two investigated sectors, it becomes apparent, that the buying, selling and licensing of patents is much more important in biotechnology than in automotive. As one expert for the automotive industry in Baden-Württemberg explains, the trade of patents in automotive only has a marginal function. For the firms in the sector it would be more important to apply for own patents and to develop own products out of it (Interview BW 6: 15). In biotechnology, on the contrary, the trading and licensing patents would be essential (Interviews BW 4: 19; BW 5: 18f.). In this, and confirming again the sectoral specialisms, this mechanism of knowledge transfer would be particularly relevant in the later stages of product development, as the executive director of the *Bio-Region RND e. V.* emphasizes:

"No, this is actually very decisive. This is connected to the fact that a drug-developing biotech-firm cannot cope with the development of a drug to the point of marketability. You also would not find an investor who would finance you all that. Therefore, it is about developing a drug far enough to extract a benefit out of it, if possible. And normally that is the so called *proof of concept*, i.e. if it turns out that a certain drug agent has shown some effect on a diseased patient in the first clinical study. Then you can sell it well, probably very well. [T]here are only very few firms that have the ambition to do what the early American biotech firms could do, that they basically developed into a pharmaceutical firm and did everything by themselves, up to the market. Most firms here are far away from that." (Interview BW 5: 18f.; translation: MO)

5.2.2.6 Community Based Mechanisms of Knowledge Transfer

To cover the community dimension of knowledge transfer processes, the interviewed experts have been asked for their estimation on the importance of three mechanisms of knowledge transfer: Informal contacts, regional platforms and additional (regional) institutions, which allow for establishing trust-based relationships and base on a common regional identity, culture and tradition. Building on the discussion on the theoretical concepts VoC and NIS, it has been assumed that these community based mechanisms of knowledge transfer should exercise a very important function within the coordinated framework of the German national market economy, since it relies strongly on non-market coordination of economic activities and less on strictly formalized and contract based coordination via the market.

Informal Contacts and Regional Platforms

In general, the interviewed experts confirmed these assumptions. *Informal contacts* were reported not only to be an important channel of knowledge transfer in the region, but also a central precondition for establishing further, then more formalized knowledge transfer activities. This distinction has been highlighted by the majority of the interview partners, since e.g. concrete R&D-projects and the transfer of IP-protected knowledge would always be formalized and base on fixed contracts. Nevertheless, the experts stated that it is a particular strength of Baden-Württemberg to offer a large variety of *regional platforms* on which the different actors can get to know and interact with each other. Apart from the already discussed cluster and network initiatives and the associations and chambers, also the metropolitan region initiatives in the Rhine-Neckar-Triangle, the Stuttgart and the Upper-Rhine region have been mentioned in this regard as platforms on the sub-regional level, on which actors from the business sector, politics, science and civil society would closely interact with each other in different formal and informal ways (a.o. Interviews BW 3; BW 6: 11ff.).

If a strong and common *regional identity* is present in Baden-Württemberg and if it helps to establish these informal contacts, cooperations and, thus, knowledge transfer processes, could not be judged clearly by the interview partners. As one interview partner explains, above all the estimated competences of the other regional actors as well as spatial proximity would be decisive in this connection, especially for the smaller firms:

> "The regional identity definitely plays a very important role for the SMEs. There, a regional cooperation partner is preferred clearly. These firms remain among each other, not least because one trusts in the competence of the regional partners. Moreover, one appreciates the short distances and the frequent opportunities to meet each other and for mutual exchange. Naturally, for the MNCs this is barely important, even if they exceptionally appreciate their regional partners in many cases. That the regional identity in Baden-Württemberg would be particularly strong in this, is hard to judge. There are also other regions in Germany with strong regional identities, take e.g. Bavaria or some regions in eastern Germany." (Interview BW 6: 11f.; translation: MO)

According to the executive director of the *Metropolitan Region Rhine-Neckar-Triangle GmbH* one reason that a common regional identity in Baden-Württemberg is not too strong would be that the *Land* it is still perceived as an artificial construct by many regional actors that emerged out of the two historical regions Baden and Württemberg-Hohenzollern after World War II. Stronger regional identities would therefore exist on these two sub-regional spaces, as well as, for example, in the Metropolitan Region Rhine-Neckar-Triangle, which covers large parts of the territory of the historical region of Electoral Palatinate (Interview BW 3: 10).

Differentiated by sectors, the picture would be similar in automotive, with the SMEs and supplier firms relying more on informal contacts and regional partners and platforms as the OEMs, which would have a more global focus (Interview BW6: 11f.). In biotechnology informal contacts would be less important, and most interaction would be formalized with fixed contracts. The reason given for this is the strong reliance on R&D-activities and the dependence on IPRs protection in the sector. Nevertheless, regional platforms, especially those provided by the respective cluster organizations, would play an important role in the sense that they would offer the opportunity to exchange information about the framework conditions of business activities, as, e.g., public funding, availability of seed- and venture capital or the provision of research facilities and instruments. Regional identity and culture would, however not play strong roles for interaction with other partners. Although particularly larger, but also smaller biotech firms would maintain intensive and stable, often long-lasting relations with other firms or scientific institutions in the region and, as one representative of a large biotech firm accentuates, appreciate their competence, expertise and spatial proximity, a science and research driven sector like biotechnology would always have a global focus and cooperate with the best partners available worldwide (Interviews BW 4: 5; BW 5: 8).

Additional (Regional) Institutions

Almost all interviewed experts brought up as *additional (regional) institutions* relevant for knowledge transfer processes in Baden-Württemberg the *Steinbeis Network*, independently of the sector. Apart from that, a broad variety of other public and private institutions, clubs and associations, from the *Lions Club* to the regional and communal economic development (*regionale und kommunale Wirtschaftsförderung*) have been mentioned. Altogether, it can be supposed that it is this variety, multitude and density of actors, platforms and institutions with a specific focus on the region that is a distinctive characteristic and strength of Baden-Württemberg's regional innovation and knowledge transfer system (Interviews BW 3: 4; BW 6: 12f.; BW 1: 1ff.).

5.2.3 Summary

In sum, this discussion demonstrated that to a large extent the theoretical assumptions derived from the literature on VoC and NIS for knowledge transfer processes in Germany's CME have been confirmed when looking at the case of Baden-Württemberg. As expected, network based mechanisms of knowledge transfer play a central role in Baden-Württemberg's regional innovation system. In this, horizontal and vertical inter-firm cooperations and networks, as well as inter-firm cooperations in R&D have been identified as crucial channels of knowledge transfer in the region. As has been shown, the informal dimension of these networks has been evaluated as less significant by the interviewed experts than has been assumed for an institutional environment of a CME, while institutionally stabilized networks, cooperations and clusters, on the other hand, were reported to be the central forms these cooperations would take. This illustrates, that in Baden-Württemberg strategic coordination of economic activities in the field of knowledge transfer finds its expression less in informal interaction, but, above all, in institutionally stabilized networks, cooperations, initiatives and clusters. The extraordinarily high number and broad variety of these institutionalized forms of cooperation can be regarded as one of the region's most distinctive characteristics in the field of knowledge transfer. In addition, cooperations between firms and the scientific world have been evaluated as more important than assumed. Both, a high demand generated by research intensive sectors in the center of the region's industrial structure requiring an intensive cooperation with scientific institutions, and a strong supply of a singular scientific infrastructure with highly renowned universities and research institutes explain this result and constitute another distinctive feature of the region's knowledge transfer system.

Also the association based mechanisms of knowledge transfer, and in particular employers-, business-, and trade associations, as well as the chambers of industry and commerce, have been attributed a decisive function. As has been shown, an historically grown tradition of chambers and also many of the employers-, business-, and trade associations in Baden-Württemberg to take a the strong focus on the region as a whole, as e.g. through the lead management principle, resulted to bring them into a highly valuable intermediary position in terms of knowledge transfer, something which strongly characterizes the region's knowledge transfer system.

Larger deviations from the theoretical assumptions could be shown for the organization based mechanisms of knowledge transfer, like the transfer of technical and scientific staff, purposeful headhunting, and business start-ups, which were assumed to not play a role in the regional innovation systems of the CMEs like Germany. Here, the empirical results for Baden-Württemberg indicate at least a medium range importance of these mechanisms. The reasons given by the interviewed experts are manifold: Concerning the transfer of technical and scientific staff and purposeful headhunting, they accentuated that a reduction of supply for qualified staff on the labor market and the beginning effects of the process of demographic change would not only open up more opportunities for highly qualified staff from the scientific world to transfer to the business sector, but also enhance the firms' recruiting activities in the scientific sector. Moreover, it is to be expected that these effects are particularly strong in Baden-Württemberg, since its industrial structure is strongly shaped by research intensive and highly innovative sectors, requiring highly qualified staff. With regard to business start-ups and the provision of seed- and venture capital, above all, institutional support, e.g. by regional government or the *Steinbeis* network, public funding and investments by private financiers have been identified as distinguishing elements for knowledge transfer processes in the region.

Concerning the state induced mechanisms of knowledge transfer (national or regional public programs and measures, (larger) state-induced innovation projects, creation of scientific parks and technological centers) the results of the case study confirmed the theoretical assumptions and accentuate the role of the state as an enabler of innovation and knowledge transfer processes in Baden-Württemberg. Only the importance of technology transfer offices or companies at the universities and public research institutes has been estimated lower than expected. As an explication most of these institutions were reported to be personally and financially underequipped, inefficient and misdirected with regard to their mission.

Furthermore, also the expected low significance of the market based mechanisms of knowledge transfer (external consultancies; mergers, acquisitions and hostile takeovers; buying, selling and licensing of patents) has been confirmed by the results of the case study. Only the trade of patents and licenses has been evaluated slightly more important than assumed, what has been explained, among other things, with the recent efforts of the public universities and research institutes to intensify the commercialization of their IPRs. Finally, as assumed, the community based mechanisms of knowledge transfer, like informal contacts, regional platforms and other (regional) institutions, resulted to have a central function in Baden-Württemberg's knowledge transfer system, in particular as important precondition for establishing further, then more formalized knowledge transfer activities. In this, the broad variety of regional platforms Baden-Württemberg offers to establish these informal and formal contacts in the form of clusters, network initiatives, chambers, associations, clubs, and metropolitan region initiatives etc., was identified as a distinctive feature of the region's knowledge transfer system.

In addition, and as expected, the results also revealed sectoral specificities as important differentiating factors. While Baden-Württemberg's automotive industry shows quite similar structures and mechanisms of knowledge transfer compared to the region as a whole, biotechnology noticeably differs from these results. The automotive sector in Baden-Württemberg displays almost all assumed features of a traditional, mature and established industrial sector in a CME-context: Network-, association- and community based mechanisms of knowledge transfer are the vital channels of knowledge transfer in this sector, while state induced and organization based mechanisms have a medium range importance and market based mechanisms only play a marginal role, except external and especially internal consultancies, which were reported to be slightly more important in this sector. In this, the sector's business structure with some large, renowned multinational OEMs and suppliers, as well as a high number and broad variety of small and medium sized suppliers, complemented by a pointed and historically grown sectoral and regional self identification, especially in case of the SMEs, constitute some of its most characteristic features, allowing this strongly network-, association and community based structures of knowledge transfer.

Baden-Württemberg's biotechnology sector, on the other hand, demonstrates characteristics of a young, strongly research based and venture-capital driven industry: Its connections to universities and research institutes as channels of knowledge transfer are strong, while other forms of networks as well as association and community based mechanisms only play an average role. Decisively important for knowledge transfer processes in the sector are state induced mechanisms, like national or regional cluster programs and initiatives, cluster management organizations, scientific parks and the provision of public funding, not least for business start-ups and academic spin-offs. Not only the emergence, but also the further development and growth of the sector in Baden-Württemberg have been strongly dependent on public action and support, which apparently is also reflected in the structures and mechanisms of knowledge transfer in the sector. In addition, and as was to expect for a new economy sector characterized by more radical modes of innovation, market based mechanisms of knowledge transfer are considerably more important than in automotive. Finally, two further factors influence strongly knowledge transfer processes in Baden-Württemberg's biotechnology sector: First, its business structure with a vast majority of small firms, only a few larger biotech companies and the absence of large pharmaceutical firms in the region constrain the use of certain channels of knowledge transfer, like mergers and acquisitions, hostile takeovers or purposeful headhunting. Second, also the strong dependence on IPRs protection impedes some forms of knowledge transfer, above all inter-firm cooperations, joint research activities and the involvement of business associations and chambers in knowledge transfer activities. Table 4 provides a summarizing overview on these results.

Table 4: Qualitative Relevance of Knowledge Transfer Mechanisms in Baden-Württemberg

MECHANISMS	GER*	BW	Auto	Bio
NETWORK BASED MECHANISMS				
business networks	+	+	+	○
informal networks	+	○	○	○
institutionally stabilized networks	○	+	+	○
customers and suppliers	+	+	+	○
cooperative inter-firm R&D-projects	+	+	+	○
cooperative R&D-projects with scientific institutions and universities	○	○	○	+
ASSOCIATION BASED MECHANISMS				
employers-, business-, and trade associations	+	+	+	○
chambers of industry and commerce	+	+	+	○
binding and common technical standards	+	○	+	+
ORGANIZATION BASED MECHANISMS				
transfer of technical and scientific staff	–	○	○	○
purposeful headhunting	–	○	○	○
business start-ups	–	○	–	○
STATE INDUCED MECHANISMS				
national government measures	○	○	○	+
regional government measures	○	○	○	+
(larger) state-induced innovation projects	–	–	○	–
scientific parks	○	○	○	+
technological centers	○	+	???	???
technology transfer agencies/companies	+	–	–	○
MARKET BASED MECHANISMS				
external consultancies	–	–	○	○
mergers and acquisitions, hostile takeovers	–	–	–	○
buying, selling and licensing of patents	–	○	–	+
COMMUNITY BASED MECHANISMS				
informal contacts	+	+	+	○
regional platforms	+	+	+	+
additional (regional) institutions	+	+	+	+

+ *Reported (*Hypothesized) as crucial mechanism of knowledge transfer*

○ *Reported (*Hypothesized) as important but not decisive mechanism*

– *Reported (*Hypothesized) as not important at all*

??? Not reported by the interview partners

= *Important deviation from hypotheses*

= *Modest deviation from hypotheses*

= *No deviation from hypotheses*

5.3 The Case of the Metropolitan Region Hannover Braunschweig Göttingen Wolfsburg

5.3.1 Locating the Metropolitan Region HBGW in the German Context

5.3.1.1 Selected Facts on the Metropolitan Region's Economic Structure

The second German region investigated for this study is the Metropolitan Region Hannover Braunschweig Göttingen Wolfsburg (MR HBGW) in the south-east of Lower Saxony. This region can be labeled a "young spatial entity", which traces back to a resolution of the *Ministerial Conference for Regional Planning* (MKRO) in the year 2005, when the metropolitan area of Hannover, Braunschweig, Göttingen and Wolfsburg was incorporated into the group of *European Metropolitan Regions.* After a reorganization and a "new start" in the year 2009, the Metropolitan Region found its present organizational form (Metropolregion 2010: 2ff.). The aim behind these efforts to become a German metropolitan region with European relevance was twofold: On the one hand, it rooted on an intensification of locational competition among the urban areas in Europe through the internationalization of markets and the accelerated European integration process. On the other hand, it resulted from the specific problems that the German urban areas faced due to the particularly decentralized structure of the German urban system, compared to the rest of Europe. Therefore, the acknowledgement as a metropolitan region was to secure the region a better position in the competition among European areas, as well as to strengthen its "cartographic presence" (Priebs 2004: 104) and improve its image on the EU-level (Priebs 2004: 102ff.).

In this connection, the MR HBGW can be described as a *polycentric region*, which is not only characterized by the four mentioned urban areas, but also by the respective rural areas in between. In this regard, the project of the MR HBGW is not exclusively to be seen as an amalgamation of communal authorities, but also as a cooperation project including a broad variety of other participating regional actors, as is also stated in its (preliminary) statute, which defines the Metropolitan Region as "a voluntary association of communal authorities and other communal associations, public and private scientific and cultural institutions, as well as business firms of national and international weight." (Priebs 2004: 109; translation: MO) Therefore, the MR HBGW must be regarded as a *nascent* spatial entity, in which historically grown regional structures and a commonly shared regional identity can be expected to be, if anything, rudimental or limited to the four constituting sub-spaces. However, it is the explicit aim of the MR HBGW to enhance cooperation among these sub-spaces (Ortiz 2008). In the following, some selected data on the region's economic profile will be briefly discussed.

Essential parts of the MR HBGW's industrial structure trace back to the national socialist industry formations in the 1930s in the automotive industry (*Volkswagenwerk* in Wolfsburg) and in the iron and steel industry (*Reichswerke "Hermann Göring"* in Salzgitter). In the post-war decades these two sectors emerged as the region's most important growth drivers. Independently of these two industrial complexes, and basing on the increasing capitalization and mechanization of agriculture, a food processing industry (sugar and canning industry) developed, which, again, gave important impulses to (agricultural-) mechanical engineering, which grew, as a consequence, in a highly dynamic way. In the present, and building on this historical industrial trajectory, the MR HBGW is still characterized by large scale industry structures and, compared to Germany as a whole, by a high share of the industrial manufacturing sector in the total value added (Bezirksregierung Braunschweig et al. 2001: 36; Stühring 2006: 90ff.).

Today (2009), the MR HBGW has about 3.9 million inhabitants and covers about one third of Lower Saxony's territory. The region's total workforce is about 1.86 million, representing about 51% of Lower Saxony's workforce[11]. With that, the MR HBGW is the economic heartland of Lower Saxony and, behind the Metropolitan Region of Hamburg, the strongest economic region in Northern Germany (BBSR 2010: 10ff.; NIW/Nord/LB 2005: 108; Priebs 2004: 101). This is also underlined by the Metropolitan Region's GDP in the year 2008, which was about 112.4 billion Euros, representing 52.6%[12] of Lower Saxony's, and 4.5%[13] of Germany's GDP (BBSR 2010: 18; Destatis 2009: 5; Eurostat 2011b).

In this, the MR HBGW is still characterized by a large share of industrial employment, with about 584,000 employees in manufacturing in 2008, representing 31.4% of the regions' total workforce[14]. This comes close to Baden-Württemberg (32.6%) and ranges considerably above the German

11 Own calculations basing on LSKN 2010: 68; Destatis 2011 and BBSR 2010: 10ff.

12 Own calculations basing on BBSR 2010: 18; Destatis 2009: 5; Eurostat 2011b

13 Own calculations basing on BBSR 2010: 18; Destatis 2009: 5; Eurostat 2011b

14 Own Calculations and estimations basing on BBSR 2010: 16ff.; LSKN 2010: 68; SÄdL 2009

average with 25.3% of industrial employment and the value for Lower Saxony (24.2%), also illustrating the region's strong industrial position in the *Land* (BBSR 2010: 16ff.; LSKN 2010: 68; SÄdL 2011). The service sector, on the other hand, is comparatively weak for a metropolitan region: With a share of 67.5% of employment in the service sector it ranges clearly below the German (72.5%) and the Lower Saxon (72.6%) average, but still above the value for Baden-Württemberg (65.6%) (BBSR 2010: 16ff.; SÄdL 2011).

Its dominant position within the Lower Saxon economy is also demonstrated by the fact that about two thirds of the hundred largest firms in Lower Saxony in terms of total sales volume are located in the MR HBGW (Nord/LB 2010: 15ff.). The location of these large firms' headquarters in the Metropolitan Region also illustrates the region's present sectoral structure: The *Volkswagen AG* in Wolfsburg and the *Continental AG* in Hannover are not only the region's two largest companies, but also represent automotive as the region's core industrial sector. Moreover, also iron and steel is an important sector in the region's economic structure, above all represented by the *Salzgitter AG* in Salzgitter. In addition, also the service sector in the Metropolitan Region has considerable strengths, in particular in the fields of tourism with the *TUI AG* in Hannover and energy supply with the *E.ON Avacon AG* in Helmstedt. A specific sectoral strength in the Metropolitan Region is the insurance industry, of which the *Talanx AG* with the *Hannoversche Rückversicherung AG*, the *VHV Vereinigte Hannoversche Versicherung AG* and the *Versicherungsgruppe Hannover VGH* (all in Hannover) are the most prominent representatives. Finally, also the financial sector is represented in the region, above all by the *Norddeutsche Landesbank* (Nord/LB) in Hannover and the *BHW Holding AG* in Hameln (Nord/LB 2010: 3ff.). However, when comparing the MR HBGW with other German Metropolitan Regions, its overall equipment with business headquarters is comparatively low (NIW/Nord/ LB 2005: 108).

Nevertheless, this already highlights that within the MR HBGW the regional capital city of Hannover is the main location of business activities in which a large part of the region's business headquarters are concentrated. In the year 2009, 58 of the largest companies in Lower Saxony were located in the MR HBGW, of which 28 were located in the city of Hannover or its surroundings, predominantly in the fields of automotive and its supply industries, energy supply, chemicals, electrical engineering, trade and (in particular insurance and financial) services. The other constituting subspaces of the Metropolitan Region are only locations of a low number of large companies (Braunschweig: 4; Göttingen: 2, Wolfsburg: 3), with a strong sectoral focus on metalworking and automotive, but also with strengths in electrical engineering, energy supply and food processing. Therefore, only the Hannover area can be described as a concentration of business headquarters. However, when considering the number of employees that is administered by the respective company headquarters, Wolfsburg with the *VW Group* and its more than 480,000 employees worldwide takes an outstanding position. This also points to the enormous weight and dominance the *VW Group* has within the MR HBGW, which had a share of about 42% (105.2 billion Euros) in the total sales volume of the hundred largest companies in Lower Saxony (Nord/LB 2010: 8ff.; Nord/LB 2007; NIW/Nord/LB 2005: 108).

Finally, and pointing to the gateway-functions of metropolitan regions in the sense of platforms, on which regional and extra-regional actors can come into contact and exchange knowledge and information, the MR HBGW is, on the one hand, particularly well connected to national and European traffic infrastructures, above all through the *Hannover International Airport*. On the other hand, the region is characterized by a highly intensive and above average access to international markets through the professional trade fairs of the *Deutsche Messe AG* in Hannover, which is Germany's leading trade fair organizer, making Hannover the world's most important trade fair location of the capital goods industry, bringing together about two million visitors and 23,000 exhibitors every year (Deutsche Messe AG 2010; Schäfer et al. 2004: 24; NIW/Nord/LB 2005: 104ff.).

Looking at some structural indicators describing the MR HBGW's knowledge generating capacity, an extraordinarily high R&D-intensity is revealed. Measured as the share of R&D-personnel in universities, research institutes and firms in all employees of the manufacturing sector, the R&D-intensity in the Metropolitan Region was of 8.7% in 2001, while the values for Lower Saxony and Germany as a whole were of 5.8% and 6.3%, respectively (NIW/Nord/LB 2005: 112). Especially the business R&D-activities are an important indicator for a region's industrial innovation potential, since they demonstrate the firms' ambition to further develop products, processes and services. In this connection, also the business R&D-intensity, measured as the share of the firms' R&D-personnel in all employees of the manufacturing sector, was, with a value of about 5.7% in 2005, considerably higher in the MR HBGW than in Lower Saxony (4.4%) and in Germany (5.3%) (Nord/LB 2007: 40ff.).

Within the Metropolitan Region, these business R&D-activities are mostly concentrated in the sub-spaces of Braunschweig and Wolfsburg, which is connected to the intensive R&D-activities of the automotive sector, including railway vehicle manufacturing, rail technology, logistics, traffic control, drive engineering and the rubber and plastics industry. More than half of the business R&D-activities in Lower Saxony are conducted in automotive, and the largest share of these activities is concentrated within the Metropolitan Region, with a focal point on Wolfsburg and the *VW Group*. Further important business R&D-activities in the Metropolitan Region, although on a lower level, are conducted in the fields of laser technology, aeronautical and measurement engineering, biotechnology, genetic engineering, pharmaceuticals, human and veterinary medicine, seed production and environmental protection (Legler/Schasse 2004: 10ff.; NIW/ Nord/LB 2005: 112f.; Legler 2004: 18ff.).

In addition, also the R&D-intensity at universities and scientific institutions is an important indicator for a region's knowledge production and innovation potential. With a value of 3.3% the Metropolitan Region ranks, again, clearly above the values of Lower Saxony and Germany as a whole, with 2.2%, respectively. Within the Metropolitan Region, university R&D-activities are further concentrated in the sub-spaces of Göttingen and Hannover. Moreover, also the R&D-intensity of the research institutes is highest in Göttingen. In this, the locations of universities and research institutes are concentrated in terms of number and variety in the three urban areas of Hannover, Braunschweig and Göttingen (NIW/ Nord/LB 2005: 113). With that, the Metropolitan Region also represents the center of Lower Saxony's university landscape. In the year 2008, 16 higher education institutions were located in the region, in which 91.662 students were matriculated, representing almost two thirds of all students at universities in Lower Saxony. The region's focus of university research activities lies in the fields of medical science as well as natural and engineering sciences (BBSR 2010: 20f.; NMWK 2010; NIW/Nord/LB 2005: 113).

Apart from the higher education institutions, some important extra-university research institutes complement the research landscape in the Metropolitan Region: Seven institutes of the *Max Planck Society* with a focus on basic research, three institutes of the *Fraunhofer Society* with a focus on applied research, six facilities of the *Leibniz Association*, as well as three facilities of the *Helmholtz Association*, providing large scale equipment for regional, national and international groups of researchers and constituting an important link to national and international research activities for the regional actors are located in the region. The most renowned of these institutes are the *Max Planck Institutes for Experimental Endocrinology* and *Experimental Medicine* in Hannover and Göttingen, as well as the *German Aerospace Center* and the *Helmholtz Centre for Infection Research* in Braunschweig. In addition, a broad variety of federal and regional facilities with R&D-missions are located in the region. These research institutes have a focus on the fields of geosciences, aerospace, biology and medical science. Most of these research institutes are part of one of the six regional initiatives (Landesinitiativen), 14 competence centers and 35 competence networks in the region, which intend to connect the R&D-activities of firms, universities and research institutes. These competence centers and networks mainly reflect the already mentioned foci of research, in particular in the fields of automotive, biotechnology and medical technology (NMWK 2011; Nord/LB 2007: 63ff.; NIW/Nord/LB 2005: 118f.).

5.3.1.2 Automotive Sector: The Mobility Competence Region HBGW

When looking at the automotive sector, only a few data are available explicitly focusing on the sector in the MR HBGW, but, rather, on Lower Saxony as a whole. Nevertheless, since automotive in Lower Saxony is strongly concentrated within the Metropolitan Region, the following remarks describe to the greatest possible extent the structural conditions of automotive in the Metropolitan Region, although most the data might refer to Lower Saxony as a whole. Automotive is the most important industrial sector in the MR HBGW, in which about 100,000 persons are employed[15], representing 23% of all employees in the manufacturing sector in Lower Saxony and almost 14% of all employees in German automotive in 2009[16]. With a total sales volume of about 71 billion Euros in 2008, the Lower Saxon automotive sector represented 38.6% of the total sales volume of manufacturing in Lower Saxony and 21.1% of the total sales volume of German automotive (MW NDS 2011a).

The automotive sector in Lower Saxony consists of about 700 firms, representing around 30% of the firms in German automotive.[17] In contrast to Baden-Württemberg, automotive in the MR HBGW is characterized by

15 Own estimation basing on VDA 2010, VDA 2011 and MW NDS 2011a
16 Own estimation basing on VDA 2010, VDA 2011 and MW NDS 2011a
17 Own estimation basing on NGlobal 2009a; BW-Invest 2011a

a strongly centralized structure, which is dominated clearly by the *VW Group* in Wolfsburg as Europe's largest OEM. Apart from the *VW Group*, also *MAN SE* as an OEM for utility vehicles maintains an important production location in Salzgitter. In addition, a high number and variety of supplier firms are located in the Metropolitan Region, of which the *Continental AG* in Hannover is the largest and most renowned (NGlobal 2009a; MW NDS 2011a; VDA 2010; VDA 2011; BW-Invest 2011b).

Like its counterpart in Baden-Württemberg, also the automotive sector in the MR HBGW is highly innovative, and constitutes a focal point of R&D-activities within German automotive. Its R&D-intensity in the year 2005, measured as the share of R&D-personnel in all employees in the sector in Lower Saxony, is 9.2%, and, therefore, slightly below the German average of 10.6%. 12.7% of all R&D-personnel in German automotive are employed in Lower Saxony, and automotive alone employs 56.9% of all R&D-personnel in Lower Saxony (in Germany: 33.2%), demonstrating the extraordinary weight of the sector for the research landscape in the *Land.* This is also underlined by the fact, that about 60% of the total R&D-expenditures in Lower Saxony's business sector are conducted by the automotive industry (Germany: 30%; Baden-Württemberg: 46%). In comparison, this strong dependence of the Metropolitan Region on automotive reveals a certain industrial monostructure in the region, constituting a decisive difference to Baden-Württemberg, which is characterized by a more diverse industry structure, above all because of a highly variegated medium size business sector (NIW 2008: 20ff.; VDA 2011; Legler et al. 2009: 90).

Apart from the sector's internal R&D-activities, research in automotive-relevant fields is conducted in 17 research institutes within the Metropolitan Region, of which most are also connected to the region's four universities or other higher education facilities (*Fachhochschulen*) with sector relevant research in this field. The most renowned of these institutes are the *Automotive Research Center Lower Saxony* and the *German Aerospace Center* in Braunschweig, as well as the *Center for Production Technology*, the *Mechatronics Center* and the *Competence Center Automotive* in Hannover. Further sector relevant R&D-activities in the MR HBGW are conducted within the frameworks of the numerous cooperation projects as well as cluster and network initiatives, which mostly have a focus on interdisciplinary topics within the field of automotive. Important examples would be the newly founded *Automotive Cluster* of the Metropolitan Region, the *Logistikinitiative Niedersachsen*, the *Landesinitiative Telematik*, the *Competence Network Adaptron-*

ics, the *Mobility Forum Lower Saxony*, the *Competence Network Production and Logistics*, as well as the *Landesinitiative Brennstoffzelle und Elektromobilität* (NGlobal 2009a; NIW/Nord/LB 2005: 118).

5.3.1.3 Biotechnology in the Metropolitan Region—Towards the BioRegioN

The biotechnology sector in the MR HBGW has a strong focus on red biotechnology, especially in the fields of functional genome research, intelligent implants, as well as infection,- neuro- and stem cell biology. Moreover, also strengths in the fields of biomedical technology and instruments, as well as green biotechnology can be found in the region. Substantial impulses for the development of the sector were given in the year 2001, when the region was among the winners in the *BioProfile* competition of the federal government and was awarded the "*BioProfile Functional Genome Analysis*" (MW NDS 2011b; BMBF 2005: 12f.).

Biotechnology in the MR HBGW is strongly concentrated within the three urban areas of Hannover, Braunschweig and Göttingen with its renowned sector relevant universities and research institutes. Together with actors from the fields of healthcare and life sciences it is integrated into a common regional cluster, the regional initiative (*Landesinitiative*) *BioRegioN*, into which, in the meanwhile, also the *BioProfile Functional Genome Analysis* merged. This regional initiative is promoted by the Lower Saxon Government and coordinates the activities of the different actors via its cluster management organization, a function that is currently exercised by the *BiomeTI e.V.* In all of the three urban centers of the biotech sector in the region scientific parks or technological centers have been established (Medical Park Hannover, BioTec Incubator Braunschweig, Science Park Göttingen), in which research institutes, biotech firms, supporting institutions and service providers have the chance to closely interact with each other and business start-ups are supported (biotechnologie.de 2010; Ernst & Young 2010; NGlobal 2009b; MW NDS 2011b; BioRegioN 2010: 4).

In 2009, the biotech business sector in the MR HBGW consisted of 30 companies[18], representing 5.6% of all biotech companies in Germany (531)[19]. In comparison of the German *Länder*, the region had considerably less biotech companies than Baden-Württemberg (84). These 30 companies had around 420 employees, representing only 2.8% of all employees in

18 Own estimation based on biotechnologie.de 2010

19 Own calculation based on biotechnologie.de 2010

the German biotech sector[20], and being one of the smallest contingents among the larger German *Länder* (Figures 5.13 and 5.14; biotechnologie.de 2010; Ernst & Young 2010; SÄdL 2011). As in Baden-Württemberg and in the rest of Germany, also biotechnology in the MR HBGW is characterized by a majority of small businesses: The average number of employees in the sector was of 14 in 2008 (BioRegioN 2009: 5).

Apart from their internal R&D-activities, research in the field of biotechnology is conducted in a large variety of research institutions such as universities, other higher education institutions, or research institutes. In 2009, 16 of these research institutions conducting research in biotechnology were located in the MR HBGW, representing a share of 7.9% of these institutions in Germany (202).[21] In the Metropolitan Region, these institutions employed around 2,300 persons involved in corresponding research activities, representing a share of almost 8.6% of the total number of employees in this field in Germany[22] (Figures 5.15 and 5.16; biotechnologie.de 2010). The fact that this share of 8.6% is considerably higher than the region's business sector's share of employees (2.8%) in German biotechnology indicates a certain disequilibrium between strong academic research and relatively weak business activities in the Metropolitan Region's biotechnology sector. The most renowned of these institutes are, apart from the *Universities of Human and Veterinary Medicine* in Hannover, the *Fraunhofer Institute of Toxicology and Experimental Medicine* (ITEM), the *International Neuroscience Institute* and the *Centre for Biomedical Technology and Innovation* (BiomeTI e.V.) in Hannover, the *Helmholtz Centre for Infection Research* in Braunschweig, as well as the *European Neuroscience Institute* and the *Max Planck Institutes for Biophysical Chemistry* and *Experimental Medicine* in Göttingen (NGlobal 2009b).

5.3.2 Structures and Mechanisms of Knowledge Transfer in the Metropolitan Region HBGW: Six Governance Dimensions

5.3.2.1 Networks as Mechanisms of Knowledge Transfer

Building on this brief discussion of some main structural economic facts, in the following section, the results of the empirical qualitative case study

20 Own estimation based on biotechnologie.de 2010

21 Own estimations and calculations based on biotechnologie.de 2010: 7

22 Own estimations and calculations based on biotechnologie.de 2010: 20

of the MR HBGW's knowledge transfer system will be discussed. Like for Baden-Württemberg, also in the case of the MR HBGW the theoretical assumptions derived from the literature on VoC and NIS will be confronted with empirical evidence. Starting with the network-based mechanisms of knowledge transfer first, unlike for Baden-Württemberg, these theoretical assumptions have only partly been confirmed by the results of the case study. Although all interview partners reported that networks of actors would have at least some significance for knowledge transfer activities in the region, the different dimensions of knowledge transfer via networks were attributed to have distinct impacts and were, overall, evaluated as less important as expected.

Horizontal Networks

With regard to *horizontal cooperations between competitors* as mechanisms of knowledge transfer, the interview partners drew a differentiated picture. Where these horizontal cooperations would happen, they would be an intensive and important channel of knowledge transfer. But, as most of the interview partners emphasized, from their perspective as practitioners in the fields of innovation and knowledge transfer, the firms in the Metropolitan Region would not cooperate often enough among each other (Interviews HBGW 1: 4f.; HBGW 2: 7ff.; HBGW 3: 2ff.; HBGW 4: 4ff.). With that, they also broadly confirmed the results of a recently conducted qualitative network study by Brandt et al. (2008) on knowledge networks in the MR HBGW, which finds out that horizontal cooperations among competitors are the weakest dimension of network activity in the region (Brandt et al. 2008: 145ff.; Figure 20).

One reason for this would be the region's business structure, with a share of more than 80% of SMEs, which were identified by the interviewees as cooperating less likely with other firms, since they would often lack the necessary organizational, human and financial resources to initiate and manage inter-firm cooperations. Moreover, a lack of knowledge about the available competences of other firms in the region has been mentioned in this connection, as well as a strong competitive attitude towards other firms, frequently driven by the fear of losing knowledge in cooperative arrangements. Not least, an insufficient regional coherence among the different sub-spaces of the Metropolitan Region has been noted, which would enforce this competitive attitude, above all among the different economic sub-spaces, as one interviewed expert for innovation processes in the MR HBGW formulates:

"This moment of competence does not come as a surprise. It is not based on a bad will, but, rather, on comprehensible self interest. The distinct sub-spaces just have different points of view and thinking, which are traditionally grown and emerged in decades. [...] Now, at least, there are common contexts for discussions and discourses, which are also institutionalized, in which the different actors from the fields of economic development and knowledge transfer can meet and interact, what was not the case before. Insofar, this is a progress, but it is not more than that. [...] Here and there it might be the case that a sensibility exists to also use the competences in the Metropolitan Region, if possible, or to even search in the closer environment and to not immediately look towards Düsseldorf and Munich. But I do not believe that this is the normal culture." (Interview HBGW 3: 4f.; translation: MO)

As one central form these cooperations would take, almost all experts mentioned the *regional initiatives* (*Landesinitiativen*) promoted by the regional government, in which firms would cooperate in specific technological fields with other firms, but also with scientific institutions. These regional initiatives would be *institutionally stabilized* forms of cooperation with a professionalized network management organization coordinating the activities of its actors and would generally be limited to a duration of about three years, but can be prolonged after an evaluation of its performance through the *Innovationszentrum Niedersachsen* (Innovationszentrum Niedersachsen 2005: 8). Furthermore, one interview partner mentions examples of predominantly business driven networks in the region, but also points to more project-related and short term forms of cooperation:

"On the one hand, there are really institutionalized networks. Some are organized like club goods, like e.g. the *Measurement Valley* in Göttingen, in which the firms can cooperate with each other in a deinstitutionalized way. Or take the project *KIM*, the cooperation initiative mechanical engineering in the Braunschweig area, in which presently two networks exist, which interact with each other in a very formalized way, e.g. in joint purchasing activities, formation and training activities, and innovation and market development activities etc. [...] Apart from that, they can also take the form of incidental and contingent networks [...]." (Interview HBGW 3: 6; translation: MO)

Fig. 20: Types of Cooperation in the Metropolitan Region HBGW

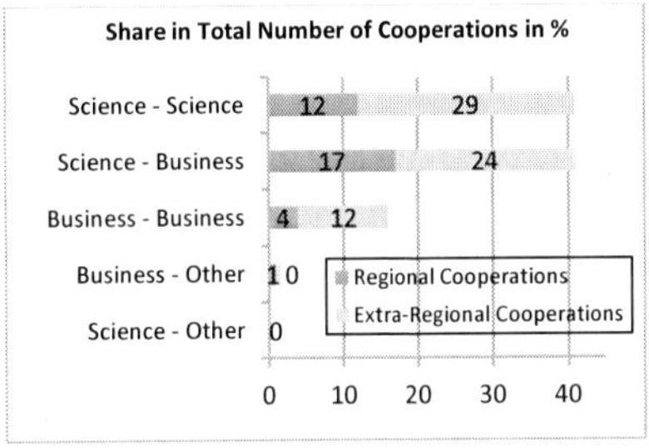

Source: Nord/LB 2007: 95

Fig. 21: R&D-Intensity: Automotive & Manufacturing, Selected Countries

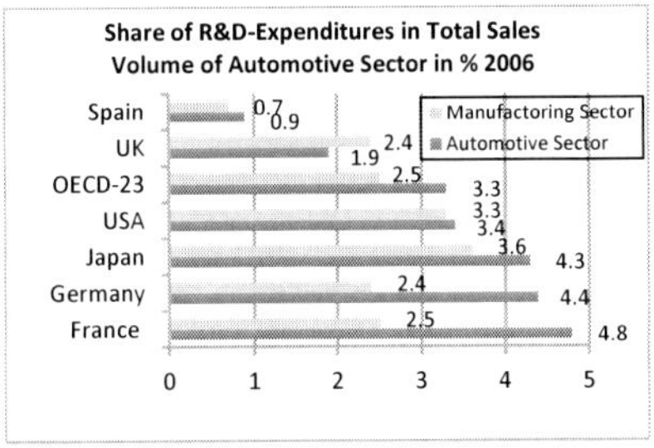

Source: Legler et al. 2009: 90ff.; data for 2006

Vertical Networks

A similarly differentiated picture arises when looking at *vertical cooperations between firms and their customers and suppliers.* Here, again, the interview partners accentuate that these cooperations would not be an economy wide characteristic feature of the region's knowledge transfer system, as it would be the case in Baden-Württemberg, but that the importance of these customer-supplier-networks would particularly show strong sectoral differences (Interviews HBGW 1: 5f.; HBGW 2: 9ff.; HBGW 4: 5ff.; HBGW 6: 5ff.). In many sectors, both vertical *and* horizontal cooperations would be less important channels of knowledge transfer, and many firms in these sectors would not cooperate with other firms at all, as one of the authors of the already mentioned network study on the Metropolitan Region explains:

"The study we conducted indeed led to some results, particularly with regard to the question in which fields of competence networking is stronger developed in comparison to other fields. E.g., the networking and the coherence is stronger in the field of production engineering than in the field of life sciences and, in particular, in biotechnology. But of course we could not compare this to other Metropolitan Regions at that time. […] Therefore, we cannot definitely state if the degree of networking within the Metropolitan Region or its respective fields of competence shows relevant deficits, since we have no reference point. […] However, these considerable differences between the fields of competence must be reflected, especially the fact that in many of them a high number of isolated actors exists, or that only bilateral cooperations exist, which is particularly obvious in the field of life sciences." (Interview HBGW 3: 2f.; translation: MO)

Also when focusing on the two investigated sectors in the study at hand, clear differences can be revealed. On the one hand, the automotive industry in the Metropolitan Region has been described as a highly networked sector (see also Brandt et al. 2008: 94ff.). In this regard, the dominant position of the *VW Group* in the sector would constitute a differentiating feature of automotive in the Metropolitan Region, also in comparison to Baden-Württemberg (Interviews HBGW 2: 9; HBGW 6: 5ff.). First, because of the lack of further OEMs in the region, cooperations between OEMs would only play a marginal role for knowledge transfer in the region. Horizontal cooperations would, therefore, predominantly occur among the supplier firms. Second, vertical cooperations would to the largest possible extent be oriented towards and dominated by the *VW Group*, turning it into the central actor and nodal point of reference in these networks. Third, this dominant and central position of only one large OEM is even reinforced by a certain industrial monostructure in the region, forcing a large part of the regional SMEs into networks and cooperations with the OEM and its suppliers. This strong industrial focus on the automotive industry would also be one major difference in comparison to Baden-Württemberg, as a representative of the *Innovationsnetzwerk Niedersachsen* explains:

"Lower Saxony is an *Autoland*, which is even stronger specialized in automotive than Baden-Württemberg. There you still have a diversified medium size business sector, something that does not exist in Lower Saxony in this way. That means that everybody who is active in automotive here in the region, is part of a network, and that is the network of *VW* [...]. [E]verybody knows everyone here since we have only one OEM in Lower Saxony. Even if I would be at the fourth level in the value added chain, I would know where my component is built in." (Interview HBGW 2: 9; translation: MO)

These vertical supply chain networks and horizontal cooperations among suppliers can, therefore, be regarded as a central dimension of knowledge transfer in the Metropolitan Region's automotive sector, although many technological developments would also be initiated top-down by the OEM, as an interviewed expert for the automotive industry in the region illustrates:

"There is the challenge and the clear goal by the OEM, to continuously improve every product coming out in comparison to previous models. And if we observe where the innovation and potentials for added value in automotive lie, then they lie in the field of the suppliers. Of course, central developments are operated by the

OEM itself. *VW* has got 12,000 persons working in the field of R&D here [...]. Moreover, there is a broad variety of suppliers, which approach the OEM and have product improvements and further developments. This is a continuous exchange process between the OEM and suppliers on the one hand, but also between supplier and supplier on the other hand. I.e., we do not only have relations between OEM and its suppliers, but also on the horizontal level, between the supplier and its sub-suppliers. Therefore, there are also a number of small and fine graded interlinks, that contribute to the improvement of a new generation of vehicles." (Interview HBGW 6: 8; translation: MO)

The networking between the OEM and its suppliers, as well as among the suppliers themselves, is actively coordinated among other things by the *Wolfsburg AG*, which, as a public private partnership of the *VW Group* and the city of Wolfsburg, acts as a regional development organization in the Metropolitan Region. In this, it could be interpreted as the counterpart of the diverse cluster management organizations in Baden-Württemberg, not least reflecting to a certain degree the strong dominance of the *VW Group* in the region. On the one hand, and through its project *LieferantenAnsiedlung* (supplier settlement), it would secure a sufficient settlement of suppliers in the region, in terms of number, quality, and variety. On the other hand, it would actively initiate and promote cooperation and networks in the sector, e.g. through promoting applications for publicly funded projects or through the organization of international supplier forums (*Internationale Zulieferbörse*) or so called *MobilityTalks*, in which especially suppliers from the different sub-spaces of the Metropolitan Region would be brought into contact, as one representative of the *Wolfsburg AG* points out:

"We initiated a broad variety of network activities, through which we purposefully bring firms into contact, so that they can do business with each other. That is the decisive point and also differentiates our events from others. It is the explicit mission of a regional development facility like the *Wolfsburg AG*, to purposefully initiate these processes. We do this for the location of Wolfsburg, but also for the region, i.e. the former greater area of Braunschweig and, as the next higher level, also the Metropolitan Region. [...] The actors here in Wolfsburg sometimes do not even know what kind of potentials do exist there [in Braunschweig, Göttingen and Hannover; MO]. With these events we can achieve a bit more transparency in these fields." (Interview HBGW 6: 5f.; translation: MO)

Concerning the forms these cooperations would take, the interview partner mentions in the first place publicly initiated and/or funded projects, like, e.g., the *Regional Initiative Fuel Cell and Electro Mobility* by the regional government, and the newly created (01.01.2011) *Automotive Cluster* by the MR

HBGW, the *Land* and the *MobilitätsWirtschaft* of the *Wolfsburg AG*. In addition, also short term project-related cooperations and the foundation of a new company have been reported as possible forms, although they were regarded as less important. In any case, these cooperations would be formalized and build on fixed contracts, to secure the use of the results and the respective IPRs, since, despite all cooperation, the relation among firms within a centralized value chain network would be highly competitive (Interview HBGW 6: 6ff.). Overall, and even stronger as in Baden-Württemberg, it can be stated that inter-firm cooperations in the Metropolitan Region's automotive sector are predominantly initiated top-down, by the EU, the federal and the regional government or by the OEM, and less bottom-up by the respective firms, what might also be explained by the high degree of centralization in the sector and the industrial monostructure in the region.

Biotechnology in the Metropolitan Region, on the other hand, can be described as a much less networked sector. One of the authors of the already mentioned network study on the MR HBGW highlights this low degree of interconnectedness, the small size and low maturity of the sector, and also points to the predominantly international orientation of cooperations in this sector, in case cooperations are realized:

> "[I]n many fields of competence a high number of isolated actors, or […] only bilateral cooperations exist, which is quite obvious in the field of life sciences. This is a very young sector, with firms that are spin-offs from academic institutes and that still have umbilical connections to their former institute. But apart from that, the interconnectedness is weak, or rather internationally oriented. This can be explained very well with the relatively small size of the national market, requiring these connections to the international community, if one wants to have a certain specialization within the field of biotechnology. The firms then only have two or three potential cooperation partners in Germany, and in the international community they might have 50. That is the explanation. And in this point we are indeed convinced that one could enhance network- and knowledge management activities and intensify the interconnectedness to detect potential cooperation partners who were previously unknown." (Interview HBGW 3: 3; translation: MO)

The few inter-firm cooperations that exist in the sector, would then predominantly take place within the framework of publicly funded and promoted networks, like, above all, the regional initiative *BioRegioN* (Interview HBGW 3: 5f.). But even in this network, concrete cooperative projects between firms would be rare, as an expert for the biotechnology industry in the Lower Saxon Ministry of Economics points out:

"If we look at the *BioRegioN*, this is a very complicated topic. There is the internet, there is a file in which all the network partners are listed, and they meet each other, sometimes. But in the last years there were only very few joint projects. This is something we have to work on. In general, we did a lot of marketing for a network, which actually only represents individual actors who meet each other here and there, but not the *BioRegioN* in a narrow sense." (Interview HBGW 5: 6; translation: MO)

The main use this network provides to the firms, would then be support regarding the framework conditions of business activities, as, e.g., in the field of legal restrictions, since the predominantly small firms in the sector would often lack the time and the resources to deal intensively with these topics (Interview HBGW 5: 6). In addition, two other networks recently emerged, the *Gesundheitsregion Hannover Braunschweig* in the field of healthcare and the *BiomeTI e.V.*, the *Center for Biomedical Technology and Innovation*, in which cooperation would, according to the interview partner, be more intense (Interview HBGW 5: 7f.). Their recent inclusion into the *BioRegioN*-network can also be interpreted as a general shift of focus away from dedicated biotechnology to healthcare and biomedical technologies in the region.

With regard to vertical cooperations between firms and their customers and suppliers, like in the case of Baden-Württemberg, the lack of larger pharmaceutical firms in the region has been reported to impede a strong customer-supplier relationship between biotech firms and the pharmaceutical sector in the field of drug development. Then these cooperations would, if anything, only be relevant with regard to firms outside the region. On the other hand, in the field of biotechnological laboratory and process technology, e.g., these vertical cooperations would be more important, not least because of the fact that with the *Sartorius AG* in Göttingen a world-wide leading company in this field is located in the region (Interviews HBGW 2: 10; HBGW 5: 10f.).

Inter-Firm R&D-Cooperations

Also when examining the importance of *inter-firm cooperations in the field of R&D* as mechanisms of knowledge transfer, the interview partners draw a differentiated picture for the region as a whole. First, they identify strong sectoral differences concerning the importance of joint R&D-activities in the region, which would only be important in those sectors that would be R&D-intensive anyway. Second, and confirming in this point the findings on Baden-Württemberg, they point out that firm size would matter in this context. While larger firms would often have the resources to conduct their own R&D-activities, and would be less likely to cooperate with other firms, for smaller firms cooperative R&D-projects would be an efficient strategy to compensate their limited resources (Interviews HBGW 2: 7ff.; HBGW 3: 2ff.; HBGW 4: 4ff.). These evaluations by the interview partners also broadly confirm the results of a quantitative study of the NIW (2008), which not only finds out that the propensity of firms in the region to conduct their own R&D-activities rises with firm size (NIW 2008: 32ff.), but also that, overall, other firms are less likely used as cooperation partners in R&D- and innovation-related activities in Lower Saxony than in the other Western German *Länder* (NIW 2008: 61ff.). While the former point can be explained by the region's sectoral structure with a strong dominance of automotive and its related industries, markedly concentrating R&D-activities within the few larger focal firms, the latter point could be interpreted as a generally more distanced business culture in the region, hindering stronger cooperations and network-building among firms. In this context, both aspects constitute important differentiating factors in comparison to the case of Baden-Württemberg.

Referring to the forms these cooperative inter-firm R&D-activities take, all interviewees agreed that they would always be formalized and fixed with contracts, to set up the mutual obligations and access rights on the respective outcomes (Interviews HBGW 2: 8f.; HBGW 3: 5ff.; HBGW 4: 7). One example is the industrial collective research (*Industrielle Gemeinschaftsforschung*), which would be joint and sector-wide research activities of especially SMEs at a pre-competitive stage (AiF 2010). Overall, these results make clear that inter-firm R&D-cooperations have some significance as mechanisms of knowledge transfer in the MR HBGW, but that their contribution to the knowledge transfer activities in the region is considerably limited by the factors discussed before.

These results are also reflected when focusing on automotive. Although, overall, this sector can be described as extraordinarily research intensive, and inter-firm cooperations in R&D were reported as being an important mechanism of knowledge transfer, a large share of R&D-activities would be conducted internally by the OEM and the large suppliers like the *Continental AG*, or with cooperation partners from the scientific world like universities and research institutes (Interview HBGW 6: 6ff.). The predominant forms these cooperations would take would be concrete

R&D-projects around future technologies like, e.g., fuel cell technologies, hybrid drives or advanced diesel technologies, while some of these joint R&D-projects would also be publicly supported (Interview HBGW 6: 9). In biotechnology, on the contrary, joint R&D-activities were reported to not play a role at all. As one interviewed expert for biotechnology from the Lower Saxon Ministry of Economics highlights, there would currently be no example of this kind of cooperation in the region. The reason for this would be the small size of both the sector and most of the firms. The interview partner recognized a disequilibrium in this connection, between very strong academic R&D-activities, which would, then, not be reflected in the region's biotech business sector (Interview HBGW 5: 9).

Cooperations between Firms and Universities or Research Institutes

As a more important mechanism of knowledge transfer in the MR HBGW as a whole the interview partners regarded *cooperations between firms and universities or research institutes.* With that, like in the case of Baden-Württemberg, the significance of this mechanism is evaluated clearly stronger by the experts than it would be expected from the theoretical assumptions derived from the literature on VoC and NIS. Also for the case of the Metropolitan Region the interviewed experts provide both demand and supply dependent reasons for the use of R&D-cooperations with scientific institutions as mechanisms of knowledge transfer: On the one hand, in the center of the region's industrial model would stand the highly research intensive automotive sector with its related industries, which require intensive channels of cooperation to academia, despite of its strong internal R&D-activities. On the other hand, the region would offer to the business sector an internationally competitive scientific infrastructure with highly renowned and specialized universities and research institutes (Interviews HBGW 1: 9ff.; HBGW 2: 7ff.; HBGW 3: 15ff.; HBGW 4: 10). Nevertheless, and in contrast to Baden-Württemberg, it must be stated that the strong focus of the region's research landscape on the broader field of health care and life sciences is not fully reflected in the regional business structure and that, therefore, cooperations of these research institutions might rather be internationally than regionally oriented.

As the most important forms these cooperations between firms and the scientific world would take, the interview partners mentioned, above all, the large technological centers in the region, which would be characterized by a strong focus on applied research, like e.g. the *Hannover Centre for Production Technology* or the *Automotive Research Centre Niedersachsen* in Braunschweig and Wolfsburg (Interview HBGW 3: 15). As one expert from the department of regional economy of the Nord/LB accentuates, these technological centers would fit particularly well to the needs of the region's core industries, also highlighting the importance of spatial proximity for the success of these cooperations:

> "These structures emerge strongly out of academia, the professors are involved and also the regional government, sometimes also the firms, like e.g. in the *Automotive Research Centre Niedersachsen* in Braunschweig and Wolfsburg, which is located in the building of Wolfsburg's *AutoUniversity*. Also the top-researchers teaching in Braunschweig have their offices and their respective laboratories there. [...] I consider this form of cooperation between firms and science as extremely important, since in innovation-oriented cooperations spatial proximity plays a big role. This is also connected to the fact that spill-over are first of all bound to specific breeding grounds, where a certain buzz is generated, showing a local density. Then you are present, or you are not. On the other hand, cooperation is always connected to social interaction and trust, [...] it is a social process requiring face-to-face. Everything you can read about this in books, is correct in this case. It is remarkable, that this is also confirmed strongly by the engineers at the universities or in the firms, that they would need this intensive exchange strongly, particularly on site and in close proximity." (Interview HBGW 3: 15; translation: MO)

Like in the case of Baden-Württemberg, both the formal and the informal dimension of these cooperations have been evaluated as important. As the interview partner specifies:

> "At least there is a strategic commitment. [...] When there is an innovation project, in which something new is developed, and one expects to arise an added value from it, then there will be agreements on what will happen with the results, which will be formal in nature. [...] On the other hand, there might also be topics about when is an exchange of knowledge, but its utility cannot be specified, so that everything remains on the informal level. Therefore, this is a modern and differentiated structure of sometimes stronger and sometimes weaker ties, which can mutually complement each other." (Interview HBGW 3: 16f.; translation: MO)

These findings were also confirmed by the interviewed experts for automotive, who further added (publicly funded) project-related cooperations in R&D as an important form in this connection (Interview HBGW 6: 6ff.). For the highly research intensive biotechnology sector, on the other hand, cooperations with universities and research institutes would be vital. Its strong dependence on basic research activities and results would require intense interaction with the scientific world by almost all types of biotech firms (biotechnologie.de 2010: 14). This dependence was noted to be particularly strong in the case of the biotech firms in the Metropolitan Region, first, because its biotech sector would be very small in size and, therefore, the possibilities to interact with other biotech firms would be rare. Second, also most of the firms would be very small, so that they would lack the necessary resources to conduct certain R&D-activities by themselves. Third, the overall immatureness of the sector would also be reflected in the fact that many firms would still have strong connections to the academic institutes they once spun-off (Interviews HBGW 3: 3; HBGW 5: 6ff.).

5.3.2.2 Association Based Mechanisms of Knowledge Transfer

With regard to the association based mechanisms of knowledge transfer, and again contrasting the case of Baden-Württemberg, the theoretical assumptions derived from the literature on VoC and NIS have only partly been confirmed by the interviewed experts when looking at the MR HBGW as a whole.

Employers-, Business-, and Trade Associations

As expected, the services provided by *employers-, business-, and trade associations* were reported to be frequently used as mechanisms of knowledge transfer, although their importance would be restricted depending on the sector they represent and their degree of specialization (a.o. Interviews HBGW 1: 6ff.; HBGW 3: 8f.; HBGW 4: 9). Also firm size would be a differentiating factor in this connection, with larger firms being less dependent on associations when initiating and organizing cooperation and knowledge transfer than SMEs (Interview HBGW 3: 8). Apart from the standard services that business associations would provide in the fields of innovation and knowledge transfer, especially their engagement in the business pillar of the Metropolitan Region has been highlighted. Through their roof organization they would contribute to networking and knowledge transfer activities within the region, as one interview partner notes:

"The *Unternehmerverband Niedersachsen* is, for example, actively participating in the Metropolitan Region and organizes the *WirtschaftsBank*. Also the *Verein Wirtschaft*, that has been founded to include the business sector into the Metropolitan Region, is directly connected to the business association. And that was explicitly intended by the business association, since they did not want to lose their active position." (Interview HBGW 3: 9; translation: MO)

Chambers of Industry and Commerce

In contrast to the business associations, the *chambers of industry and commerce* were reported to only play a very limited role for knowledge transfer processes in the MR HBGW (Interviews HBGW 1: 6f.; HBGW 3: 8ff.). Neither would they be used intensively by the regional actors, nor would they play an important role for the networking and knowledge transfer activities in the organizational structures of the Metropolitan Region. The reasons given are manifold: First, a lack of coherence in their activities would lower the demand for their services in the fields of innovation and knowledge transfer. Second, the chambers would lack a focus on the region as a whole, and, third, specific historical reasons, would make particularly the Chamber of Industry and Commerce Hannover a less frequently used cooperation partner, as one of the interviewees remarks:

> "Here in this region it is different than e.g. in the Metropolitan Regions of Hamburg or Bremen-Oldenburg, where the chambers play a dominant role. That is not the case here. […] The chambers are not so well organized, also because of internal contradictions among each other. That is the first site. The second is that particularly the Hannover Chamber is not permanently approached as a cooperation partner, also not by politics, because of a specific incident in the past […]. Insofar, the motivation to include the chambers was not too high. The willingness to approach the business association was considerably higher." (Interview HBGW 3: 9; translation: MO)

These structural weaknesses of the chambers in the Metropolitan Region can be traced back to the fact that the single local chambers in the region have a more individualistic perspective. Unlike their counterparts in Baden-Württemberg, the chambers of Hannover and Braunschweig are not members of the Federation of the Lower Saxon Chambers of Industry and Commerce (*Niedersächsischer Industrie- und Handelskammertag*). Their focus on the region (or the Land) as a whole can, therefore, be supposed to be less developed, not least because they are not organized according to the principle of lead management (*Federführungsprinzip*), through which the other local chambers in Lower Saxony, and also the chambers in Baden-Württemberg, coordinate their activities in specific fields on the regional level, taking into account region-wide developments and perspectives, what also allows them to articulate their interests in a concerted way towards the regional government. As one of the interview partners explains, the chambers in the Metropolitan Region historically never exercised this role of taking responsibility for the whole region or the *Land*, not least because of a historically grown rivalry among the different sub-spaces of the Metropolitan Region. Even if also the two chambers of Hannover and Braunschweig, which cover most of the territory of the Metropolitan Region, would intent to coordinate their activities through the *Niedersächsische IHK-Arbeitsgemeinschaft Hannover-Braunschweig*, and although they would also be members of the *Innovationsnetzwerk Niedersachsen*, this coordination is viewed critically by the interview partner:

> "It must be mentioned, that the Metropolitan Region not only has one chamber, but two, Braunschweig-Wolfsburg and Hannover-Hildesheim-Göttingen. I do not know exactly how good the culture of cooperation between the local chambers is, but I have got hints that there were severe problems in the past. Mostly, that not only has historical reasons, but is also rooted in the participating persons, the management directors etc." (Interview HBGW 3: 8; translation: MO)

However, as one representative of the *Innovationszentrum Niedersachsen* highlights, through their participation in the region-wide *Innovationsnetzwerk Niedersachsen* their services would be made available for all actors in the region and would constitute helpful tools herein. Not least, their role in defining binding and common quality standards for professional training programs would be valuable in this context (Interview HBGW 2: 11ff.).

This interview partner also points to the role of the regional trade unions in innovation and knowledge transfer processes, as a third pillar of corporatist actors, apart from chambers and associations (Interview HBGW 2: 11). This factor has not been noted by the interviewed experts for Baden-Württemberg, what could be interpreted in the sense that the active role of trade unions in knowledge transfer and innovation processes might constitute a specific feature of the MR HBGW, as has also been stated elsewhere (a.o. Ortiz 2008: 106ff.; Widuckel 2004: 61; Prätorius 2003a: 3). In the words of the interview partner:

> "[T]hese are different pillars through which firms can be accessed. The question would also be, who do you want to approach within the firms. Take e.g. the employers association, which addresses the management directors and the plant managers. But are they the only representatives of the firm? No! The representatives of the firm are multilayered, from the shift foreman to the trade union, which prefers to abdicate a part of the wage, if, in return, there will be investments in a new development. Knowledge and technology transfer is, therefore, not only of interest

for the management, which you would predominantly reach via the associations and chambers. But you also have to reach the designing engineer or the sales manager, who needs a new marketing method which he would like to learn at a university." (Interview HBGW 2: 11; translation: MO)

For the automotive sector, business associations were attributed an active and positive, but not dominant role in the field of knowledge transfer, particularly through the services they would provide in form of internet platforms, newsletters, publications, conferences and, most important, cluster and networking forums, like e.g. the *Unternehmerforum* (Interview HBGW 6: 10). The strong position of the OEM can be expected to lower the influence of the business associations in the region. Unlike for the Metropolitan Region as a whole, also the role of the chambers, especially in initiating and promoting networking activities and creating joint events and conferences, has been estimated as relevant in this connection (Interview HBGW 6: 10ff.). This might among other things be explained with the specific position the *Braunschweig Chamber* has in the automotive cluster around Wolfsburg and Braunschweig.

In the biotechnology sector, on the other hand, both business associations and chambers would not play a role at all, as one interview partner details:

"In fact, they do not play a role, except of the *VCI*, the *German Chemical Industry Association*, which also has an own biotech section. Here in Lower Saxony we have e.g. the *Biotech Stammtisch*, which is held by the *VCI* twice a year. [...] This would be one activity, but apart from this the associations do not play a role at all. And that would also hold true for the chambers. It [biotechnology; MO] might be more or less a topic of interest for them, and actors might meet each other from time to time at the *Life-Science-Day*, but apart from that, actually not." (Interview HBGW 5: 11; translation: MO)

As a reason for this low importance of chambers and associations the interview partner points, again, to the small size of most biotech firms in the region. For them, even to pay a membership fee would be a considerable barrier not to engage themselves in associations and chambers. Moreover, the services these organizations provide would not be of interest for the majority of the biotech firms in the region (Interview HBGW 5: 11f.).

Binding and Common Technical Standards and Norms

Finally, the importance of *binding and common technical standards and norms* fostered by associations and chambers has been evaluated as less significant by the interviewed experts than one could have assumed from the theory. Although some experts indeed attributed binding and common technical standards a positive and reinforcing effect on cooperation and knowledge transfer, its significance for the success of knowledge transfer processes in the region as a whole has been described as only average (Interviews HBGW 2: 13f.; HBGW 3: 12), like in Baden-Württemberg. The main advantage these common standards would have, would be their relieving effect in contracting processes, as one interview partner explains:

> "Above all, I think that through these standards cultures of trust can emerge, so that contractual gaps, which you will always have in cooperation contracts, can be accepted more easily. Especially in the context of cooperation not everything can be specified, otherwise you could just take a standard work contract." (Interview HBGW 3: 13; translation: MO)

A similar picture has also been revealed for automotive. As an expert for this sector in the MR HBGW comments, in many cases these standards would be a central precondition allowing cooperation in the first place. Nevertheless, he also points to the ambivalence of standards, which could also constrain knowledge transfer among actors, and that, therefore, coordination via liberal markets would have at least a similar importance in the sector (Interview HBGW 6: 13). Like in the case of Baden-Württemberg, also the interviewed experts for biotechnology in the Metropolitan Region did not report on binding and common technical standards and norms. Yet, also for biotechnology in the Metropolitan Region it can be assumed that large parts of interaction and knowledge transfer processes would be influenced by common technical (as well as medical and chemical) standards, norms, regulations and clinical test requirements, as it would be e.g. *DIN- and ISO-Norms*, among many other things.

5.3.2.3 Organization Based Mechanisms of Knowledge Transfer

Like for Baden-Württemberg, also in the MR HBGW the assumptions on the organization based mechanisms of knowledge transfer, i.e. transfer of technical and scientific staff, purposeful headhunting and business startups, have basically not been confirmed by the interview partners. Although all three organization based mechanisms were not attributed a decisive importance, they were all evaluated by the experts to have a medium range significance for knowledge transfer processes in the Metropolitan Region.

Transfer of Technical and Scientific Staff

Looking at the *transfer of technical and scientific staff* first, most of the interview partners identified this as a naturally used, although not dominant mechanism of knowledge transfer. Above all, they pointed to an increasing academization of services and production processes in this connection. More difficulties to transfer knowledge through this channel would exist in some SMEs, where actually only a few academics would be employed, so that certain barriers to contact and recruit highly qualified staff would be to observe, as one expert points out:

"But in those cases, in which this personnel does not exist, this [transfer of staff; MO] is just not happening, because the inhibitions are too strong and the communicative barriers are too high, etc. But I would not interpret this in a way that there are different systemic worlds which would be mutually exclusive. They can be compatible in the moment in which the respective internal preconditions are fulfilled. One should think again about starting respective programs in the field of promoting the employment of R&D-staff in SMEs." (Interview HBGW 3: 17f.; translation: MO)

Specifically important in this regard would therefore be the *Innovationsassistenten Programm* of the Lower Saxon Ministry of Economics, supporting the transfer of academic staff to SMEs, as another interview partner explains (Interview HBGW 4: 10f.). In addition, the interview partners also emphasized other region specific factors impeding a stronger transfer of staff. Above all, they mentioned that the region might be a comparatively less attractive place of residence for highly qualified staff and top managers, as the managing director of the *Innovationszentrum Niedersachsen* illustrates:

"There are firms in the Metropolitan Region, like *Sennheiser* […] and *Bosch-Blaupunkt* […], which just have difficulties to get qualified staff. At *Bosch-Blaupunkt* in Hildesheim, headquarters are located close to Stuttgart, the top-managers get more money in Hildesheim than in Stuttgart, although the living costs would suggest it would be vice versa. The reason is that they have got the problem that the spouses do not want to go with them. That is a very practical problem. The spouses just prefer to stay in the sunny Southern Germany than in Hildesheim." (Interview HBGW 2: 14f.; translation: MO)

Also the less positive image of the Metropolitan Region in comparison to other German regions would cause problems to find adequate staff, as the interview partner continues:

> "Another problem has *Sennheiser*. If you are thinking about high-end consumer electronics, in this case headphones, and ask people where they would think that this firm is located, then they would mostly assume Bavaria or Baden-Württemberg. That has something to do with image. And since we recognized that, we started the Lower Saxon image campaign. With that, we aim to adapt the perception to the evidence. People think that *Sennheiser* is located in Baden-Württemberg and look for job announcements in the *Stuttgarter Zeitung*, so that the firm is just not found." (Interview HBGW 2: 15; translation: MO)

As the managing director of the *Metropolregion GmbH* adds in this connection, the organizational bodies of the Metropolitan Region would try to address this problem with a project on talent recruitment, to make the region more attractive for the "creative class" and improve its image (Interview HBGW 1: 10f.).

Purposeful Headhunting

Purposeful headhunting of staff from other firms would mostly be restricted to the larger firms. Since many recruitments of qualified staff would be rooted in cooperations in the field of innovation, and since most of these cooperations would be institutionalized in nature, one interview partner remarks that headhunting activities in the region would not be too intensive because of the fear to endanger these cooperations. Overall, this mechanism would indeed used by regional firms, but it would not play a dominant role in their knowledge transfer activities (Interviews HBGW 2: 11; HBGW 3: 17; HBGW 6: 14f.).

From a sectoral perspective, transfers of staff were reported to be particularly important for automotive, and, unlike in Baden-Württemberg, not only for the larger firms and the OEM, but also for the supplier firms. Because of the limited availability of qualified staff on the labor market and the beginning effects of the process of demographic change also the firms in the MR HBGW's automotive industry would find themselves more and more in a competition on these highly qualified persons, and especially on engineers. One approach of recruiting qualified staff is described by a representative of a large automotive firm in the region:

"In the field of automotive many suppliers have systematic cooperations with universities and technical colleges and found scholarships to recruit new engineers for their firms. In concrete terms, this means that there is one cooperation between some suppliers we have here in this location and the technical college and the TU Braunschweig, in which scholarships are granted in the field of automotive technology and automotive research and students get financial support. These students then commit themselves to work for two or three years in the firms, as soon as they reached their engineering degree. […] It is not a secret that in the field of automotive many hundreds of engineer-posts are not occupied, because of the lack of qualified personnel." (Interview HBGW 6: 14; translation: MO)

Moreover, purposeful headhunting would also play a role, although only for the highest levels of qualification. This would be true for transfers among firms, as well as for transfers from academic institutions to firms, and, although less often, also for transfers from firms to academic institutions (Interview HBGW 6: 14f.).

For biotechnology, transfers of technical and scientific staff would have a medium range importance as mechanism of knowledge transfer in the MR HBGW. The predominantly small and not dynamically growing firms in the sector would not invest much effort in their recruiting activities, and many scientists would prefer to create their own start-up firm instead of changing into one of the existing biotech firms (Interview HBGW 5: 12f.). The relevance of purposeful headhunting in biotechnology could not be estimated by the interviewed experts.

Business Start-Ups

A quite similar picture is revealed when focusing on *business start-ups* as mechanisms of knowledge transfer: Looking at the Metropolitan Region as a whole, business start-ups are realized, but only to a limited extend, and, as almost all interview partners add, less as would be desirable and necessary from their point of view. However, they state that in the last years the propensity to create a business start-up has risen, not least because of an intensive public funding for entrepreneurs by regional government, above all through programs like the *GründerCampus Niedersachsen* and the *technology and business incubators* at many higher education facilities in the region, or institutional support, as e.g. through the *Innovationsnetzwerk Niedersachsen*, the *NBank* or the numerous regional economic development facilities in the Metropolitan Region, like *HannoverImpuls*, *Projekt Region Braunschweig* or the *Wolfsburg AG* (Interviews HBGW 1: 15f.; HBGW 2: 19f.; HBGW 3: 18ff.; HBGW 4: 11ff.).

Nevertheless, this indicates that also in the MR HBGW business start-ups are still strongly dependent on external funding and support and that only to a limited extent they emerge out of market dynamics. Like in the case of Baden-Württemberg, the interview partners see, above all, cultural differences at work and emphasize distinct mentalities with regard to entrepreneurship and risk aversion when comparing the Metropolitan Region with the UK or the USA (Interviews HBGW 2: 19f.; HBGW 3: 18ff.; HBGW 4: 12). The managing director of the *Innovationszentrum Niedersachsen* explains this as follows:

> "The decisive difference is rather sociological and cultural-historical. What is an American? It is a European who had the heart to move towards new horizons. In Germany, we have another culture. [...] We are the country that invented the reinsurance companies. You would only do that, if as a nation you were anxious and insecure, and then you are not a founder. There were founders in the region, like *Siemens* and *Telefunken*, but they experienced twice a total destruction in the world wars and that is a psychological momentum that we do not have in the Anglo-Saxon world." (Interview HBGW 2: 19; translation: MO)

In particular, a different mentality in case of failure has been highlighted in this connection, since a failure of an entrepreneurial project would, unlike in the Anglo-Saxon world, lead to damage of reputation and block, above all, the access to new seed- or venture capital (Interview HBGW 3: 18f.). However, in general, most of the interviewed experts see a sufficient provision of seed- and venture capital in the Metropolitan Region, although especially the experiences in the New Economy crisis at the end of the 1990s would have reduced the motivation to provide this kind of capital, as a representative of the *Nord/LB* explains:

> "I can only ascertain, also from our own inquiries, that bottlenecks on the capital side only play a subordinated role. [...] On the one hand, there are people saying that somebody who founds a company and has a business idea and a convincing business plan will get money here. We do have the saving banks, i.e. different institutions than in the Anglo-Saxon world, which understand themselves as capital providers for SMEs and business start-ups. But also in this point one has to consider that the crisis of the New Economy led to a lower disposition to provide capital in the seed phase, than it was the case in the 1990s. Therefore, at this point processes of institutional learning can also be counterproductive for innovation processes." (Interview HBGW 3: 19; translation: MO)

Overall, the interview partners ascribe the regional savings banks and the *Volksbanken*, as well as the *NBank*, which administrates among other things the funding for public start-up programs by the regional government and coordinates a network of business angels, a central role in the provision with seed- and venture capital. And also the *Nord/LB* as a federal state bank would contribute through its engagement in the *S-Finanzgruppe* to the funding of regional business start-ups. As the interview partner from the Nord/LB estimates, around 75% of the start-up activities in the Metropolitan Region would be financed by the regional savings banks (Interviews HBGW 3: 20; HBGW 1: 15f.; HBGW 2: 19f.; HBGW 4: 16). With that, the role of these regional banks for business start-ups is stronger emphasized than in Baden-Württemberg, what could be traced back to the fact that the financial service sector in general plays a more important role in the MR HBGW, not least in terms of seed- and venture capital provision as well as business angel support.

In the Metropolitan Region's automotive sector, in contrast to Baden-Württemberg, business start-ups were reported to be important, although not decisive, mechanisms of knowledge transfer. The main reason provided by the interviewed experts is the effort by the public and private actors, and especially the *VW Group*, to raise the number, the quality and the variety of suppliers in the region. Therefore, even in a mature and consolidated sector like the Metropolitan Region's automotive industry business start-ups would be frequently realized and used as mechanisms of knowledge transfer. Not least, the active promotion of entrepreneurs in the sector through the *Wolfsburg AG* and its *InnovationCampus* would be decisive for these start-up activities. As an expert for automotive in the region accentuates, this promotion would already start in very early stages, namely in schools and universities:

> "There are two competitions of the *Wolfsburg AG*'s *InnovationCampus*, which try to promote the disposition to found a company in a very early stage, through the competitions *PromotionSchool* and *PromotionStudent*, which have the aim to animate start-up activities by students. This early stage of start-up activities is attributed to a specific importance. And it is also about creating an awareness for the fact that somebody who once founded a company without success is not a failed person, but that he will get specific promotion, like in other countries as well." (Interview HBGW 6: 16f.; translation: MO)

Apart from that, the *Wolfsburg AG*, but also the *NBank* and some regional development facilities, like *Projekt Region Braunschweig* or *HannoverImpuls*, would exercise an important function in assessing entrepreneurs in the sector and bringing them into contact with business angels, venture capitalists and private equity firms (Interview HBGW 6: 22).

In biotechnology as a sector relying strongly on analytic knowledge, business start-ups and, above all, academic spin-offs from scientific institutions should play a decisive role for knowledge transfer from academia to the market. However, the foundation of new biotech firms in the MR HBGW would predominantly be connected to public programs like *Bio-Profile* or the regional initiative *BioRegioN* and would only in a limited number of cases occur out of market dynamics. As an extraordinarily research intensive sector, in which firms often spend up to ten years or more only on conducting R&D-activities without producing any marketable product, funding would play a crucial role in this connection. In this, the Metropolitan Region's biotechnology sector would still be strongly dependent on public funding by national and regional government, and would lack a sufficient provision of seed- and venture-capital via the market, a fact that would effect strongly entrepreneurial activities (Ernst & Young 2009; Ernst & Young 2010; Interview HBGW 5: 13f.). Apart from this, one interviewed expert from the Lower Saxon Ministry of Economics emphasizes that many of the excellent researchers in the field of biotechnology would refrain to create a spin-off, since they would lack the necessary business and managerial skills and would fear the respective risks. This would also be one major explication for the pronounced disequilibrium between a strong research landscape in biotechnology and a comparatively weakly developed business sector, which is characteristic for biotechnology in the Metropolitan Region (Interview HBGW 5: 13f., 19f.).

5.3.2.4 State Induced Mechanisms of Knowledge Transfer

With regard to the state induced mechanisms of knowledge transfer, the results of the qualitative case study on the MR HBGW only confirmed some of the theoretical assumptions, while others were not approved.

National and Regional Government Measures

Looking at the single state induced mechanisms in detail, *national and regional government measures* were estimated by the interviewed experts to play a central role for knowledge transfer in the Metropolitan Region, and, therefore, as more important than assumed. In particular, the role of the *Lower Saxon Regional Government* has been highlighted by all interviewed experts. Its measures and programs in the fields of innovation and knowledge transfer, like e.g. the regional initiatives, the competence networks, technology and business incubators, programs to foster start-up activities or the *Innovationsförderprogramm* for SMEs were evaluated as important mechanisms of knowledge transfer in the Metropolitan Region (Interviews HBGW 2: 2ff.; HBGW 3: 2ff.; HBGW 5: 2ff.).

One explaining factor for this central role of the regional government would be the specific centralized and focused structure of the Lower Saxon innovation policy, with the *Innovationszentrum Niedersachsen* coordinating and evaluating the activities of the different regional ministries and the *NBank GmbH* (*Investitions- und Förderbank Niedersachsen*) implementing and controlling regional policy measures, supported by the broad variety of intermediary actors organized within the *Innovationsnetzwerk Niedersachsen* (Interviews HBGW 2: 3ff.; HBGW 4: 3f.; Ortiz 2008: 75ff.). Although this structure is not specifically tailored for the MR HBGW, but rather for Lower Saxony as a whole, it strongly shapes the structures and mechanisms of knowledge transfer in the Metropolitan Region (Ortiz 2008) and was reported to be used intensively by the regional actors. Especially the *Innovationsnetzwerk Niedersachsen*, which is directed by the *Innovationszentrum Niedersachsen GmbH* and represents about 250 intermediary institutions and more than 380 contact persons in the fields of innovation and knowledge transfer, would be the central interlocutor in the Metropolitan Region's innovation and knowledge transfer system (Interview HBGW 2: 4ff.).

Divided by sectors, automotive would not deviate much from the conditions described for the region as a whole (Interview HBGW 6). Public programs and measures, as e.g. the *Regional Initiative Fuel Cell and Electro Mobility* by the Lower Saxon Government, or other innovation programs promoted by the federal government or the EU, as well as the new *Automotive Cluster* initiated by the MR HBGW, would play an important role for knowledge transfer processes in the sector. Moreover, for automotive, also the activities of the local policy levels were highlighted, especially through their engagement in regional economic development institutions, like the *Wolfsburg AG*, *Projekt Region Braunschweig* or *HannoverImpuls*. Overall, as one representative of the *Wolfsburg AG* summarizes, there would be a good mix of top-down action by the state and bottom-up action by the economic actors in the knowledge transfer activities of the Metropolitan Region's automotive sector (Interview HBGW 6: 7).

The biotech sector in the Metropolitan Region, on the other hand, would depend much more on public measures and programs in the field of knowledge transfer than automotive. One decisive difference to the biotech sector in Baden-Württemberg would be that it got much less support by national government, since it only was awarded a *BioProfile*, but has not been funded by the *Gesundheitsregion*, the *BioRegio* and the *Leading Edge Cluster Competitions*. Therefore, in the MR HBGW the measures of the Lower Saxon Government would be the most important impulse givers for knowledge transfer processes in biotechnology, especially through its engagement in the regional initiative *BioRegioN*. In addition, also the already described core structures of the Lower Saxon innovation policy, with the interplay of ministries, *Innovationszentrum*, *Innovationsnetzwerk* and *NBank*, would be important in promoting knowledge transfer processes in the sector, above all through the provision of seed- and venture capital for business start-ups and academic spin-offs, as well as the coordination of a business angels network (Interviews HBGW 5: 2ff.; HBGW 2: 3ff.).

Scientific Parks and Technology Centers

Scientific parks have been evaluated as important mechanisms of knowledge transfer where they exist, but there would be too few of them operating in the region (four; see Nord/LB 2007: XXXI) to ascribe them a decisive role for the Metropolitan Region's knowledge transfer system as a whole (Interview HBGW 5). *Technology and start-up centers*, on the other hand, are more numerous, with about 17 of them operating in the region (Nord/LB 2007: XXXI; ADT 2011). They were seen as important, but not central mechanisms of knowledge transfer in the MR HBGW. Their particular strength would lie in providing a physical space for collaborative research and close face-to-face interaction between actors from business and academia, as well as in their role as business incubators (a.o. Interviews HBGW 2: 17f.; HBGW 3: 15f.; HBGW 4: 12).

Differentiated by sectors, the picture is slightly different. In the Metropolitan Region's automotive sector, science parks can be supposed to only play a minor role in the overall knowledge transfer activities. Technology centers, on the other hand, were reported to play an important role in the regional knowledge transfer activities, especially in the cluster around

Wolfsburg with the *InnovationCampus*, the *SE-Center* and the *Automotive Research Center Niedersachsen* (Interviews HBGW 6: 3ff.; HBGW 2: 17f.; HBGW 3: 15f.).

In biotechnology, science parks and technology centers were both evaluated to be important mechanisms of knowledge transfer by the interviewed expert from the Lower Saxon Ministry of Economics. The sector's strong dependence on basic research and product development activities would require a close connection of the firms to the region's renowned universities and research institutes, what would be particularly well to achieve within the framework of these installations. Almost all important biotechnological research institutions in the region would be connected to one of these parks or centers (Interview HBGW 5: 8; biotechnologie.de 2010; Ernst & Young 2010).

Technology Transfer Offices

Furthermore, with regard to the role of the *technology transfer offices or companies* at the universities and research institutes, most of the interviewed experts did not confirm the high importance these institutions have been ascribed for knowledge transfer processes in the region. All experts consistently stated that they would consider technology transfer offices, if anything, as mechanisms of a medium range importance, and that a large part of them would not frequently be used by the regional actors, and especially not by the firms (a.o. Interviews HBGW 2: 17f.; HBGW 3: 21f.; HBGW 4: 12f.). The reasons given are manifold, and similar to the ones provided by the experts for Baden-Württemberg. First, the majority of these offices would lack sufficient human, organizational and financial resources to play an active and efficient role in transferring knowledge between academia and the business world. Second, their focus would be more on commercializing marketable knowledge from the universities and research institutes, and less on bringing the firms and researchers together to resolve specific requests, initiate concrete projects, and, therefore, foster knowledge transfer through these channels.

Concordantly, the experts stated that from a firms' perspective there would be a great need to grade up the technology transfer facilities by a *switchboard function*, so that they could provide a competent overview on the current research activities in their respective institutes and could ease the access to the universities, research institutes and single researchers. This would especially be useful for smaller firms, which would often lack the knowledge how to access experts at universities and research institutes and would also have general reservations to interact with the scientific world. In addition, the interviewed experts identified a general preference of firms to directly interact with individual researchers in the research facilities, without having to choose the way through intermediate institutions like the technology transfer offices, which would frequently be perceived as additional bureaucratic barrier (Interviews HBGW 2: 17f.; HBGW 3: 21f.; HBGW 4: 12f.).

Nevertheless, these shortcomings would partly be compensated through the participation of a vast majority of the region's technology transfer facilities in the *Innovationsnetzwerk Niedersachsen* (Interviews HBGW 2: 2ff.; HBGW 4: 9). This would connect them to the other intermediary institutions in the field of knowledge and technology transfer in the region and, therefore, also give them access to their respective contacts and networks. This unique structure of the *Innovationsnetzwerk Niedersachsen* would constitute a distinctive feature of the Metropolitan Region's knowledge transfer and innovation system, and would also represent the future organizational form of knowledge and technology transfer in the region, as the executive director of the *Innovationszentrum Niedersachsen* predicts:

> "I think that these technology transfer facilities are in a flux. They were necessary at a time, when technology transfer at the universities was not desired. In my eyes, these technology transfer agencies are, in the way they were operated until short time ago, a relict and will not exist anymore in ten years. The universities themselves recognize that knowledge and technology transfer has to be done in a decentralized way, and not just via one single person. The world and the requirements have become much more complex and the only thing that those central facilities can do is to organize events and open days, but not much more. And I can only disadvise everybody to always and only operate technology transfer through these technology transfer facilities." (Interview HBGW 2: 18; translation: MO)

A distinct picture resulted for automotive, not least because of the generally more intensive connections between the sector and the scientific world. As one interviewed expert for the sector accentuates, there would be an active exchange between the firms and the technology transfer facilities, although dependent of the participating individuals:

> "There is an active exchange between the firms here in the region and these transfer facilities. Indeed, they are used, simply because many firms cultivate intensive contacts to higher education institutions. On the one hand, they want to recruit

qualified staff, and on the other hand they also actively teach at the higher education institutions. Therefore, they have an easier access to these technology transfer facilities than somebody who is completely inexperienced in approaching a university and still has to inform himself. So there is a very intensive exchange of information. But this is also dependent on persons, i.e. it must be an active entrepreneur approaching the universities and initiating cooperation." (Interview HBGW 6: 17; translation: MO)

In biotechnology, technology transfer offices would only be one mechanism of knowledge transfer among others. Since most of the biotech firms would be academic spin-offs and would maintain intensive connections to the higher education institutions anyway, other channels of interaction without an intermediary actor in between would be used more intensively (Interviews HBGW 5: 15; HBGW 3: 3).

Finally, none of the interview partners could estimate the role of *(larger) state induced innovation projects* as mechanisms of knowledge transfer in the Metropolitan Region.

5.3.2.5 Market Based Mechanisms of Knowledge Transfer

Concerning the market based mechanisms the interviewed experts have been asked for the role of external consultancies, mergers, acquisitions and hostile takeovers as well as the buying, selling and licensing of patents in knowledge transfer processes, and did basically not agree with the theoretical assumptions derived from the literature on VoC and NIS.

External Consultancies

With regard to the use of *external consultancies* the interview partners indeed reported that service providers would be used by the regional actors in questions of knowledge transfer, although they would not be decisive in this context. Therefore, overall they have been ascribed a medium range importance for knowledge transfer processes in the region. In particular, these external consultancies would be used for applications for public funding or for installing professional innovation management departments within firms or other organizations (Interviews HBGW 1: 14; HBGW 2: 21; HBGW 3: 22; HBGW 4: 15; HBGW 6: 19f.). In contrast to the case of Baden-Württemberg, an insufficient supply with consultancy services in the region has not been remarked in this connection. Thus, it can be assumed that the Metropolitan Region's position as one of Northern Germany's major service centers might also explain the more intensive use of external consultancy services from the supply side.

Above all, smaller and specialized consultancy firms would be engaged in this regard, but the services of the *Steinbeis Beratunsgzentren*, some regional banks or even public consultancy services would also play a role (Interview HBGW 3: 22). In addition, for the Metropolitan Region as a whole, in the past the large consultancy firms like *McKinsey* would have played a specific role on the meta-level, since they were decisively contributing to the development and implementation of the previously described regional growth concepts, like *AutoVision* or *HannoverImpuls*, which aim to promote cooperation and knowledge transfer in the different sub-spaces of the Metropolitan Region (Interview HBGW 6: 20).

For automotive, these results were not confirmed. Here, external consultancies would be less important, as one interviewed expert for the sector in the MR HBGW explains:

> "There are firms promoting knowledge transfer, but from my point of view these services are used less intensively than in the USA and the UK. I think this is related to the entrepreneurial mentality we have here in Germany, which is more cautious and reserved with regard to consultancy services. In the background, it is assumed that also the consultant has the aim to do business, and that is approached with a certain reservation." (Interview HBGW 6: 19f.; translation: MO)

Like for the large automotive OEMs in Baden-Württemberg, also for the *VW-Group* its internal consultancy, the *Volkswagen Consulting*, would be more actively used in the field of knowledge transfer (Interview HBGW 6: 20).

In the biotechnology sector, like in the Metropolitan Region as a whole, external consultancies would have a medium range importance for knowledge transfer processes. In case they were used, these would then predominantly be small and specialized consultancy firms, and less likely the larger consultancies (Interview HBGW 5: 18).

Mergers, Acquisitions and Hostile Takeovers

Also *mergers, acquisitions and hostile takeovers* as mechanisms of knowledge transfer would play a certain, although not decisive role. On the one hand, one interview partner emphasized that in the German business culture, in which cooperation and fairness would be highly valuable pillars, mergers, acquisitions and hostile takeovers would be less likely than in the Anglo-

Saxon world. Legal restrictions, however, would not be decisive barriers in this connection (Interview HBGW 3: 23). On the other hand, other interview partners pointed out that nevertheless, in a comparison of the German *Länder*, Lower Saxony would be one of the most active regions in terms of mergers, acquisitions and hostile takeovers, as the managing director of the *Metropolitan Region GmbH* remarks:

> "The classic example would be the *Continental AG* [and its hostile takeover battle with the *Schaeffler Group*; MO]. Within Germany, Lower Saxony is almost the classic *Land* for that, and also our two largest proprietors, *VW* and *Continental* have been involved in these kind of activities. Obviously, *VW* won against the hostile takeover by *Porsche*. That really formed us and in this regard we are a prime example among the German *Länder*." (Interview HBGW 1: 14; translation: MO)

One representative of the Lower Saxon Ministry of Economics also points to the fact that many firms in the region would be owned by Anglo-Saxon investors, what would lead to an increasing diffusion of a business culture in which mergers, acquisitions and hostile takeovers would be natural entrepreneurial operations:

> "I think that we will experience this more often, simply because among the larger firms there are only a few left which are purely Lower Saxon. In many fields we have firms which are owned by larger business groups, and in which the decision takers are located in the Anglo-Saxon countries. Therefore, we are not so far away from that anymore. Many former family-owned firms in Lower Saxony, e.g. in the chemical industry, are bought at a certain point. We experience it more and more that decisions on business headquarters are not taken in Lower Saxony anymore." (Interview HBGW 4: 15f.; translation: MO)

A similar picture has been drawn by the interviewed experts for the automotive industry in the Metropolitan Region. Apart from the already mentioned intensive activities in the field of mergers, acquisitions and hostile takeovers by the *VW Group*, it would be, in particular, concentration processes among the suppliers that give these mechanisms of knowledge transfer a specific weight in the region:

> "There are two options, if one wants to open up a market: Either you built something up, what would require a lot of effort, time and costs, and you need adequate staff, or you buy the know-how by acquiring a competitor or a small firm, and also acquire the know-how. In the field of the suppliers this indeed is a usual practice, simply because we experience a concentration process right now, what means that in 10 to 15 years we will have considerably less firms on the market than before.

The concentration process towards highly specialized firms will even be further enforced in this field and continue." (Interview HBGW 6: 20f.; translation: MO)

In the Metropolitan Region's biotechnology sector this situation would be different: In general, mergers and acquisitions would be crucial mechanisms of knowledge transfer in the sector, but only in the advanced phases of product development and clinical trials, when e.g. a smaller biotech firm might be bought by a large pharmaceutical firm to bring a new drug to the market, as the managing director of the *Innovationszentrum Niedersachsen* explains:

"Especially with regard to the younger and smaller firms it is often the case that they have an idea, which they bring to a certain matureness, and finally they have the strategy to be bought by a large firm which would then take care of the commercialization. Especially in the field of biotechnology this is common practice, above all in drug development, which has to pass four clinical trials and can take between 12 and 16 years. These days, you will not find any financier who would finance you for 16 years, with a probability of one out of hundred that a drug agent is approved and without knowing if it will be a blockbuster. In this field it would be more likely to define in advance a stage in which a result could be sold. And in many cases these results are not sold at the end of product development, but at certain previously defined milestones in between." (Interview HBGW 2: 21; translation: MO)

However, since in the MR HBGW there would only be a few larger pharmaceutical firms that could exercise this function, and since a great majority of the region's biotech firms would still be very young, with a low share of firms in advanced phases of product development (see also Ernst & Young 2009), mergers and acquisitions would still not play an important role for knowledge transfer in the region's biotech sector. If anything, smaller firms in the region would be acquired by foreign companies (Interview HBGW 5: 18f.).

Buying, Selling and Licensing of Patents

Finally, the *buying, selling and licensing of patents* were attributed a medium range importance as mechanisms of knowledge transfer, although strongly dependent on the sector. On the one hand, the interview partners emphasized that in the last years the importance of trading patents and licenses has risen, not least because of the stronger dependence of economic processes on highly specific knowledge (Interview HBGW 3: 23f.). Above all, also the active promotion of these activities by the regional government, e.g. through the assessment by the *Erfinderzentrum Niedersachsen* and the technology and start-up centers would have caused positive effects in this connection (Interview HBGW 4: 18). On the other hand, many firms would explicitly refuse to apply for or trade patents and licenses, but rather invest more energy in new developments (see also Interview HBGW 3: 23f.), as the managing director of the *Innovationszentrum Niedersachsen* explains:

> "Here you have to look at the different sectors. The *VDMA*, the German *Engineering Federation*, e.g., in the meanwhile officially recommends its members to not apply for patents anymore. The background is that with your patents you create a description of your innovation and you would point your competitor to it. If you now take into account that the life span of machines is declining, and that everybody copies, if you have got a patent or not, then you must rather bring a new generation of your products to the markets. Therefore, they explicitly recommend not to apply for patents. In the biotechnology sector, on the contrary, this is absolutely necessary, since you have extremely long phases of development." (Interview HBGW 2: 22; translation: MO)

This already points to strong sectoral differences in this regard. When looking at the two investigated sectors, it becomes obvious, that the buying, selling and licensing of patents is much more important in biotechnology than in automotive. As one expert for automotive in the MR HBGW remarks, the trade of patents would only have a marginal function in the sector, and would be clearly less developed than in the Anglo-Saxon countries. For the firms in the sector and especially for the OEM it would be more important to apply for own patents and to develop own products out of it (Interview HBGW 6: 22f.). In biotechnology, on the contrary, the trading or licensing of patents was evaluated as essential by the interview partners, above all in the later stages of product development. Because of the long lasting development processes in the sector, only at these later stages it could be secured that the running time of a patent would still be long enough to successfully commercialize a product out of it. Therefore, the trade of patents would still only be relevant for a limited number of the very young biotechnology firms in the region (Interviews HBGW 5: 20f.; HBGW 2: 22f.).

5.3.2.6 Community Based Mechanisms of Knowledge Transfer

To cover the community dimension of knowledge transfer processes, the interviewed experts have been asked for their estimation on the importance of three mechanisms: Informal contacts, regional platforms and additional (regional) institutions, which allow for establishing trust-based relationships and base on a common regional identity, culture and tradition. The theoretical assumptions on these mechanisms have only partly been confirmed by the interviewed experts.

Informal Contacts

Overall, *informal contacts* were reported to be an important, but not decisive channel of knowledge transfer in the region (Interviews HBGW 1: 5; HBGW 3: 16ff.; HBGW 4: 7f.; HBGW 5: 16f.). Their most significant use would be to work as an important precondition for establishing further, then more formalized knowledge transfer activities. This distinction has been highlighted by the majority of the interview partners, since e.g. concrete R&D-projects and the transfer of IP-protected knowledge would always be formalized and base on fixed contracts.

Regional Platforms

With regard to *regional platforms*, the interviewed experts drew an ambivalent conclusion. Although also the MR HBGW would offer a broad variety of regional platforms on which the different actors could get to know and interact with each other, their efficiency and acceptance by the actors has been questioned by some of the interview partners (a.o. Interviews HBGW 1: 6f.; HBGW 3: 8ff.; HBGW 5: 6ff.). Some of the already mentioned regional cluster and network initiatives would not exercise their function as a regional platform in a satisfying way, and this would also be true for some business associations and, above all, the chambers of industry and commerce, whose lack of coherence and focus on the region as a whole has been criticized, as discussed above. And also the project of the MR HBGW itself would only exercise to a limited extent a function as a regional platform (Interview HBGW 3: 24). In this, the lack of a common and coherent *regional identity* in the Metropolitan Region has been identified as one of its major shortcomings, sometimes impeding knowledge transfer activities. These common identities would only be found on the sub-regional levels, and the relations between the different constituting parts of the Metropolitan Region would still be highly competitive instead of cooperative, as one interview partner stresses:

> "It may be much easier in other metropolitan regions. This is a polycentric space and it is historically grown. There are very different regional sub-identities, which in parts are also highly oppositional, like e.g. between Hannover and Braunschweig. Even in 50 years there will be no 'we-feeling' in the MR HBGW. The decisive point is that the creation of the Metropolitan Region is a form of spatial adjustment, which arises from purely functional aspects, [...] and this functional perspective definitively leads, from my point of view, to the right results, but not necessarily to a good feeling. That is the problem." (Interview HBGW 3: 24; translation: MO)

As an explanation for this lack of a common and strong regional identity in the Metropolitan Region HBGW the interview partner especially highlights the region's past as a Prussian province, when he continues:

> "One major historical reason for this situation is certainly the particularism (*Kleinstaaterei*). Lower Saxony has been Prussian territory, and Prussia was big and its sub-spaces and provinces have not been advocated so much in their self identification. It was surely more important to create the large scale connection in the first place." (Interview HBGW 3: 24; translation: MO)

Differentiated by sectors, this picture concerning informal contacts and regional platforms would be similar in automotive. Regional platforms, like e.g. the *SuppliersDay* in Wolfsburg, would constitute important knowledge transfer mechanisms, and would also allow for establishing informal contacts, above all as a basis for then more formalized interaction. Like for the region as a whole, also the interviewed experts for automotive could not identify a strong regional identity as a precondition for knowledge transfer activities (Interview HBGW 6: 19). Nevertheless, the strong connection of the *VW Group* towards the Metropolitan Region and also towards Lower Saxony as a whole should be taken into account in this regard. Not least because of the fractional ownership that the Land holds on the *VW Group*, a specific responsibility for and focus on the region as a whole would exist, leading to intensified cooperation and knowledge transfer activities within the region (Interview HBGW 6: 1ff.).

For the biotech sector, informal contacts would be less important. Most interaction would be formalized with fixed contracts. The reason given for this is the strong reliance on R&D-activities and the dependence of IPRs protection in the sector. Shortcomings in the structure, organization and the management of the sector's central regional platform, the regional initiative *BioRegioN*, would, in addition, also impede informal (as well as formal) interaction and knowledge transfer via this channel. Moreover, also regional identity and culture would not play a strong role for interaction with other partners, since a science and research driven sector like biotechnology would have a more global focus and cooperate with the best partners available worldwide (Interviews HBGW 5: 16ff.; HBGW 3: 3).

Additional (Regional) Institutions

As *additional (regional) institutions* relevant for knowledge transfer processes in the MR HBGW the majority of the interviewed experts mentioned, independently of the sector, above all, the so called *regional growth concepts* (*Regionale Wachstumskonzepte*) in the different sub-spaces (Interviews HBGW 1: 9ff.; HBGW 2: 16; HBGW 3: 13ff.; HBGW 6: 1ff.). These regional growth concepts are part of the initiative *Regionale Wachstumskonzepte Niedersachsen* by the Lower Saxon Government, and have the explicit aim to connect firms, political actors and the scientific world and to promote cooperation and knowledge transfer between them. Most of them are operated as public private partnerships between local authorities, academic institutes and firms (see also Revilla Diez 2004: 73; Krumbein 2003: 30ff.; Prätorius 2003b: 22ff.; Fromhold-Eisebith Eisebith 2005: 1265f.).

Within the Metropolitan Region, the regional growth concepts *HannoverImpuls GmbH*, the *Weserbergland AG*, the *Projekt Region Braunschweig GmbH*, the *Regionalverband Südniedersachsen e. V.*, as well as the *Wolfsburg AG,* would exercise a decisive and well accepted role in bringing the actors together, offering regional platforms, and promoting knowledge transfer (Interviews HBGW 1: 9ff.; HBGW 2: 16; HBGW 3: 13ff.; HBGW 5: 12; HBGW 6: 1ff.). A stronger coordination of these different regional growth concepts, which would have a more competitive attitude towards each other, at the level of the Metropolitan Region would be pretended, but not yet reached, as the managing director of the Metropolitan Region GmbH states (Interview HBGW 1: 9ff.) Finally, also some facilities of the *Steinbeis Network* in the Metropolitan Region would actively be used in connection with knowledge transfer activities (Interview HBGW 3: 14).

5.3.3 Summary

Summarizing these results, it becomes obvious that in the case of the MR HBGW many of the theoretical assumptions derived from the literature on VoC and NIS on knowledge transfer processes in Germany's CME have not been confirmed. In this, above all, the Metropolitan Region's polycentric structure, its young and "constructed" history, the antagonism between its constituting sub-spaces and its strong dependence on the automotive industry and the VW Group have been identified as central explanatory factors for these results, which also represent major differentiating features when comparing the Metropolitan Region with the case of Baden-Württemberg.

When looking at the single dimensions of knowledge transfer, network based mechanisms have been evaluated as less significant as expected. Horizontal and vertical business networks, as well as cooperative inter-firm R&D-projects have been attributed only a medium range significance for knowledge transfer in the Metropolitan Region. A generally lower propensity of firms to cooperate, a more distanced business culture and a strong competitive attitude of actors basing on a lack of coherence among the different regional sub-spaces explain these results and constitute main differentiating characteristics of the Metropolitan Region's knowledge transfer system. Not least, they also explain the limited importance of informal networks in this context. Networking activities in the region are, then, actively fostered by regional government, above all through the publicly promoted and institutionally stabilized regional initiatives and competence networks. This illustrates, that in the MR HBGW strategic coordination of economic activities in the field of knowledge transfer finds its expression less in informal interaction and bottom-up action by the economic actors, but, above all, in institutionally stabilized networks, cooperations and initiatives, with an important role of the state in initiating and promoting these networks top-down. This stronger role of the state in the regional networking activities constitutes a main difference to Baden-Württemberg, where the state has been described as equally active, but not in initiating cooperation top-down, but in promoting as a bystander bottom-up networking activities by the economic actors.

In addition, cooperations between firms and the scientific world have been clearly evaluated as more important than assumed. Apart from a generally increasing dependence on specialized scientific knowledge in the region's core industries, both a high demand generated by research intensive industrial sectors, and in particular the automotive industry, requiring intensive cooperation with scientific institutions, and a strong supply of an excellent scientific infrastructure with highly renowned universities and research institutes explain this result and constitute another distinctive feature of the region's knowledge transfer system. Nevertheless, the strong academic research focus on human and veterinary medicine, medical technology and biotechnology is not fully reflected in the region's industrial structure. Therefore, the "perfect" fit of scientific infrastructure and industrial structure that has been identified for the case of Baden-Württemberg is not been found again in the MR HBGW, so that it can be assumed that this is a rather specific feature of Baden-Württemberg's innovation and knowledge transfer system.

Also the role of association based mechanisms of knowledge transfer in the MR HBWG resulted to be weaker than assumed. Apart of some of the business associations, above all the chambers of industry and commerce in the region do not fulfill the central role they are ascribed in the theory. Organizational weaknesses, competitive rather than cooperative relations between the different local chambers in the region, as well as a lack of tradition in taking responsibility for the Metropolitan Region as a whole are the main reasons provided for that. This also constitutes a major difference in comparison to Baden-Württemberg, where the strong focus of corporatist actors on the region as a whole has been identified as central feature and enforcing element in innovation and knowledge transfer processes. It can be assumed, that the comparatively low propensity of firms in the Metropolitan Region to cooperate among each other can inter alia also be traced back to this lack of intermediary moderation by chambers and associations.

Like in the case of Baden-Württemberg, also in the MR HBGW the significance of the organizational dimension of knowledge transfer has been evaluated as higher than assumed. The respective mechanisms like transfers of staff, purposeful headhunting and business-start-ups/ academic spin-offs have all been attributed a medium range importance for knowledge transfer processes in the region. Concerning the transfer of technical and scientific staff and purposeful headhunting, it resulted that a reduction of supply for qualified staff on the labor market and the beginning effects of the process of demographic change would not only open up more opportunities for highly qualified staff to transfer from academia to the business sector, but also enhance the firms' recruiting activities in the scientific sector. Specific drivers of this development are the lack of qualified staff in the region's automotive sector, but also public programs to foster transfer activities in the region. However, with the low attractiveness of the region as a place of residence for highly qualified persons and top managers and it's comparatively less positive image also factors hindering stronger transfer and recruitment activities have been discussed as major differences in comparison to Baden-Württemberg. With regard to business start-ups, the strong support by regional government and other regional institutions, but also the active role of the regional banks in providing seed- and venture capital as well as business angel support constitute specific features of the Metropolitan Region's knowledge transfer system. For this point, the metropolitan concentration of financial service providers in the region can be assumed to be a major explanation. Contrariwise, private maecens compensating for some shortcomings in the provision of seed- and venture capital like in Baden-Württemberg did not result to play a role in the Metropolitan Region.

Many of these arguments already demonstrate that national and especially regional government measures are a key element in the Metropolitan Region's knowledge transfer system. In this, the state plays a considerably more active role in the region's knowledge transfer activities than has been assumed for the German case, and also than could be shown from the case study on Baden-Württemberg. The major explaining factor for this resulting from the analysis above is the centralized and focused structure of the Lower Saxon innovation policy with the specific interplay of regional ministries, the *Innovationszentrum Niedersachsen* and the *NBank*. Above all, the coordination of intermediary actors in the fields of knowledge transfer and innovation within the *Innovationsnetzwerk Niedersachsen* and its role as central interlocutor in the Metropolitan Region's innovation and knowledge transfer system have been highlighted in this connection. In comparison, Baden-Württemberg's innovation policy resulted to follow a more decentralized and bottom-up approach, leaving more initiative to the economic and corporatist actors, while the analysis revealed a stronger top-down orientation of the Lower Saxon innovation policy. While the former showed to being able to rely more on the respective coordination capabilities of the economic and corporatist actors, the latter has to assume more responsibility in initiating and coordinating the activities in the fields of innovation and knowledge transfer. Nevertheless, these more centralized coordination efforts, above all through the participation of most of the chambers, associations and other intermediary actors in the *Innovationsnetzwerk Niedersach-*

sen, in fact resulted to be evaluated as positive and efficient by the interviewed practitioners, and were estimated to at least partly compensate the organizational shortcomings characterizing many of these actors per se.

In addition, like in the case of Baden-Württemberg, also in the MR HBGW the importance of technology transfer offices or companies at the universities and public research institutes has been estimated considerably lower than expected. As an explication, also in this region most of these institutions were reported to be personally and financially underequipped, inefficient and misdirected with regard to their mission. Apparently, these shortcomings in the organization of technology transfer agencies are an overarching problem in the German regions. However, the analysis of the MR HBGW's innovation and knowledge transfer system revealed that these shortcomings might be overcome through the integration of technology transfer offices into wider networks of intermediary actors, like, e.g., the *Innovationsnetzwerk Niedersachsen*.

Furthermore, the expected low significance of the market based mechanisms of knowledge transfer (external consultancies; mergers, acquisitions and hostile takeovers; buying, selling and licensing of patents) has not been confirmed by the results of the case study. The intensive use of external consultancies for knowledge transfer processes and also their role in elaborating and implementing innovation and knowledge transfer oriented regional growth concepts are major reasons and differentiating factors in this regard. In addition, also intensified activities in the field of mergers, acquisitions and hostile takeovers in some industrial sectors characterize the knowledge transfer system in the Metropolitan Region.

Finally, the community based mechanisms of knowledge transfer resulted to be less significant in the MR HBGW than expected. Informal contacts and regional platforms were only attributed a medium range significance. Here, the region's polycentric structure, a competitive attitude among the different sub-spaces, as well as the missing of a historically grown, common and strong regional identity have been identified as central reasons. At the level of these different regional sub-spaces, however, the respective regional growth concepts have been acknowledged as central actors and differentiating features of the MR HBGW's knowledge transfer system. These regional growth concepts can be seen as the counterparts of the large variety of cluster initiatives in Baden-Württemberg in assuming the respective functions in the fields of innovation and knowledge transfer. The difference is that the former are organized according to a "sub-spatial logic", focusing on many central sectors within one specific regional sub-space, while the latter are mostly organized around only one specific value added chain. This sub-spatial, rather than sectoral organization of knowledge transfer activities, providing more general and less sector specific institutional support, can be regarded as another specific feature of the Metropolitan Region's knowledge transfer and innovation system.

As expected, the results of this case study demonstrated sectoral specificities as important differentiating factors. Like in the case of Baden-Württemberg, the MR HBGW's biotechnology sector deviates much stronger from the assumed structures and mechanisms derived from the theory than automotive. Moreover, both sectors also noticeably differ to the situation identified for the region as a whole. In this, automotive in the MR HBGW displays some of the assumed features of a traditional, mature and established industrial sector in a CME context: Network- and association based mechanisms are central channels of knowledge transfer, while organization- and community based mechanisms have a medium range importance. Both, state induced and market based mechanisms, on the other hand, play a surprisingly important role, except of the trade of patents. Most of these features have been explained by the sector's strongly centralized business structure with only one large, multinational OEM, only a few larger and many small suppliers, but without a differentiated medium sized firm sector. Moreover, also the dominant position of automotive within the region's industrial model and the resulting monostructural tendencies strongly shape the knowledge transfer activities in this sector.

Like its counterpart in Baden-Württemberg, the Metropolitan Region's biotechnology sector also shows characteristics of a young, strongly research based and venture-capital driven industry: Its connections to universities and research institutes as channels of knowledge transfer are strong, while other forms of networks as well as association and community based mechanisms only play an average or minor role. Decisively important for knowledge transfer processes in the sector are state induced mechanisms, like national and, especially regional cluster programs and initiatives, cluster management organizations, scientific parks, technological centers and the provision of public funding, not least for business start-ups and academic spin-offs. Not only the emergence, but also the further development and growth of the sector in the MR HBGW resulted to be highly dependent on public action and support. Above all, the centralized coordination of the sector's innovation and knowledge transfer activities

Table 5: Qualitative Relevance of Knowledge Transfer Mechanisms in the Metropolitan Region HBGW

MECHANISMS	GER*	MR	Auto	Bio
NETWORK BASED MECHANISMS				
business networks	+	O	+	–
informal networks	+	–	O	–
institutionally stabilized networks	O	O	+	O
customers and suppliers	+	O	+	–
cooperative inter-firm R&D-projects	+	O	O	–
cooperative R&D-projects with scientific institutions and universities	O	+	+	+
ASSOCIATION BASED MECHANISMS				
employers-, business-, and trade associations	+	+	O	–
chambers of industry and commerce	+	O	+	–
binding and common technical standards	+	O	+	+
ORGANIZATION BASED MECHANISMS				
transfer of technical and scientific staff	–	O	+	O
purposeful headhunting	–	O	O	???
business start-ups	–	O	O	O
STATE INDUCED MECHANISMS				
national government measures	O	+	+	O
regional government measures	O	+	+	+
(larger) state-induced innovation projects	–	???	???	???
scientific parks	O	O	–	+
technological centers	O	O	+	+
technology transfer agencies/companies	+	O	O	O
MARKET BASED MECHANISMS				
external consultancies	–	+	O	O
mergers and acquisitions, hostile takeovers	–	O	+	O
buying, selling and licensing of patents	–	O	–	+
COMMUNITY BASED MECHANISMS				
informal contacts	+	O	O	–
regional platforms	+	O	O	–
additional (regional) institutions	+	+	+	+

+ *Reported (*Hypothesized) as crucial mechanism of knowledge transfer*

O *Reported (*Hypothesized) as important but not decisive mechanism*

– *Reported (*Hypothesized) as not important at all*

??? Not reported by the interview partners

[dark grey box] = *Important deviation from hypotheses*

[light grey box] = *Modest deviation from hypotheses*

[white box] = *No deviation from hypotheses*

through the regional initiative *BioRegioN* has to be highlighted in this connection. In addition, and as was to expect for a New Economy sector characterized by more radical modes of innovation, market based mechanisms of knowledge transfer are considerably more important than in the region as a whole.

Finally, two further factors strongly shape knowledge transfer processes in the Metropolitan Region's biotechnology sector: First, its business structure with a vast majority of small firms, only a few larger biotech companies, and the absence of large pharmaceutical firms in the region constrain the use of certain channels of knowledge transfer, like mergers and acquisitions, hostile takeovers or purposeful headhunting. Also the overall sectoral structure with a low absolute number of firms and a very young business demography exercises this effect. Second, also the strong dependence on IPRs protection impedes some forms of knowledge transfer, above all inter-firm cooperations, joint research activities and the involvement of business associations and chambers in knowledge transfer activities. Table 5 provides a summarizing overview on these results.

5.4 Lessons from the German Case

With regard to the two research questions guiding this study, the qualitative analysis of the knowledge transfer systems of the two German regions of Baden-Württemberg and Metropolitan Region HBGW revealed considerable differences between the two cases. For the German case, this indicates that the German system of innovation and knowledge transfer is by no means homogeneously structured across its constituting regional innovation and knowledge transfer systems. Rather, the analysis exposed a number of specific factors explaining this inter-regional variation that will be discussed in the following section.

The regional innovation and knowledge transfer system of Baden-Württemberg showed most of the elements that have been assumed for a region within a CME environment: Network-, association-, and community based mechanisms have been identified as central channels of knowledge transfer, state induced mechanisms play a medium range role, especially in enabling coordination among the economic actors, and market based mechanisms are no more than marginally important. Only organization based mechanisms resulted to be more relevant than assumed. As has been demonstrated, a common and sustainable regional identity, a cooperative and proactive business culture, regional self-identification of large parts of the business sector, corporatist actors with a strong sense of responsibility and focus on the region as a whole, and a regional government structuring and reinforcing bottom-up action by the economic actors are main factors explaining this fit in the case of Baden-Württemberg. In addition, also the region's business structure, with a strong and diversified medium sized business sector, and an industrial structure with traditional sectors like automotive, mechanical engineering, chemicals and electrical engineering with their respective incremental modes of innovation in its center are major explanatory factors (Figure 22).

However, against the assumption of broad parts of the literature on VoC and NIS, the analysis of the MR HBGW demonstrated clearly that Baden-Württemberg is not to be seen as a typical, but most probable as a specific case within the German context in fulfilling most of the conditions discussed above. The qualitative case study on the MR HBGW illustrated that within the German context also other regional models of knowledge transfer systems exist, which are the result of and base on different structural conditions, as well as of distinct adaptation processes to these conditions. In this, the knowledge transfer system in the MR HBGW resulted to rely less on network-, association-, and community based, but more on state induced mechanisms of knowledge transfer. Moreover, organization and market based mechanisms play a medium range role (Figure 23). As central explanatory factors for this specific shape of the Metropolitan Region's knowledge transfer system its polycentric structure, its young and "constructed" history, the absence of a common regional identity, the antagonism between its constituting sub-spaces, its more individualistic and competitive business culture, its centralized business structure dominated by a few larger firms, its strong dependence on the automotive industry and the *VW Group*, the proactive and centralized Lower Saxon innovation policy and the coordination of intermediary activities within the *Innovationsnetzwerk Niedersachsen* have been identified.

Overall, this provides first empirical evidence that in the German case institutional coherence of national systems of innovation and knowledge transfer is less strong than assumed by large parts of the literature on VoC and NIS. Rather, it demonstrates that regional systems within national institutional models deviate to a considerable degree from their respective national institutional frameworks, at least with regard to their structures and mechanisms of knowledge transfer. Two specific explanatory dimensions for these inter-regional differences in the field of knowledge transfer have received particular attention in the two German case studies: The traditional industrial region vs. metropolitan region dimension, and the sectoral dimension. These will briefly be discussed in the following.

As expected, the case studies of the two German regions revealed important insights regarding the differences rooted in the antagonism between traditional industrial regions and metropolitan regions. As has been shown, Baden-Württemberg shows almost all characteristics that have been defined for a traditional industrial region: Historical path-dependent and industrial trajectories leading to established practices, a business structure dominated by SMEs and complemented by MNEs standing at the head of dense networks of customers and suppliers, institutional frameworks specifically tailored for the needs and the support of the particular industrial focus, entrepreneurial mentalities of a "commercial community", a commonly shared socio-cultural identity, and historically grown and formalized networks. These characteristics have been demonstrated to influence and structure the region's knowledge transfer system strongly, with network-, association- and community based mechanisms as central channels of knowledge transfer (see also Figures 22 and 23).

In contrast to that, the knowledge transfer system of the MR HBGW is shaped by many characteristics that have been defined as typical for metropolitan regions: A pronounced gateway-function with strong inter-regional and international channels of knowledge transfer, a control-function with a high concentration of headquarters of larger multinational companies and central financial service institutions, as well as a learning-function with a dense infrastructure of higher education and research institutions. Above all, the comparatively high significance of organization- and market based mechanisms of knowledge transfer has been traced back to these metropolitan functions. Moreover, the region's specific structure as a polycentric and "constructed" metropolitan region could be identified as central explanatory factors for the comparatively lower importance of network-, association- and community based mechanisms of knowledge transfer. Overall, this indicates, that traditional industrial regions and metropolitan regions are characterized by specific and distinct structural conditions, which influence the shape of their respective knowledge transfer systems strongly. These distinct structural conditions can be, thus, regarded as important differentiating factors for regional knowledge transfer systems within specific national institutional frameworks.

Fig. 22: Regional and Sectoral Variation the German Coordinated Model: Baden-Württemberg[23]

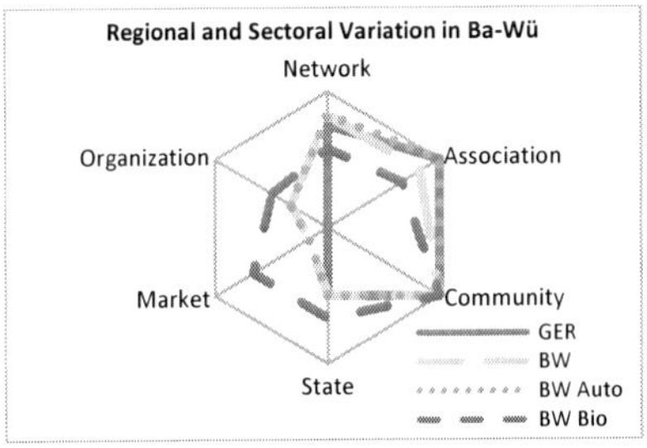

Source: Own Illustration

Fig. 23: Regional and Sectoral Variation from the German Coordinated Model: The MR HBGW

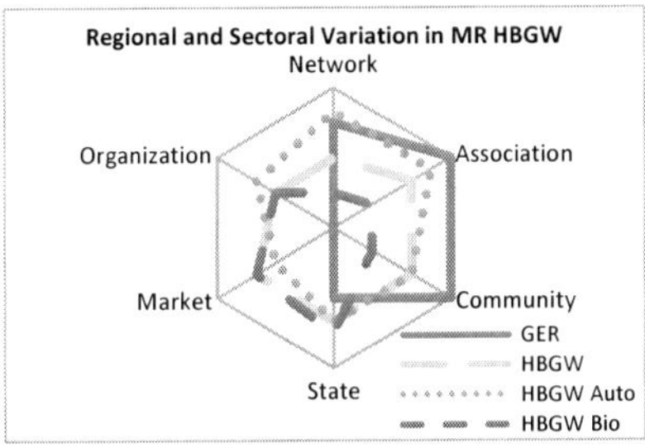

Source: Own Illustration

Furthermore, it has also been shown that *sectoral* differences decisively shape regional knowledge transfer systems. The analysis of the two German cases demonstrated clearly that there are "sectoral knowledge transfer systems", analogous to the notion of sectoral innovation systems (e.g. Malerba 2004; Casper/Soskice 2004; Coriat/Weinstein 2004). In this, the special focus of this study on automotive as a traditional industrial sector and biotechnology as a young high-tech sector representing the New Economy disclosed a broad number of specialisms in the knowledge transfer activities of both sectors (Figures 22 and 23). On the one hand, in automotive network-, association- and community based mechanisms of knowledge transfer are central. In particular, vertical supply chain networks between firms and their customers and suppliers are decisive, not least be-

23 This and all following radar-charts intend to further aggregate the results of the qualitative case studies in a comparative perspective. Parting from the ordinal scale differentiating between crucial (**+**), important but not decisive (**O**), and those mechanisms which have been evaluated (hypothesized) as not important at all (**–**), it ascribes the three categories numerical values of (**+**) = 1, (**O**) = .5, and (**–**) = 0 according to the fsQCA scores (see section 4.1.6), which are then summated for each of the six dimensions of governance and divided by the number of mechanisms constituting each dimension (not reported values (???) are estimated where possible). The resulting values are the data on which these radar charts base. This proceeding is exclusively motivated by visualization purposes, and does not intend a quantification of purely qualitative data, as has already been discussed in detail in section 4.1.6.

cause of its incremental mode of innovation and the increasing importance of modularized production processes. Institutionalized cluster and network initiatives are predominant forms of cooperation in this sector. Business associations and chambers moderate this cooperation and enable and foster a long term orientation of the sectoral actors in incremental innovation processes. Especially cooperative inter-firm R&D-projects and cooperations with scientific institutions, but also transfers of qualified personnel, take account of the increasing dependence on highly specific knowledge in the sector's production processes, while its consolidated character impedes intensive start-up activities.

On the other hand, the biotechnology sector in the German regions relies more on state induced, as well as organization- and market based mechanisms of knowledge transfer. Above all, the sector's emergence and large parts of its growth can be traced back to public action and support programs, giving public institutions a central role for knowledge transfer in this sector. Moreover, the sector is strongly research driven, making it necessary to maintain intensive channels of knowledge transfer to the scientific world. Long product development phases create a strong dependence on seed- and venture capital supply, often impeding more intensive start-up and spin-off activities as mechanisms of knowledge transfer in the German regions. In addition, the sector's strong awareness on IPRs protection creates a barrier to cooperations and networks with competing firms, but also for a strong and moderating role of intermediary actors like business associations and chambers. Finally, the specific business model of a large part of dedicated biotechnology firms (especially in the field of red biotechnology), explains the comparatively high relevance of market based mechanisms like the trade of patents and licenses, as well as mergers, acquisitions and hostile takeovers.

Overall, this indicates, that also the sectoral dimension must be taken into account when looking at regional knowledge transfer systems. Different industrial sectors are characterized by specific and distinct structural conditions, business models, knowledge bases, innovation modes and functional needs, which influence the shape, logic, structure and the applied mechanisms of knowledge transfer activities within these sectors strongly. In addition, they give the different sectors distinct capacities to efficiently use the available structural conditions and to find adequate support structures in their respective regional environment. The specific sectoral structure of a region can be, therefore, regarded as an important differentiating factor for regional knowledge transfer systems within specific national institutional frameworks.

In sum, these results demonstrate that in contrast to the coherence of national market economy models assumed by large parts of the literature on VoC and NIS, the Germany coordinated market economy should rather be seen as a regionally variegated set of regional governance models in the fields of innovation and knowledge transfer. Major drivers for this regional variation in the German case are regional historical industrial trajectories, particular industrial and sectoral structures, the business structure, the matureness of industrial sectors and institutional structures, business cultures, diverging policy approaches, the degree of regional coherence, regional identity, the organizational capabilities and the coherence of corporatist actors, as well as the importance of metropolitan functions (supply of advanced business services, equipment with business headquarters, openness and intensity of cross-regional interaction).

This suggests that the approaches of VoC and NIS not only overestimate the coherence of national systems of innovation and production and are not sufficiently aware of these driving factors of regional variation, but also that they rely too much on single regional cases to define their idealized models as wholes. In the concrete case of Germany, the too strong focus on Baden-Württemberg when describing the German coordinated model as a whole seems questionable in the light of the analysis above, since other regional variations of the coordinated model do exist. This study particularly focused on the case of metropolitan regions, which are under-reflected in the theory-building process on VoC and NIS so far, as well as on specific functional and structural needs generated by sectoral specializations which can be found in other regions as major instances accounting for regional variation. Based on this, the need to amplify the approaches of VoC and innovation systems towards a more regional-varieties-oriented research agenda must be derived, as will be discussed in detail in the concluding section 8. Table 6 summarizes the results for the German case.

Table 6: Qualitative Relevance of Knowledge Transfer Mechanisms in two German Regions and Sectors: Baden-Württemberg and Metropolitan Region HBGW

MECHANISMS	GER*	BW	Aut	Bio	MR	Aut	Bio
NETWORK BASED MECHANISMS							
business networks	+	+	+	O	O	+	–
informal networks	+	O	O	O	–	O	–
institutionally stabilized networks	O	+	+	O	O	+	O
customers and suppliers	+	+	+	O	O	+	–
cooperative inter-firm R&D-projects	+	+	+	O	O	O	–
cooperative R&D-projects with scientific institutions and universities	O	O	O	+	+	+	+
ASSOCIATION BASED MECHANISMS							
employers-, business-, trade associations	+	+	+	O	+	O	–
chambers of industry and commerce	+	+	+	O	O	+	–
binding and common technical standards	+	O	+	+	O	+	+
ORGANIZATION BASED MECHANISMS							
transfer of technical and scientific staff	–	O	O	O	O	+	O
purposeful headhunting	–	O	O	O	O	O	???
business start-ups	–	O	–	O	O	O	O
STATE INDUCED MECHANISMS							
national government measures	O	O	O	+	+	+	O
regional government measures	O	O	O	+	+	+	+
(larger) state-induced innovation projects	–	–	O	–	???	???	???
scientific parks	O	O	O	+	O	–	+
technological centers	O	+	???	???	O	+	+
technology transfer agencies/companies	+	–	–	O	O	O	O
MARKET BASED MECHANISMS							
external consultancies	–	–	O	O	+	O	O
mergers and acquisitions, hostile takeovers	–	–	–	O	O	+	O
buying, selling and licensing of patents	–	O	–	+	O	–	+
COMMUNITY BASED MECHANISMS							
informal contacts	+	+	+	O	O	O	–
regional platforms	+	+	+	+	O	O	–
additional (regional) institutions	+	+	+	+	+	+	+

+ *Reported (*Hypothesized) as crucial mechanism of knowledge transfer*

O *Reported (*Hypothesized) as important but not decisive mechanism*

– *Reported (*Hypothesized) as not important at all*

??? Not reported by the interview partners

[dark grey box] = *Important deviation from hypotheses*

[light grey box] = *Modest deviation from hypotheses*

[white box] = *No deviation from hypotheses*

6. Spain—Knowledge Transfer Systems in a Mediterranean Economy

6.1 The Spanish Economy—Some Structural Facts

In section 2.3.1.2, Spain has been described as a MME according to the VoC-approach (Hall/Soskice 2001a). Building on this discussion, Spain has been selected as an example of a MME in Europe in the case study at hand. This section will give a brief overview on some fundamental quantitative empirical indicators to describe Spain's current economic structure.

With 44.3 million inhabitants in 2008 Spain is the fifth largest country in the EU (OECD 2010b), and also Europe's fifth largest economy, with a GDP of 1.01 trillion Euros, representing 8.75% of the EU's GDP, and less than half of Germany's GDP (2.46 trillion Euros) (Eurostat 2010: 98). As in the other EU-countries, also Spain's economic structure is mainly dominated by the service sector, with a share of 68.4% in the total GVA. With a share of 2.8% in the total GVA, its primary sector is comparatively large, and more than triples the values for Germany (0.9%) and the UK (0.8%). The Spanish industrial sector (without construction), on the contrary, with a share of 17.3%, contributes considerably less to the total GVA than is the case in Germany (25.6%), but almost equals the value for the UK (17.6%). A specific feature of Spain's economic structure is the strong position of the construction sector: With a share of 11.6% in the total GVA the country ranks among the EU countries only behind Romania, and almost doubles the value for the UK (6.1%) and even almost triples the value for Germany (4.2%) (Figure 6; Eurostat 2010: 100).

Fig. 24: Spain's Business Structure I

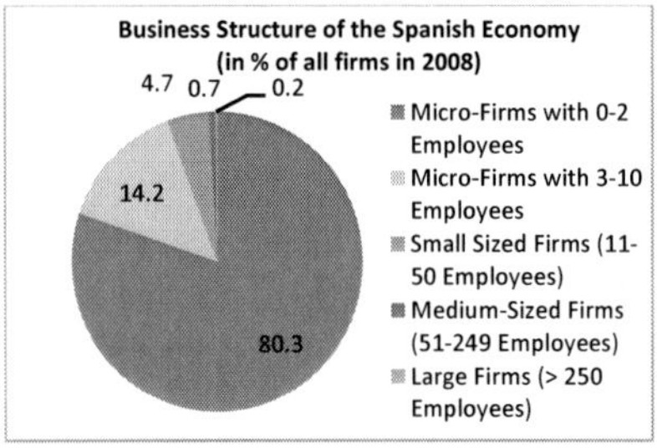

Source: Own Calculations and Illustration based on MITC (2010): 8

Fig. 25: Spain's Business Structure II

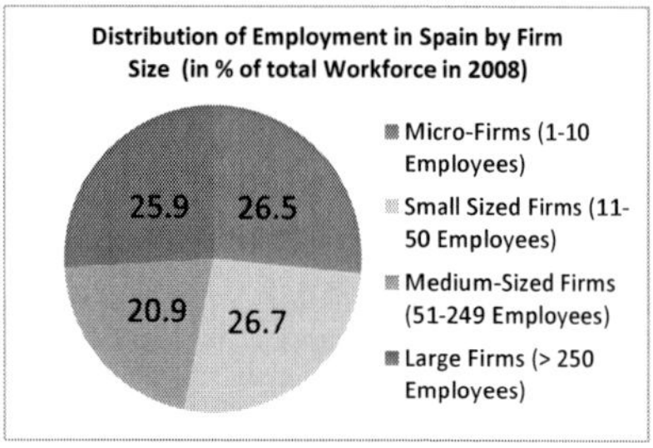

Source: Own Illustration based on MITC (2010): 9

When looking at Spain's business structure, the strong dominance of (often family-owned) small and especially micro-firms must receive special attention. In 2008, 94.5% of all Spanish firms were micro-firms with less than ten employees. Small firms with ten to 49 employees represented 4.7% of all Spanish firms, medium sized firms from 50 to 199 employees 0.7% and large firms with more than 200 employees only 0.2%. Within the group of micro-firms those firms stand out that have in between zero and two persons employed: They represent about 85% of all micro-firms in the country. In this, the micro-firms only employed 26.5% of the country's total workforce, while small, medium and large firms employed 26.5%, 26.7%, and 25.9% of Spain's total workforce, respectively (Figures 24 and 25; MITC 2010: 8f.).

Fig. 26: Sectoral Structure of Spain, Catalonia and the Madrid Community

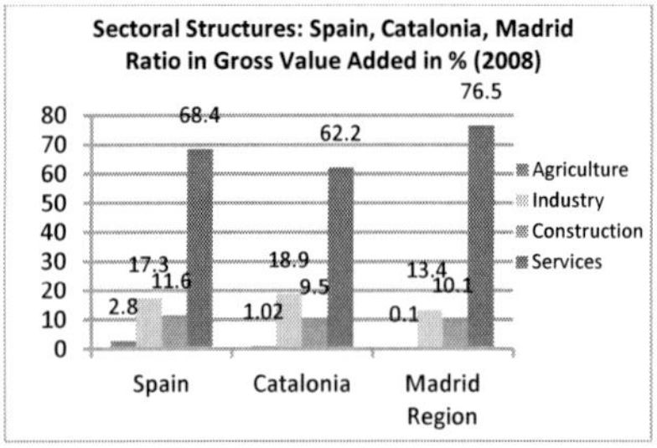

Source: Own Calculations and Illustration based on Eurostat (2010), IESTADIS (2011) and Idescat (2011)

Fig. 27: Spain's Industrial Structure

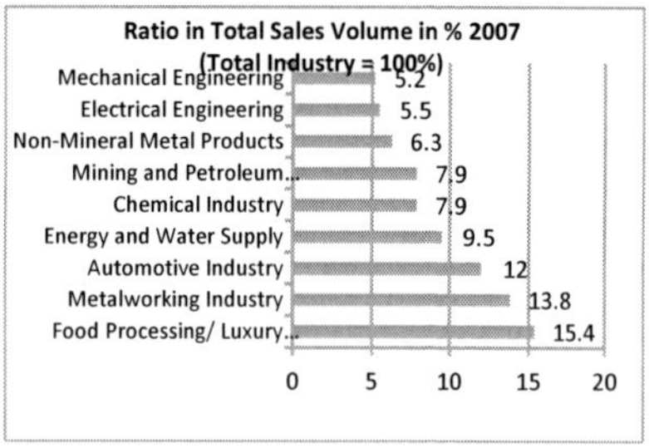

Source: Own Illustration based on INE (2010): 144

The sectoral structure of Spain's industry is mainly shaped by medium-high and medium-low technology sectors. The Spanish industry is characterized by a high sectoral diversity and, therefore, by a low degree of specialization, with no industrial sector accounting for much more than 15% of the industry's total sales volume (Figure 27). The most important manufacturing sector is food processing and luxury foods, with a share of 15.4% in the total sales volume of the industrial sector, followed by the metalworking industry (13.8%), automotive (12.0%), energy and water supply (9.5%), the chemicals (7.9%), and mining and petroleum (7.9%) (Figure 27; INE 2010a: 144f.).

Fig. 28: Sectoral R&D-Expenditures in the Spanish Industry

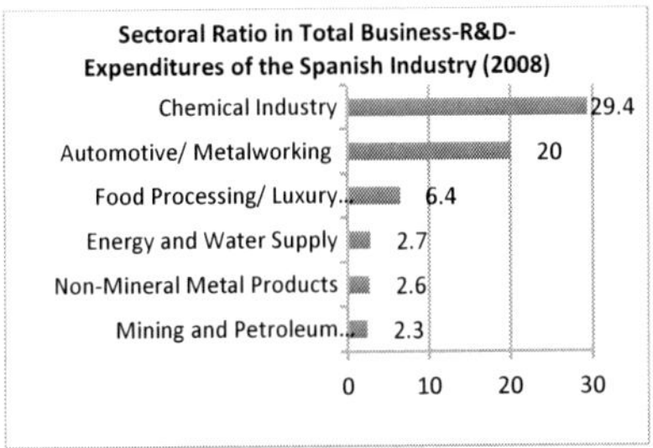

Source: Own Calculations and Illustration based on COTEC (2010): 279

Fig. 29: R&D-Intensity: Spanish Automotive Industry in Int. Comparison

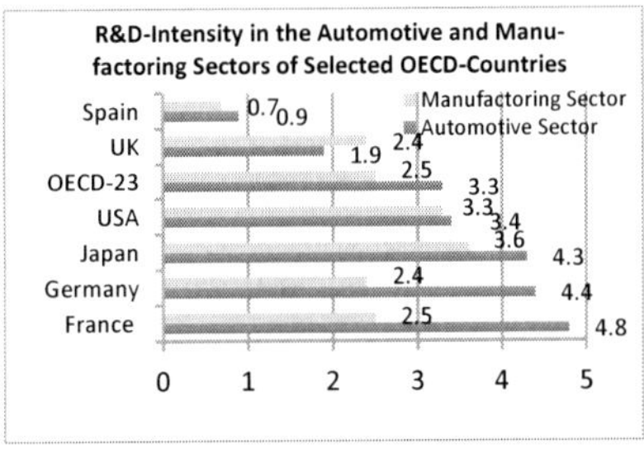

*Source: Own Illustration based on Legler et al. (2009): 94 *OECD-23: GER; F; UK; IT; BEL; NL; DK; IR; GR; E; P; SW; FIN; A; N: PL; H; CZ; CAN; USA; JP; KOR; AUS*

To describe Spain's knowledge base, some structural indicators of the country's R&D-activities will briefly be discussed. In 2008, 215,676 persons were employed full time in the field of R&D in Spain (COTEC 2010: 17), representing 1.06% of the country's total workforce of 20.3 million (INE 2011). Spain's total expenditures for R&D-activities were about 14.7 billion Euros in the same year, representing 1.35% of the Spanish GDP (GERD for Germany: 2.68%; UK: 1.77; EU: 1.92%), and of which 54.9% were covered by the business sector and 46.1% by the public sector plus other non-commercial public and private research institutions (COTEC 2010: 246). When differentiating again in this context between SMEs (< 250 employees) and large firms (> 250 employees), the SMEs had a share of 48.9% in the total R&D-expenditures of the business sector, while the large firms covered 52.1% (OECD 2009a). Given Spain's overall business structure, this points to a strong concentration of the country's R&D-ac-

tivities within the large firms. A sectoral perspective reveals that most of Spain's key industries are also on top in terms of their R&D-activities: Measured by their share in total R&D-expenditures of the industrial sector, chemicals is Spain's most R&D-intensive sector (29.4%), followed by mechanical engineering and automotive (together 19.9%), food processing and luxury foods (6.4%), and energy and water supply (2.7%) (Figure 28; COTEC 2010: 279).

One of the two sectors of interest for this study is automotive, which is Spain's third most important industrial sector. In 2007, the sector's total sales volume was of 75.4 billion Euros, representing a share of 12% in the total sales volume of the Spanish industry. In the same year, Spanish automotive employed 211,888 persons, representing about 8% of the total workforce of the Spanish industry. The sector consisted of 3,395 firms, representing about 2% of all industrial firms in Spain. 86.0% of these firms are small firms with less than fifty employees, while 10.6% have between fifty and 249 employees. 3.4% of the firms in Spanish automotive have more than 250 employees. This demonstrates that in Spanish automotive medium sized and large firms have a considerably higher weight than in the Spanish economy as a whole. When looking at the persons employed in the distinct firm-size classes, this is further underlined: The large firms employ 64.9% of all employees in the sector, the medium sized firms 19.6% and the small firms 15.4%. This strong position of the large firms in the sector is also illustrated by their share in the sector's total sales volume of 80.8%, while the medium sized firms only contribute with a share of 13.6% to the sector's total sales volume, and the small firms with 5.6% (INE 2010a: 138f.).

A specific feature of Spanish automotive is the lack of own large OEMs. Spain must therefore be considered as a pure production location for many of the large MNEs in the sector operating worldwide (AXESOR 2009: 28). Important strategic decisions in the sector are, therefore, in most cases not taken in Spain, but in the OEMs' headquarters abroad, and will, moreover, depend strongly on foreign capital. The most important of these production facilities in terms of sales volume are those of the *Peugeot Citroën Group*, *Fasa Renault*, *Ford*, SEAT, *General Motors España*, *Volkswagen Group*, *Nissan Motor Ibérica* and *Daimler España* (Fira de Barcelona 2009: 5; AXESOR 2009: 16f.). The regional distribution of business activities in automotive in terms of number of firms, share in the sector's total sales volume and number of employees is mostly concentrated within the Autonomous Regions of Catalonia and Madrid, but also in Aragon, Castile and Leon, and Galicia (AXESOR 2009: 34f.).

This also explains the fact that Spanish automotive is a predominantly supply oriented industrial sector, in which the dedicated vehicle producing firms only represent about 13.0% of all the firms (but 46% of all employees), while the suppliers to these production plants represent 87% of the firms (but only 54% of the employees) (AXESOR 2009: 22f.). Also in Spanish automotive, production plants and supplying firms (components, materials, services) are interconnected intensively with each other as customers and suppliers along the value added chain. More than 70% of the sector's total production value stems from supplied advanced performances, reflecting the strong intra- and inter-sectoral division of labor. This especially implies supplies of suppliers among each other and supplies between suppliers and the production plants of the larger OEMs located in Spain (Candil Gonzalo 2010: 15).

Focusing on some R&D-oriented indicators characterizing the knowledge generation function of Spanish automotive, it becomes obvious that it is a comparatively R&D-intensive sector within the Spanish economy: In 2008, around 8.9%[24] of the total internal R&D-expenditures of the Spanish business sector and about 20% of those of the Spanish industrial sector have been realized by the automotive industry (plus mechanical engineering), representing the largest share in the total internal R&D-expenditures among the country's industries (COTEC 2010: 125f.). However, its comparatively high intra-national innovativeness is still only low when compared internationally: When looking at the sector's R&D-intensity (share of R&D-expenditures in total sales volume in 2006) of 0.9% it becomes clear that although this value is above the value for the country's manufacturing sector as a whole (0.7%), Spanish automotive is clearly less R&D-intensive than other leading automotive industries, like e.g. France (4.8%), Germany (4.4%), Japan (4.3%), UK (1.9%) and ranks even below the average value for the automotive industries of the largest OECD-countries (3.3%) (Figure 29; Legler et al 2009: 94). Not least, this might be explained by the fact that the headquarters of the large OEMs operating in the country are located abroad, and, therefore, also substantial parts of the R&D-activities exercised by these OEMs are conducted outside of Spain.

While Spanish automotive can be regarded as a traditional and mature industrial sector, which decisively influenced the industrial development of

24 Own calculation based on COTEC (2010): 125f.

the country in the last decades, biotechnology in Spain is a very young sector, of which basic parts only emerged within the last decade, and with strong support of the national and regional governments. In 2008, the Spanish biotechnology sector consisted of 942 firms active in biotechnology related activities, of which 305 were dedicated biotechnology firms, and which had a share of about 1.1% in the country's GDP (ASEBIO 2010: 34ff.). The business structure of Spanish biotechnology is dominated by small firms: 79% of the Spanish biotech firms are small companies with less than 50 employees, while 12.3% have between 50 and 250 employees, and 8.5% have more than 250 employees[25] (OECD 2009b: 21). However, the share of large biotechnology firms is considerably higher in Spain than in Germany (1.4%) and UK (3%), pointing to comparatively stronger growth dynamics in Spanish biotechnology, especially when taking into account the overall younger demography of the Spanish biotech firms. In addition, the share of small firms with less than fifty employees in Spain is slightly lower than in the other two investigated countries Germany (87.0%) and UK (84.0%[26]) (OECD 2011a). The total sales volume of the sector was of 31.1 billion Euros in 2008, of which about 78.5% have been realized by the large firms with 250 or more employees, which might more likely have products in phases of commercialization. Finally, in 2008, the whole sector employed 108.347 persons, representing not even 0.6% of the country's total workforce. 72.4% of these persons were employed in the large biotechnology firms with more than 250 employees[27] (ASEBIO 2010: 35f.).

Furthermore, the geographical distribution of biotechnology in Spain shows a considerable concentration of business activities in the two Autonomous Regions of Madrid and Catalonia, where a share of about 22.6% and 21.0% of all Spanish dedicated biotechnology firms is located, respectively. Other important regional concentrations of biotech activities can be found in Andalusia (12.5%), the Valencia Region (11.5%) and the Basque Country (9.8%) (ASEBIO 2010: 36f.).

Looking at some innovation related indicators, overall, 22.2%[28] of all employees in Spanish biotechnology were R&D-personnel (OECD 2009b: 45). When only focusing on those persons conducting research in biotech-

25 Own calculation based on OECD (2009b): 21

26 Data for 2010

27 Own calculations based on ASEBIO 2010: 35f.

28 Own calculations based on OECD 2009b: 45; data for 2006

nology relevant fields, in 2008 about 26% of these persons were employed in the large firms with more than 250 employees, and 74% in firms with less than 250 employees[29] (ASEBIO 2010: 35f.). The total expenditures in biotechnology R&D-activities represent 4.5% of the total R&D-expenditures of the Spanish business sector (Germany: 2.4%), so that R&D-activities in the field biotechnology can be regarded as a specific focus of research within the Spanish business sector (OECD 2009b: 26). In this, 31.2% of the total expenditures in R&D-activities are realized by the large biotechnology companies with more than 250 employees[30] (ASEBIO 2010: 35), and 33.4% by the small firms with less than fifty employees (OECD 2011b).

Apart from basic research, in particular in the fields of red and medical biotechnology the development of new products, and especially new pharmaceuticals, requires long clinical test phases (phase I–III) of up to ten years or more, before a new pharmaceutical is approved for commercial distribution. In total, Spanish biotech firms had 35 drug candidates in phases of clinical trials, clearly less than the UK (235), and Germany (141). Looking at the distribution of drug candidates in the three clinical phases, in 2009, 7% of the drug candidates were in phase I, 74% in phase II and only 19% in phase III[31]. With that, in comparison to the mean values of all other European countries active in biotechnology, the Spanish biotech firms have comparative strengths in drug candidates in phase II, but are behind in drug candidates in phase III, pointing to certain weaknesses of the sector to bring their R&D-efforts to the markets, but also, again, to the sector's "young business demography" (Ernst & Young 2010: 42).

29 Own calculations based on ASEBIO 2010: 35f.

30 Own calculations based on ASEBIO 2010: 35

31 Own estimations and calculations basing on Ernst & Young 2010: 42

6.2 The Case of Catalonia

6.2.1 Locating Catalonia in the Spanish Context

6.2.1.1 Selected Facts on Catalonia's Economic Structure

Catalonia is one of only a few historically grown and traditional industrial regions within Spain, together with the Ebro Valley (Aragon, Navarre and the Basque Country) and, although to a lower degree, the Autonomous Regions of Valencia and Madrid (Bacaria et al. 2004: 64f.). Its industrial development traces back to the industrialization and mechanization processes in the textile and agrarian sectors in the 19th century and converted Catalonia to the most industrialized region in Spain, also called the "Manchester of Southern Europe." Over the past 50 years, due to the transition process from autarky and inward-orientation during the Franco-Regime to internationalization and openness, Catalonia experienced a profound economic transformation, accompanied by an industrial take-off. While Catalonia's economic model in the 1960s was mainly based on the attraction of low-skilled workers from the rural regions of Spain, the acquisition of investments from other European countries, the use of raw materials in the production process and a dynamically expanding construction sector, the economic crisis of the 1970s and early 1980s deeply affected the region's competitiveness and questioned its economic model. After 1984, driven by Spain's incorporation into the European Union in 1986, Barcelona's hosting of the 1992 Olympic Games, and the accompanying investments in the communication and transport infrastructure, the Catalan economic development model became more specialized in the tertiary sector, less dependent on low-cost labor and more capital intensive (OECD 2010c: 64f.; Bacaria et al. 2004: 63ff.).

In particular, this process of economic and industrial renewal was paralleled and fostered by the processes of regional decentralization and devolution developed in the Spanish Constitution of 1978 and the Catalan Statute of Autonomy (recently renewed in 2006), deeply reforming Catalonia's public administration and society. This led to the consolidation of a specific "Catalan Governance Model", not least in the fields of R&D and innovation policies, driven by a strong political will to be differentiated, if not separated, from the Spanish polity. In this, the region's specific industrial and institutional setting in comparison to the rest of the country, and, above all, its particular socio-cultural characteristics, with a strong regional identity basing on an own linguistic and cultural heritage, constitute a decisive momentum for the specific path of development of the Catalan innovation system (Bacaria et al. 2004: 63f.).

Today (2010), with a population of about 7.5 million Catalonia is Spain's second largest region in terms of inhabitants after Andalusia, and represents around 16% of Spain's population. Catalonia is Spain's strongest economic region, with a GDP of 215.2 billion Euros in 2008, representing 19.8% of the Spanish GDP, and functions as an important commercial trading hub in the Mediterranean. Catalonia is also the center of industrial activities in Spain, whose industrial sector's GDP of 40.6 billion Euros contributed 23.9% to the GDP of the Spanish industry, and where 20% of all Spanish industrial firms are located[32] (see also Figures 6.7 and 6.8; INE 2010a: 39). Within Spain, the region's economic structure is still characterized by a comparatively large share of industry in the region's GDP (18.9%). However, like in the rest of the country, services are dominating clearly Catalonia's economic structure with a share of 62.2% in the region's GDP, while the agrarian sector has a share of 1.02% and construction 9.5%[33] (Figure 26).

Fig. 30: Regional Distribution of Industrial Firms in Spain

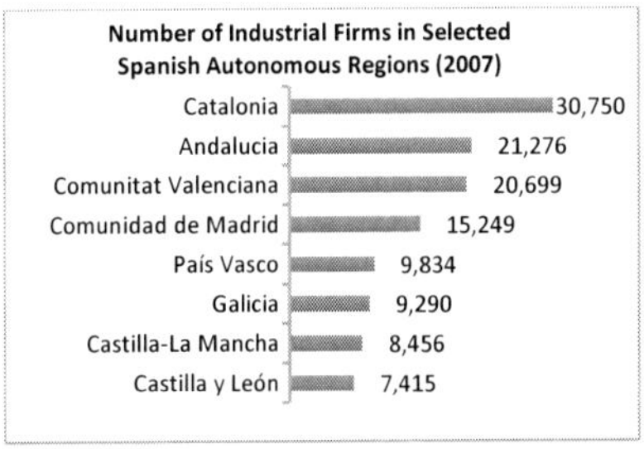

Source: Own Illustration based on INE (2010): 39

Fig. 31: Regional Distribution of Industrial Activities in Spain

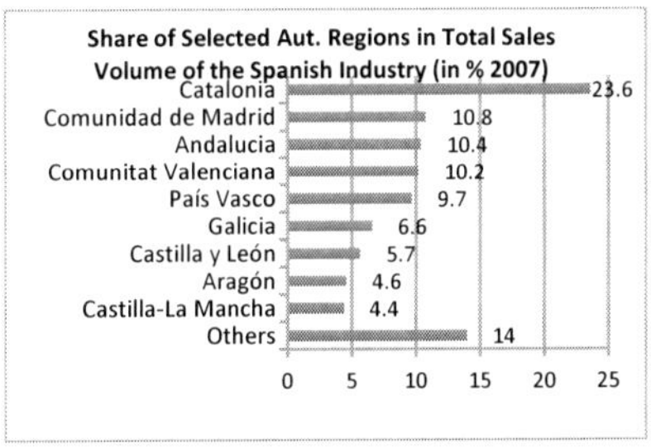

Source: Own Illustration based on INE (2010): 39

Catalonia's industrial structure is characterized by a high degree of diversity, with no industrial sector accounting for more than 17% of the Catalan industry's total sales volume. The region's most important industries are food processing and luxury foods with a share of 16.5% in the regional industry's total sales volume, chemicals (16.2%), automotive (10.9%),

32 Own calculations based on data from Idescat 2011, data for 2008

33 Own calculations based on data from Idescat 2011, data for 2008

energy and water supply (9.4%), the metalworking industry (8.4%), electrical engineering (6.9%), paper and printing (4.6%), mining and petroleum (4.2%), mechanical engineering (4.0%) and the textile industry (3.9%)[34] (Figure 32). When looking at the technology level of Catalonia's manufacturing sectors, it can be stated that it is more technology-intensive than the rest of Spain, but ranks about average when compared to the OECD regions in general. Only 4.5% of the total employment in the Catalan manufacturing sector is in high-technology, but 32.1% in medium-high technology and 63.5 % in medium-low or low technology, which predominate clearly the region's manufacturing activities (OECD 2010c: 63f.).

Fig. 32: Catalonia's Industrial Structure

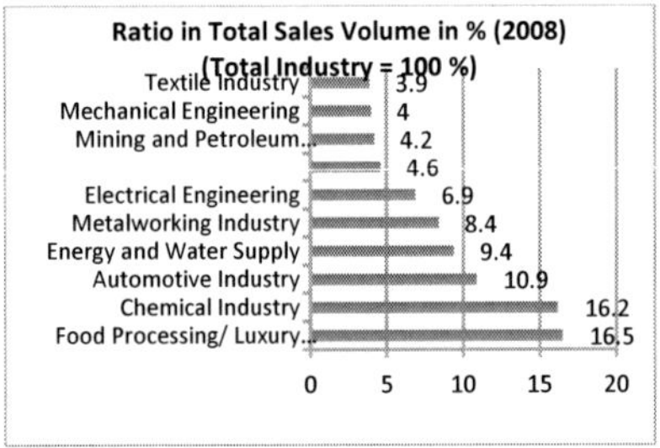

Source: Own Calculations and Illustration based on Data from Idescat 2011

Fig. 33: Regional Distribution of R&D-Expenditures in Spain

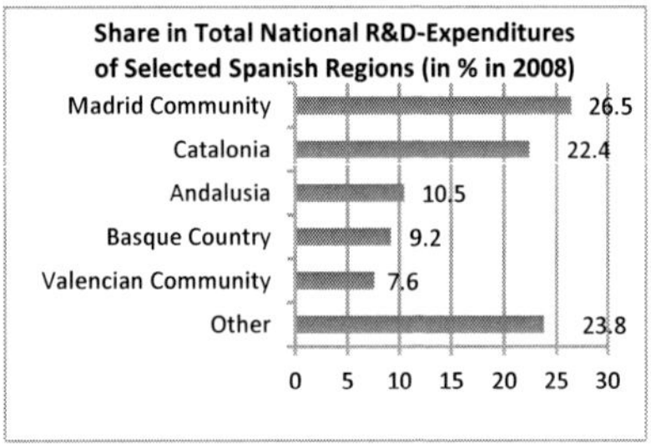

Source: Own Calculations and Illustration based on COTEC (2010): 26

Catalonia's business structure almost equals its national Spanish counterpart, especially in being strongly dominated by (mostly family-owned) micro firms with less than nine employees, which represent 93.1% of the total number of firms in the region. Within the group of micro-firms those firms stand out that have no person employed: They represent 55.9% of the micro firms in the region and 52.1% of the total number of firms in Catalonia. Small firms with ten to 49 employees represent 5.5% of all Catalan firms, medium sized firms from 50 to 199 employees 0.98%. Only the value for large firms with more than 200 employees (0.4%) almost doubles the value for Spain (0.2%)[35]. In this, firms without employees represented 10.8% of Catalonia's workforce, the micro-firms up to ten employees employed 20.0% of the region's total workforce, while small, medium (50–249

34 Own calculations based on data from Idescat 2011, data for 2008

35 Own calculations based on data from Idescat 2011, data for 2008

employees) and large firms (with more than 250 employees) employed 22.7%, 18.8%, and 27.7% of Catalonia's total workforce, respectively (Figures 34 and 35; PIMEC 2010: 19).

Fig. 34: Catalonia's Business Structure I

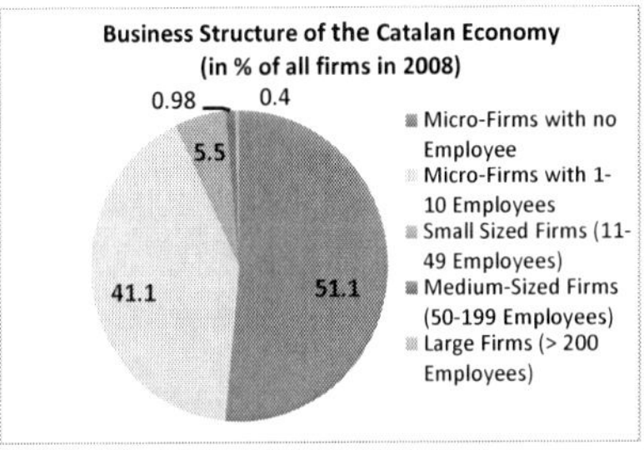

Source: Own Illustration and Calculations based on data from Idescat (2011)

Fig. 35: Catalonia's Business Structure II

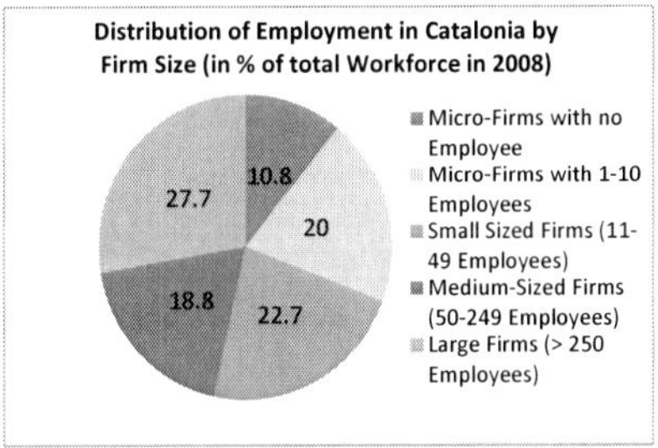

Source: Own Illustration based on PIMEC (2010): 19

Another important structural indicator describing a region's knowledge base is its R&D-activities. In 2008, Catalonia had total R&D-expenditures of about 3.3 billion Euros. This represents 22.4% of Spain's total R&D-expenditures and is the second highest contribution among the Spanish regions, after the Madrid Community (26.5%) (COTEC 2010: 26). Catalonia's R&D-intensity (total R&D expenditures as a percentage of the regional GDP) is of about 1.61%, and is above the national average (1.35%), but ranks behind the Autonomous Regions of Madrid (2.0%), Basque Country (1.96%) and Navarra (1.94%) at the fourth place (COTEC 2010: 25). 61.1% of these total Catalan R&D-expenditures were covered by the business sector and 38.9% by the public sector[36], giving the business sector's R&D-activities in Catalonia a stronger position than those at the national level (54.9%) (COTEC 2010: 246).

Basic data for a sectoral perspective on the business R&D-activities in Catalonia, as well as the distribution of these activities by firm size are not available. Nevertheless, a recent study of ACC1Ó (2010) on the R&D-activities of the region's 50 largest companies reveals that most of Catalonia's key industries are also on top in terms of their R&D-activities: Measured by their share in the total R&D-expenditures conducted by Catalonia's 50 largest companies, automotive (43.5%) and the chemical and pharmaceutical industry (39.0%) are the most research-intensive sectors in this

36 Own calculations based on data from Idescat 2011, data for 2008

sample (ACC1Ó 2010: 44), and should also be considered to be the most R&D-intensive sectors in the region (OECD 2010c: 82).

6.2.1.2 Catalonia—The Center of Spanish Automotive

A focus on Catalan the Catalan automotive industry reveals that it is the third most important industrial sector in the region, representing a share of 10.9% (15.7 billion Euros) in the Catalan industry's total sales volume in 2008[37]. Also in a national comparison, Catalonia is the largest contributor to the Spanish automotive industry among the Autonomous Communities, with a share of about 22% in the total sales volume (INE 2010a: 40). The sector consisted of 793 firms, representing about 1.6% of all industrial firms in Catalonia. 84.2% of these firms are small firms with less than fifty employees, while 10.1% have between fifty and 249 employees. 5.6% of the firms in Catalan automotive have more than 250 employees. This demonstrates that in Catalan automotive medium sized and large firms are considerably more important than in the Spanish economy as a whole. In addition, the sector employs more than 51,000 persons, representing about 24% of the employees in Spanish automotive, and 9.2% of the total industrial employment in the region.[38] When looking at the persons employed in the distinct firm-size classes, this is further underlined: The large firms employ about 65% of all employees in the sector, the medium sized firms about 20% and the small firms about 15%[39] (INE 2010a: 138f.).

Catalonia's automotive activities are mostly concentrated in and around the metropolitan area of Barcelona, which constitutes the main local production cluster sector. Like the rest of Spain, the region lacks its own OEMs, but is a production location for many internationally renowned OEMs. The most important of these production facilities in terms of sales volume and employment are those of SEAT (as part of the *Volkswagen Group*), *Nissan Motor Ibérica*, *IRISBUS* (Iveco), and *Daimler España* (GenCat 2010; Fira de Barcelona 2009: 5; AXESOR 2009: 16f.).

37 Own calculations based on data from Idescat 2011, data for 2008

38 Own calculations based on data from Idescat 2011, data for 2008

39 Own estimation based on INE (2010): 138f.

6.2.1.3 Catalonia's Young Biotechnology Sector

As in Spain, Catalan biotechnology is a very young sector, with most companies and research centers created after 2000, fostered by strong intervention of the regional and national governments. With that, Catalan biotech firms are on average also considerably younger than their counterparts in the German regions, which mostly emerged in the 1990s. In 2008, Catalonia's biotech sector consisted of 215 firms in biotechnology related fields, of which 65 were dedicated biotech firms. All Catalan biotech firms together created a total sales volume of 160.3 million Euros, representing 22.7% of the Spanish biotech industry's total turnover, and about 0.11%[40] of the total sales volume of the Catalan industry. These firms employed around 1,200 persons, representing 28.3% of the total workforce in biotechnology in Spain (BioCat 2009: 24). Also in Catalonia the business structure of biotechnology is predominantly shaped by small firms: More than 70% of the Catalan biotech firms are small companies with less than 50 employees, of which almost 60% are micro firms with less than ten workers. Only 5.5% of all firms are medium sized with 50 to 100 workers. However, a considerable number of 24% of the Catalan biotech firms have more than 100 employees[41] (BioCat 2009: 103).

Regarding the geographical distribution of biotechnology in Catalonia, one can detect clearly a strong concentration of business activities within the metropolitan area of Barcelona. Apart from that, also some biotech firms are located in Girona and Reus. In this, the Catalan biotech sector is mainly specialized in the field of red biotechnology, with a strong focus on drug discovery, drug development and commercialization activities. In addition, also the field of medical technology constitutes a main field of activity in the sector (BioCat 2009: 79ff.). Overall, a large share of the activities in Catalonia's biotech sector are coordinated by the central cluster organization *BioCat*, founded in 2006, which acts, supported by the regional government, as a sectoral management and development agency (BioCat 2011).

In order to evaluate the R&D-activities in Catalan biotechnology, and following the *BioCat 2009 Report* (BioCat 2009: 103), a measure for the R&D-personnel-intensity (number of R&D-workers as a percentage of the total) will be used. To this, a base 1 ratio is established on a biotech-firm

40 Own calculation based on data from Biocat 2009 and Idescat 2011, data for 2008

41 Own calculation based on BioCat 2009: 103

sample allowing evaluating the weight of R&D-employees in each firm-size-category. As a result, it becomes apparent that companies with less than 5 employees have a ratio of 0.95 to 1 in R&D-personnel, demonstrating that nearly all employees in these small firms carry out R&D-activities. A wider variation is found in companies with 20 to 100 employees, which have ratios of between 0.5 and 0.95. Finally, most companies with more than 100 employees have a 0.5 to 1 ratio of R&D-personnel, indicating that in these larger firms also a large share of other than R&D-activities are conducted and that these firms might be less research intensive (BioCat 2009: 103).

As already discussed before, a large share of business activities in biotechnology, and in particular in the fields of red and medical biotechnology, is related to the development of new products, and especially new pharmaceutical products, requiring long pre-clinical and clinical test phases (phase I—III) of up to ten years or more, before a new pharmaceutical is approved for commercial distribution. In this connection, and referring again to the already mentioned *BioCat 2009 Report* (BioCat 2009: 92), Catalan biotech companies have considerable strengths in the discovery and pre-clinical phases of product development, with a share of 47.1% of companies being active in these phases. Much lower are the rates of firms being active in the later phases of product development: 12.4% of the firms are active in phase-I clinical tests, while in phase-II and phase-III 10.5% of the firms are conducting clinical tests, respectively. Finally, 19.6% of the firms are active in the phase of commercializing pharmaceutical products[42]. This points to both the very young demography and the immaturity of the biotech firms in Catalonia, as well as to the comparatively strong role of larger pharmaceutical oriented biotech firms in the region (BioCat 2009: 92ff.).

6.2.2 Structures and Mechanisms of Knowledge Transfer in Catalonia: Six Governance Dimensions

6.2.2.1 Networks as Mechanisms of Knowledge Transfer

Turning to the results of the empirical qualitative case study of Catalonia's knowledge transfer system, the first dimension of knowledge transfer mechanisms that has been investigated in this connection is the network di-

42 Own calculation based on BioCat (2009): 92

mension. In this regard, the theoretical assumption derived from the literature on VoC and NIS was that in MMEs like Spain horizontal business networks, informal networks, institutionally stabilized networks like business clusters and vertical networks between firms and their customers and suppliers should be of a medium range importance for knowledge transfer activities. Moreover, both cooperative inter-firm R&D-projects and cooperative R&D-projects with scientific institutions and universities should not be important as knowledge transfer mechanisms.

Horizontal and Vertical Networks

When evaluating the expert interviews on Catalonia, these theoretical assumptions have only partly been confirmed. Business cooperations in general, and especially *horizontal cooperations* among competitors, albeit in a bilateral way or within *institutionally stabilized* consortia, were reported to be very rare in Catalonia (Interviews CAT 1: 2f.; CAT 2: 5ff.; CAT 3: 2ff.). Therefore, these cooperations would only play a very limited role in knowledge transfer processes in the region. If firms in Catalonia would cooperate, they would do this in an *informal*, unstable and project-related way, and within *vertical supply chain networks* between firms and their customers and suppliers, although also in this context only to a limited extent and strongly dependent on the sector. Automotive would be the most important example in this connection, since it is one of only a few sectors in the region in which integrated value added chains would exist, as will be discussed in more detail below. This overall low propensity to cooperate with other firms has been evaluated by all interview partners as a distinctive characteristic of the region and as a major shortcoming in Catalonia's innovation system, since it would impede flows of knowledge and larger innovation projects within the business sector. As one major reason for this unwillingness of Catalan firms to cooperate an individualistic business culture has been put forward by the interview partners. As the director of the department for technology transfer in the *Fundació Bosch y Gimpera* explains:

"There is very few cooperation. We do not have this cooperative culture here in Catalonia. It is a cultural factor that the firms enclose themselves and are more inward oriented, and that maybe cooperation is not what they are doing. It would be desirable they would do so, since we have a lot of SMEs here, but they barely cooperate. [...] No, [there are no legal barriers for that; MO], it is just the way people are and it is cultural. They are a bit uncommunicative and individualistic." (Interview CAT 1: 2; translation: MO)

Another interview partner from the *ACC1Ó* agency agrees with that and highlights the uncooperative attitude of firms in Catalonia as a specific and differentiating feature in comparison to other Spanish regions:

"There are no laws that specifically favor or impede cooperation. In Spain, e.g. in the Basque Country, people cooperate a lot, there are many cooperatives and they have a cooperative attitude. In Catalonia they do not cooperate and we have the same laws here. [...] By the way, in Madrid they do also not cooperate. Therefore, it cannot be the legal framework, since it is the same everywhere, this factor cannot be relevant. What is left are cultural factors, mentality and the business structure." (Interview CAT 2: 6; translation: MO)

Especially this last factor pointing to the atomistic business structure has been emphasized by all interview partners as an explanation for the low relevance of business networks in knowledge transfer activities in Catalonia. As the interview partner continues:

"We have a lot of SMEs here, but there are no large firms. There are no large corporations like in Germany, as *Siemens* e.g., generating a high demand of projects and provoking outsourcing activities which again promote cooperations. We do not have this. We have a highly atomized business structure, which is, moreover, not focalized on one concrete sector, but on all the sectors. This means, we have a lot of firms and they are active in all sectors. This impeded the creation of more cooperative structures. And vice versa, perhaps because of cultural factors, we have firms that do not cooperate, so that now we have a lot of small firms in all sectors that do not group together and grow." (Interview CAT 2: 6; translation: MO)

As in many other fields of the Catalan innovation policy, also in this context the regional government would intervene with specific programs to overcome market shortcomings and to enhance the creation of business networks in Catalonia (Interviews CAT 2: 2ff.; CAT 3: 3ff.; CAT 4: 3). The most important of these regional programs to foster the creation of business networks would be the *NUCLIS* program (*Nuclis d' Innovació Tecnològica*). This program also has the aim to help the predominantly small firms in the region to access funding from the national or the EU-level, which often require certain minimum volumes in terms of project sizes, which the single firms would not reach alone, as the director of the technology networks department at the *ACC1Ó* agency explains:

"We provide incentives to the firms to cooperate [...]. We have created this *NUCLIS Program*, through which we provide a certain quantity in form of a subvention for a project, half a million or one million Euros, given that the firms present projects double as large, i.e. one or two million, in which firms participate, what is obligatory. [...] We want this to be the first phase of posterior projects to help also these groups of firms to apply for European projects, like the *Seventh Framework Programme*. Since our firms have strong difficulties to cooperate, we start giving them incentives to group together in projects, which from an European point of view might be small, but from our point of view are large, so that from these nuclei they pass to further phases applying for larger European projects in the *Seventh Framework Programme* or in Madrid, which offers the *CENIT Program* (Consorcios Estratégicos Nacionales de Investigación Tecnológica; MO) which are larger projects up to 15 million Euros to 20 million." (Interview CAT 2: 5f.; translation: MO)

Joint Inter-Firm R&D-Activities/Cooperations of Firms with Academia

These innovation nuclei would be promoted in almost all sectors, including biotechnology and automotive. Moreover, the *NUCLIS* program also tries to improve cooperation among firms in *joint R&D-activities* and *cooperations of firms with scientific institutions and universities*, which were both reported to be traditionally low in Catalonia (Interviews CAT 1: 2f.; CAT 2: 2ff.; CAT 3: 3ff.). Many factors have been brought up to explain this low propensity to cooperate in R&D-activities: First, again, the low overall cooperative business culture in Catalonia also impedes cooperation in R&D-activities. Second, the Catalan economy is predominantly characterized by medium-low and low technology sectors (see 5.3.1), which are less R&D-intensive and innovation-oriented, so that the necessity to cooperate with other firms in R&D-activities, as well as the demand for cooperation with universities and research institutes would be low. Third, this comparatively low business R&D-activities in Catalonia and the low tradition of firms to cooperate in R&D-activities among each other or with the scientific world would be rooted in the historical economic and industrial development of the region, as one interview partner remarks:

"Unfortunately, Catalonia has no real tradition in innovative activities. For many decades, like the rest of Spain, Catalonia was not exposed to international market competition, due to the exclusion of the Spanish economy from international markets and its more inward oriented economy. Most of the Catalan firms only produced goods for the Spanish markets, and the lack of competition did not motivate them to invest in their technologies or processes, to innovate or to conduct own R&D-activities. If there were innovations in the Catalan economy, they have mostly been introduced from abroad, or by foreign firms using the region as their production location. Catalan firms never had the necessity to innovate, and they never had the necessity to cooperate in their innovative activities between each other. They just never learned to learn. That is also the reason why for many firms contacts to universities and research institutes are not part of their everyday business activity." (Interview CAT 1: 2f.; translation: MO)

In this connection, also the pointed separation of the Catalan business sector and the scientific world has been highlighted. Distinct logics of actions, goals and worldviews, as well as mental barriers rooted in the different socialization of the actors from the two worlds would hinder cooperation and knowledge transfer (Interviews CAT 3: 4; CAT 2: 5ff.). Especially the lack of academic personnel in the predominantly small firms in the region would constitute an important barrier to interactions between the two worlds. This would also explain the incongruence between the comparatively well performing Catalan scientific institutions and the only little innovative business sector, in which the results of academic R&D-activities are not sufficiently reflected (see also OECD 2010c; OECD 2010d). In addition, one interview partner from the *ACC1Ó* agency accentuates the structural barriers within the Catalan higher education system in this connection, which would not motivate Catalan researchers to cooperate with the business world:

"The ratio of publications in the rankings is high, we have the capacities to attract researchers from abroad to conduct research here in Catalonia. And the capacity of Catalan researchers to publish in international journals is high. However, we occupy the last position in the ranking on patents. […] The explanation of this is structural. […] 20 years ago, the main goal was to improve the quality of teaching. Spain came from a dictatorship in 1975 and the universities were opened to the masses, more and more people came to the universities and the quality of teaching was feared to decrease. Therefore, the main goal was to maintain the quality of teaching. […] This was paralleled by increasing research activities as the second mission of the Spanish universities. And the way to create a sufficiently large critical mass of teaching and investigation was to put a lot of money into projects, but also into the careers of the researchers, so that their different career steps and their professional incentives were only determined by the publications they made. […] In their careers there still are no incentives to engage in technology transfer. If these persons work with firms additionally, it has no consequences for their careers. […] We have a scientific sector which is at the same level than those of the other developed countries. The enduring challenge are the firms generating the money and the wealth, which in Catalonia are still little innovative." (Interview CAT 2: 2f.; translation: MO)

In sum, it can be stated that the barriers to knowledge transfer through cooperations between the business sector and the scientific world are located in both the demand *and* the supply side of scientific knowledge. To overcome these barriers in the Catalan innovation and knowledge transfer system, the Catalan Government has developed an innovation policy instrument named *TECNIO*. In a first step, this is a quality label awarded by the Catalan Government to those university institutes, research groups, scientific centers and institutes which already cooperate with firms or are motivated to do so, so that their visibility for the firms is increased and cooperation between the two worlds is eased. In further steps, this program tries to professionalize interaction between firms and researchers. On the one hand, the *TECNIO*-centers are provided with assessment by the Catalan Government's regional development agency *ACC1Ó* in questions concerning cooperation with firms. On the other hand, it provides funding for both a proof of concept program, and a business start-up program to foster knowledge transfer from the scientific world. Moreover, and addressing the business world, funding to engage external consultancies in initial and early stages of cooperation with the scientific world is provided to the firms (Interview CAT 2: 2ff.).

Networks in Catalan Automotive

When focusing on the network dimension of knowledge transfer mechanisms in *automotive*, most of the points discussed for Catalonia as a whole have also been confirmed for knowledge transfer processes in this specific sector. Horizontal business networks between competitors, cooperative inter-firm R&D-projects and cooperations between firms academia were reported to only play a very limited role for knowledge transfer processes in the sector (Interviews CAT 3: 2ff.; CAT 1: 7f.). In this, also for automotive a low propensity to cooperate and transfer knowledge, an individualistic business culture and mental barriers between the automotive firms and the scientific world have been named as major reasons. The facts that all major OEMs (and also many larger multinational supplier firms) in the sector conduct broad parts of their R&D-activities outside of Catalonia, and that the region is, above all, used as a production plant for constructing vehicles, reinforces this effect. In addition, also a lower congruence between scientific institutions and the business sector would represent an important differentiating factor in comparison to the German regions, and in particular to Baden-Württemberg, as an interviewed expert for the automotive industry in Catalonia remarks:

> "As a former technology assessor for the automotive industry in Catalonia at *CIDEM* [Centro de Desarrollo Empresarial; MO], I had the opportunity to visit the Stuttgart Cluster, where the automotive firms around Mercedes, and the production plants of Mercedes and Bosch are located. There I really got the impression that in this sense they are very different in comparison to us, since there you could observe that the whole scientific and transfer system was strongly oriented towards generating knowledge and technologies for the firms, which at the end are the clients of this type of research centers and universities, etc. Here in Catalonia, with regard to this orientation, we still have a long way to go." (Interview CAT 3: 4f.; translation: MO)

However, in contrast to the region as a whole, vertical cooperations between firms and their customers and suppliers were reported to be more important, since automotive would be one of only a few sectors in the region in which integrated supply chains would exist. In particular, these vertical cooperations would exist around the two large production plants of SEAT and NISSAN (Interviews CAT 3: 2ff.; CAT 1: 7f.). Nevertheless, also in this point the peculiarities of Catalan automotive sector limit the knowledge transfer activities within these vertical cooperations. Referring to the example of SEAT, the strong competition on prices imposed by the *VW Group* would reduce the incentives to cooperate within the value added chain, and would also make cooperations instable and short term oriented, what would impede continued and stable flows of knowledge. Also the lack of strategic decision making capacity in the region's automotive sector would hinder stronger cooperation in this context, as another expert for the automotive industry explains:

> "SEAT has a development center [here in Catalonia; MO], which is a small center and which depends again on Germany. This means, we do not have large firms, we have some larger production plants, whose decision making capacities are located in other parts of the world. And that has the consequence that neither the R&D-activities are concentrated within these larger production plants, nor that they are capable to structure very strong cooperation chains." (Interview CAT 2: 7; translation: MO)

Therefore, the interviewed experts explicitly highlight that in terms of stability and intensity the supply chain networks in Catalonia would be by no means comparable to their German counterparts (Interview CAT 3: 5). However, also the automotive industry is addressed by initiatives of the Catalan Government to improve cooperation and knowledge transfer activities. Especially the recent establishment of the *ACC1Ó* agency as a sectoral cluster organization has been mentioned in this connection, but also the already discussed *NUCLIS* and *TECNIO* programs would exercise positive effects on the sector (Interview CAT 3: 3ff.). Overall, these results point to a strong role of top-down action by the regional government in initiating cooperation in the Catalan automotive sector, while bottom-up cooperative activities only play a very limited role.

Networks in Catalan Biotechnology

In Catalan biotechnology, on the other hand, the network dimension of knowledge transfer is strongly shaped by the young and unconsolidated character of the sector. Most of the firms in the sector are very young and small in size, so that in many cases cooperative structures are not established yet. This would hold true for both horizontal cooperations between competitors and vertical cooperations between firms and their customers and suppliers (Interviews CAT 4: 2ff.; CAT 1: 3). If firms would cooperate in these initial phases of their entrepreneurial activity, it would be more in a project-related, short term and instable way, and, in most cases, base on personalized informal contacts and predominantly include regional cooperation partners. One expert for biotechnology in Catalonia from the *BioCat* cluster organization illustrates this and also points to the active role that *BioCat* plays in initiating cooperation in the sector:

> "No, we do not have institutionalized cooperations yet. This is emerging now, but it is not real yet. Right now these are more consortia which are created on the basis of personal knowledge and affinity between the proprietors, or which are facilitated by us. There is no established network yet. Many times we as a cluster organization exercise this function of connecting the actors, and sometimes they do it themselves, but this happens only in a punctual, and not in an organized way." (Interview CAT 4: 3; translation: MO)

Cooperative inter-firm R&D-projects would have some relevance, although their function as channels of knowledge transfer would be limited in these initial stages of the business life cycle, since these cooperations would mostly consist of the joint use of research facilities or equipment. Furthermore, also the strong IP-protection in the sector would impede an intensive exchange of knowledge in this connection. More important, however, in terms of knowledge transfer would be R&D-cooperations between biotech firms and scientific institutes and universities (Interview CAT 4: 2ff.). Most of the young firms in the sector would have a high demand for the results of research conducted in scientific institutions and would also need to use the respective research facilities and equipment. Moreover, many biotech firms in the region would be spin-offs from the Catalan universities and research institutes and would still maintain close personal linkages to their former institutes easing cooperation. Many of these cooperations would be initiated and supported by public policy measures, of both the regional and the national government. The most important regional government measures in this regard would be the assessment activities of the *ACC1Ó* agency, the support of the central cluster organization *BioCat*, as well as the already mentioned *NUCLIS* and *TECNIO* programs. In addition, the Spanish Ministry of Science and Innovation would provide support for larger innovation projects in consortia through the *CENIT* programs (Interview CAT 4: 3; see also BioCat 2009: 96).

6.2.2.2 Association Based Mechanisms of Knowledge Transfer

The second dimension of knowledge transfer processes is represented by the association based mechanisms, namely employers-, business-, and trade associations, chambers of industry and commerce, and binding and common technical standards. From the literature on VoC and NIS the theoretical assumptions have been derived that all the three of them should play a medium range role for knowledge transfer processes in the Spanish regions. These assumptions have not been confirmed by the interviewed experts when looking at Catalonia as a whole.

Employers-, Business-, and Trade Associations/Chambers of Industry and Commerce

All interview partners consistently reported that they would not attribute any significance to *employers-, business-, and trade associations*, as well as *chambers of industry and commerce* in knowledge transfer processes (Interviews CAT 1: 3f.; CAT 2: 8f.), and, therefore, evaluated them clearly as less important as assumed for the Spanish case. The reasons for this would be threefold: First, the distinct business-, employers- and trade associations would mostly be nationally organized and would often lack a specific regional focus and concentrate their activities more on interest representation towards the national government and less on innovation and knowledge transfer. Second, the different local chambers in Catalonia would not be sufficiently coordinated on the regional level and follow more individual and local interests, instead of developing a coherent perspective on Catalonia as a whole in their different fields of action. In addition, they would also often lack efficient internal organization structures and qualified staff to be more active in the fields of innovation and knowledge transfer. Third: For both regional and national business associations, like *PIMEC* and *Fomento*, and the chambers, from a historical perspective the fields of innovation and knowledge transfer would have never been part of their core activities, as one representative of the *ACC1Ó* agency explains:

> "Normally, the role exercised by the employers' associations has not been the one of innovation. Their main mission has basically been to defend their interests towards the administration. [...] Therefore, the relation between the associations and the administration mainly consists in the negotiation of norms and laws, and in collective agreements with the labor side. But they have never had the role to promote cooperative innovation, nor to promote the relations between firms and universities. The second function they exercised, and this points more probably to the chambers, was fostering exportation. They did not foster innovation, but trade fairs, congresses, etc, i.e. the searching for groups of firms to open up new markets in the world. [...] As a consequence, there has been a deficit in who would organize the firms in promoting innovation." (Interview CAT 2: 8f.; translation: MO)

Even today, associations and chambers would still not exercise an active role in the fields of innovation and knowledge transfer, making that their services like internet platforms, congresses and conferences would not focus on innovation and technology relevant topics and would therefore not be frequently used as mechanisms of knowledge transfer by the firms (Interview CAT 2: 9). Moreover, and focusing more on the demand side, one interview partner from the *Fundació Bosch y Gimpera* remarks that also the firms would not actively request the associations and chambers to be more active in these fields:

> "I think it is more likely a problem of the firms. If the firms would ask for it, the business- and employers associations would have to do it. If the business associations do not do it, it is because their members are not asking and demanding for it. If the firms themselves are not into these topics and do not have them as their priority, the others will not pick it up." (Interview CAT 1: 4f.; translation: MO)

Associations and Chambers in Catalan Automotive

One exception in this context would be automotive. In this particular sector, the (again mostly nationally organized) sector specific associations like *STA* (*Sociedad de Técnicos de Automoción*), *SERNAUTO* (*Asociación Española de Fabricantes de Equipos y Componentes para Automoción*) and *ANFAC* (*Asociación Española de Fabricantes de Automóviles y Camiones*), would play a more active role in the fields of innovation and knowledge transfer and promote cooperation, as one expert for the Catalan automotive sector illustrates:

"These associations at least try to establish some initiatives, i.e. they act as agents who try to dynamise a little this type of market. Not only the associations I already mentioned, but also other associations which are active in automotive related technologies are active in these fields. E.g., there is a technological center of an association named *Asociación Catalana de Empresas Constructoras de Moldes y Matrices.* This association is very active in the field of automotive and as an association created a technological center with a high impact in the sector." (Interview CAT 3: 6; translation: MO)

In addition, also other activities offered by these associations would be used by the sectoral actors like, e.g. internet platforms, congresses and conferences (Interview CAT 3: 6f.). Pointing to the role of the chambers in automotive, they would be less active in the fields of innovation and knowledge transfer than the associations. However, as one of the region's most important industrial sectors the chambers would have a special focus on the development of automotive, but rather in other fields than innovation and knowledge transfer, so that their influence in this regard would be, if anything, indirectly (Interview CAT 3: 7).

Associations and Chambers in Catalan Biotechnology

In biotechnology, on the other hand, business associations would only play a very limited role. The only established sectoral business organization would be *ASEBIO* (*Asociación Española de Bioempresas*) which has a national focus and is therefore perceived as distanced within the sector. On the regional level, the Catalan biotech business association *CataloniaBio* has been founded in 2007, which would still be in a process of finding its role within the sector. Both associations would offer working groups, conferences, meetings and discussion forums to the regional actors, but only to a very limited extent, not least because of the small size of the sector. More important in this regard would, therefore, be the services offered by the regional cluster organization *BioCat* (Interview CAT 4: 4). Through this cluster organization also the chambers intent to offer their services to the Catalan biotech firms. They recently (2009) established a cooperation agreement with *BioCat* and start to participate in specific projects within the sector. However, also their engagement has been evaluated as rudimental in the fields of innovation and knowledge transfer, and they would not yet have recognized the young biotechnology sector as one of their main fields of action (Interview CAT 4: 4f.).

Finally, no one of the interviewed experts reported on the role of *common and binding technical standards* in knowledge transfer processes.

6.2.2.3 Organization Based Mechanisms of Knowledge Transfer

As organization based mechanisms of knowledge transfer in regional innovation systems the transfer of technical and scientific staff, purposeful headhunting and business start-ups have been identified. Building on the literature on VoC and NIS it has been assumed that all of these mechanisms should play a medium range role for knowledge transfer processes in the regional innovation systems of Spain's MME. These assumptions have basically been confirmed by the interview partners when looking at Catalonia as a whole.

Transfer of Scientific and Technical Staff

Concerning the *transfer of scientific and technical staff*, the interview partners stated that in Catalonia there would be dynamic transfer activities in both dimensions, between firms and between firms and scientific institutions (Interviews CAT 1: 6f.; CAT 2: 13). However, these transfer activities would be strongly dependent on the sector, and would increase with the technology level the respective firms would operate in, as well as with firm size. As one interview partner accentuates, one important source for these transfers of qualified academic staff are those scientific institutes in Catalonia wearing the label of and participating in the *TECNIO* program. These institutes would be the ones maintaining continuous relations to the business sector, above all through joint R&D-projects:

"The firms which continuously work in projects with these 100 research groups and scientific institutes, […] once they have seen the technical and scientific staff, are often interested to incorporate staff which worked in their projects into their firms. This means that there is a continuous flux of personnel from these 100 *TECNIO* institutes to the firms, because these firms already know them from joint projects. […] One part of the personnel at these institutes are researchers which are doing their doctoral studies, and they come and go. They come to these institutes, spend some time there and then they go to the firms. This generates a continuous flux of technical personnel and is, therefore, one form of knowledge transfer to the market. However, this is only one fifth of the 500 research groups and institutes we have at the universities. This means, that this only happens in 20% of the research groups at Catalan universities. In the other 80% this practice does not happen in a habitual form, but, if anything, only punctually." (Interview CAT 2: 13; translation: MO)

This demonstrates again, that also in the field of the transfer of staff, public support measures like the *TECNIO* or *NUCLIS* programs are important (indirect) initializing preconditions, while purely market driven dynamics are low. Therefore, transfer activities are also explicitly supported, above all by the *Torres Quevedo* program at the national level, and by the *TALENT* program at the regional level, which both provide funding to firms employing doctoral students or other scientific and technical staff to work on innovative projects in Catalan firms (Interview CAT 1: 6f.). However, the scope of these programs would still be limited to innovative firms, as one interview partner points out:

"[These programs; MO] are only starting now, and it is hard for the traditional sectors. Right now, these programs are more requested by the technology intensive firms, the spin-offs from the universities or the technological centers. […] The traditional sectors are not yet entering in such systems, because they do not know what to do with a doctor or a technician. Since these firms do not have innovation as one of their priorities, they do not use them." (Interview CAT 1: 7; translation: MO)

Purposeful Headhunting

Purposeful headhunting, in turn, would be a naturally and frequently, although not dominantly used mechanism to broaden up the firm's knowledge base (Interviews CAT 1: 7; CAT 2: 14).

Business Start-Ups

Finally, also with regard to *business start-ups* and *academic spin-offs* as mechanisms of knowledge transfer the experts draw a differentiated conclusion

(Interviews CAT 1: 7f.; CAT 2: 14ff.). On the one hand, there would be considerable business start-up activities in the region, which is not least manifested in the large share of micro- and one person firms discussed in the previous sections. A lack of entrepreneurial mentality, like in the two investigated German regions has, therefore, not been identified for Catalonia. However, most of these business foundations would not take place in technology relevant fields, and would, therefore, not contribute to the transfer of technologically relevant knowledge to the market. On the other hand, also academic spin-offs would have become more numerous in the last years, especially because of extensive public support, but, as the director of the technology networks department at the *ACC1Ó* agency illustrates, only with low success rates:

"We created the technological business start-up program. As a consequence, between 2001 and 2008 300 firms have spun-out these 500 research groups, in which for many academics the way to start working or to enter the labor market was to found a business. But only very few of these firms grow. […] This means, they serve to give employment only to the respective founders, which are engineers or technical staff, but to only very few persons more. It is not typical that these firms would grow and expand and, therefore, would create a continuous flow of engineers and technical staff to these companies. […] The reason for this is that in many cases they do not have a concrete product, but only know-how. These people know to do something and they start as a firm offering services in something. But they do not have a technology which would allow them to create a product that grows." (Interview CAT 2: 14; translation: MO)

Apart from the regional and national governments (not least through the universities and technological centers) also private foundations like the *Fundación Bosch y Gimpera*, which is the technology transfer center of the *University of Barcelona*, would actively promote business start-ups and academic spin-offs, above all through professional assessment, business-plan development and business incubators (Interview CAT 1: 7). Decisively important in this connection would also be the provision of seed-capital by the national and regional governments through programs like *NEOTEC* and *ENISA* (national) and *GENESIS* (Catalan), which provide funding for new firms in initial phases of their business activities. In addition, also business angel networks and venture-capital firms would be active in providing early capital to entrepreneurs in the region, so that, in contrast to the German regions, in these initial phases enough capital would be available for start-up activities (Interviews CAT 1: 7f.; CAT 2: 16). However, even if the Catalan Government itself would invest and hold shares in venture capital firms through its regional development agency *ACC1Ó*, in the subsequent phases of business development especially the provision of venture capital and the investments of the banking sector would be insufficient in Catalonia, as an interview partner from *ACC1Ó* remarks:

"Referring to the venture capital firms it is not that clear if there are as much as would be needed. I.e., while the Government's funding for the initial phases finances the firms sufficiently well, it is not that clear if there are enough venture capital firms to invest in posterior phases and support the growth of these firms. Therefore, there might be a gap between business angels [and venture capital firms; MO], which have a very limited scope, and the banks, which do not take the responsibility. […] The banks do not play a role in this, simply because they did not need it. They invest in the real estate sector. And they invest in loans for already established firms, but not in the initial phases. The banking sector never played this role and is still not doing it. […] The stagnation of growth might also be a failure of the financial system. The government creates the incentives in the first phases and then there are no financial and venture capital firms investing in posterior phases." (Interview CAT 2: 15; translation: MO)

In sum, business start-up and academic spin-off activities in Catalonia would play a role as mechanisms of knowledge transfer, but they would be limited by two major obstacles: On the one hand by the human capital side, since excellent researchers would not necessarily also be the best entrepreneurs. And, on the other hand, by a gap between sufficiently available seed-funding by the government and insufficient venture-capital provision in later phases of business development by venture-capital firms and banks (Interview CAT 2: 16).

Organization Based Mechanisms in Catalan Automotive

In automotive, like in Catalonia as a whole, transfers of scientific and technical staff have been reported to be of medium range relevance as mechanisms of knowledge transfer. On the one hand, there would be a large pool of highly qualified academic and technical staff in the Catalan universities like, above all, the renowned *Catalan Polytechnic University*, the *University of Villaterra* or the *University of Barcelona*, which would serve as recruiting grounds for the automotive firms in the region. On the other hand, and because of the fact that only a few R&D-activities in the sector are conducted within Catalonia, the demand for highly qualified staff and engineers would be much lower than in the home-countries of the OEMs, like e.g. Germany. However, and again confirming the results for the region as a whole, also in automotive purposeful headhunting activities would be conducted frequently (Interview CAT 3: 9f.).

In turn, business start-ups and academic spin-offs would not play a role in the Catalan automotive sector. In a mature and consolidated industrial sector like automotive the creation and the establishment of a new business would be extremely difficult and be impeded by many sector specific barriers, as one expert for the Catalan automotive sector reports:

> "I would say that obviously many start-up initiatives exist here in Catalonia, but only very few go to automotive. [...] Automotive is a very mature sector, in which it is difficult to just generate a spin-off from university and go directly to the sector. That is simply because you would find some barriers there, since finally your customers would be the OEMs or the *tire-one*, which normally have their pool of suppliers, and for being one of their suppliers there is a process, which is a slow process and this already works as a barrier. [...] It is a problem of idiosyncrasy in automotive. If you want to be an OEM's supplier, you need an entry certification, the *QSTS* of quality. [...] Then, if you enter a project, in many cases you carry its risks, since you first have to invest and then you might get paid amortizing the sold pieces. This whole environment works as an entry barrier for recently created firms, which probably are firms with difficult technologies, but in any case with little resources, which is one of the most important factors, at least here in Catalonia. In other words: For creating a business, the main barrier is funding." (Interview CAT 3: 10f.; translation: MO)

Organization Based Mechanisms in Catalan Biotechnology

In biotechnology, the transfer of scientific and technical staff from academic institutions to the firms would mostly be limited to the start-up or spin-off phase (Interview CAT 4: 8). Transfer activities in later phases of the business life cycle would be rare, what could also explain the fact that the majority of Catalan firms remains small in size and does not grow. The main reasons mentioned by one interviewed expert for the Catalan biotech sector would be distinct mentalities between the business and the scientific world, but also regulatory barriers, which would impede the mobility of researchers to the firms. Nevertheless, purposeful headhunting activities would absolutely be undertaken, also in biotechnology (Interview CAT 4: 8). Finally, business start-ups, and in particular academic spin-offs, have been evaluated as vital channels of knowledge transfer in the Catalan biotech sector. As one of the interviewed authors of the *BioCat Report 2009* on biotechnology in Catalonia remarked, Catalonia would be the bioregion in Europe with the highest share of business start-ups among all the Euro pean bioregions (Interview CAT 4: 8f.). This might among other things be explained by the fact that the sector is still very young (younger than in many other European bioregions) and that in an emerging and expanding sector start-up activities are natural processes. In addition, the sector is not yet consolidated, so that closing tendencies like in automotive are not active yet. However, also in the highly capital intensive biotechnology sector venture capital provision in later phases of the business cycle has been reported as a major bottleneck to start-up activities and further growth dynamics of the sector. While there would be enough private and public funding for the initial phases, specialized venture-capital for subsequent phases would often miss (Interview CAT 4: 10f.).

6.2.2.4 State Induced Mechanisms of Knowledge Transfer

The state has been defined to intervene in knowledge transfer processes in regional innovation systems above all through mechanisms such as public programs, differentiated by national or regional measures, through (larger) state-induced innovation projects, the creation of scientific parks and technological centers, as well as the maintenance of technology transfer offices or companies at the universities and public research institutes. Basing on the discussion of the literature on national production and innovation systems, it has been assumed that in the Spanish regional innovation systems all these mechanisms should play a decisive role for knowledge transfer activities, according to the interventionist role exercised by the state in the Spanish MME, compensating for market failures and failures of strategic coordination among the economic actors. While the first five of these assumptions have fully been confirmed by the interviewed experts, ascribing them indeed a central significance for knowledge transfer processes in the region, the assumption concerning the important role of the technology transfer offices at the Catalan scientific institutions has not been confirmed.

National and Regional Government Programs and Measures

In detail, all interview partners consistently ascribed both *national and regional public programs and measures* in the fields of R&D and innovation a highly significant role in initializing, promoting and funding knowledge transfer activities in Catalonia (Interviews CAT 1: 4ff.; CAT 2: 10ff.; CAT

3: 7f.; CAT 4: 5f.). The legal framework for the national Spanish policy in the fields of R&D, innovation and knowledge transfer mainly builds on the *National Plan for Scientific Research, Development and Technological Innovation* (currently in its sixth version 2008–2011), defining the national strategy in these fields and allocating the respective budgets for a four-year-period. This *National Plan* is complemented by the *INGENIO 2010* initiative, which is part of the wider *National Reform Plan* to approximate Spain to Europe's most advanced countries in the fields of science, technology and innovation. The policy measures of *INGENIO 2010* include the promotion of public-private partnerships for innovation through the *CENIT* program, venture capital funds and the *CONSOLIDER, CIBER*, and *TORRES QUEVEDO* programs to increase the country's research and knowledge transfer capacity (for more details on the Spanish R&D- and innovation policy see also OECD 2010c: 217ff.).

However, and despite its actual reorientation towards stronger affords in terms of innovation and knowledge transfer, the *National Plan* would be more important in improving the framework conditions for the R&D-activities in the region, but less in giving impulses for innovation in and knowledge transfer to the firms (Interviews CAT 1: 4f.; CAT 2: 10ff.; CAT 3: 8). More important in this connection would be the measures offered by the *CDTI* (*Centre for the Development of Industrial Technology*), which is a public agency under the *Spanish Ministry of Science and Innovation*, whose mission is to help Spanish companies to increase their technological profiles and innovation activities. In this connection, it administrates some of the public programs to explicitly foster cooperation between firms, cooperation between firms and the scientific world, larger innovation projects (especially *CENIT*) and business start-up activities (*NEOTEC*), of which Catalonia is the largest recipient of funding among the Spanish regions (OECD 2010c: 217ff.).

In this connection, the Catalan interview partners consistently evaluated the *CENIT* program as the most important national measure in terms of knowledge transfer (Interviews CAT 1: 4ff.; CAT 2: 10ff.; CAT 3: 7f.; CAT 4: 5f.). This program offers substantial amounts of funding to consortia of firms (including larger and small firms) and research institutes or universities working together on larger innovative and technology oriented projects and would finance a high number of projects in different sectors in Catalonia as well. In this, Catalonia is also the largest recipient of funding from the *CENIT* program among the Spanish regions (OECD

2010c: 217ff.). Through this program, therefore, cooperation, network-building, joint R&D-activities, knowledge transfer and innovation would actively and successfully be promoted. As one decisive reason for this important function of these public programs in the fields of innovation and knowledge transfer, the interview partners consistently pointed out that both economic and scientific actors per se would have little auto-initiative to cooperate, innovate and transfer knowledge. One interview partner from the *ACC1Ó* agency formulates this as follows:

> "The *National Plan* is very successful among the researchers, the measures of the *CDTI* are very successful among the firms. But pay attention to the fact that the firms only move due to incentives! If there is no money, they do not cooperate. Therefore, this is a good way to start cooperation. […] We created these programs so that they can be one step in the ladder to cooperation in R&D-activities. And I think that internally they were pretty successful. But I am not so sure if they will be similarly successful in going one step further to access European programs." (Interview CAT 2: 10; translation: MO)

While the national level (complemented by funding from the EU-level) would be the decisive level in initiating large scale innovation projects and would provide the large amounts of funding for university-, applied and extra-university R&D-activities in Spain, the financial resources for this type of R&D- and innovation activities available on the regional level would be comparatively small (Interviews CAT 1: 4f.; CAT 2: 10ff.; CAT 3: 8). However, like all other Spanish Autonomous Regions also Catalonia has the responsibility for the higher education system and in particular for the universities, where it also provides funding for R&D-activities. Overall, about 20% of the public expenditures on R&D-activities are carried by the Catalan level (OECD 2010c: 228). The specific role of the regional government would then be more the one of financing research institutes and promoting network activities in the fields of innovation and knowledge transfer, like e.g. through the *NUCLIS*, *TECNIO* and *TALENT* programs. Anyway, these regional government measures would, like their national counterparts, exercise decisive contributions in all stages and levels of knowledge transfer activities in Catalonia (Interviews CAT 1: 4f.; CAT 2: 10ff.; CAT 3: 8).

As the decisive player in the Catalan innovation and knowledge transfer system all interview partners consistently identified the *ACC1Ó* agency, which is the public agency for business development, innovation support and external promotion of the *Catalan Ministry of Innovation* (Interviews

CAT 1: 5; CAT 2: 2ff.; CAT 3: 3ff.). This agency is the result of the recent (2009) merger of the former *Centre for Innovation and Business Development* (CIDEM) and the *Consortium for Commercial Promotion of Catalonia (COPCA)*, now coordinating and centralizing most of the activities of the Catalan Government and its single ministries directed towards innovation and knowledge transfer between the scientific and the business sector (Interview CAT 2: 21). This centralization and concertation efforts within the Catalan regional innovation system have been evaluated as very positive and efficient by all the interview partners, although the concrete effects of this centralization would only recently start to be visible (Interviews CAT 1: 5; CAT 2: 2ff.; CAT 3: 3ff.).

Moreover, like its Spanish counterpart, also the Catalan policy in the fields of innovation and knowledge transfer would show strong elements of central planning by the state. What would be the *National Plan for Scientific Research, Development and Technological Innovation* on the Spanish level, would be the *Catalan Agreement on Research and Innovation* on the regional level. The difference between both would be that the former defines and allocates concrete expenditures, while the latter would only be a strategic commitment defining clear goals in the mentioned policy areas without directly allocating expenditures (Interviews CAT 1: 5; CAT 2: 11). However, given that also the Catalan innovation and knowledge transfer policy shows a highly centralized structure, with the *ACC1Ó* agency as the central player, it can be stated that in Catalonia the state (national and regional, but also the EU-level) is the all decisive actor in a multilevel governance system of innovation and knowledge transfer, basically structuring and promoting knowledge transfer activities in the region top-down, allocating the decisive amounts of funding to it and creating the necessary structures and incentives.

(Larger) State Induced Innovation Projects

This strong and interventionist role of the state in knowledge transfer processes in Catalonia has also been confirmed by the interview partners when asked for *(larger) state induced innovation projects* as mechanisms of knowledge transfer (Interviews CAT 1: 6; CAT 2: 12). State induced innovation projects, as e.g. in the fields of defense or aerospace, in which the state as a contracting body initiates, organizes and at least partly finances innovation activities would be important occasions to interact with other actors and exchange knowledge. In addition, also the *NUCLIS* program at the regional (Interview CAT 2: 12), but above all the *CENIT* program at the national level would show similar effects, since the state would often impose its own priorities in certain sectors when deciding about the approval of funding, as the director of the department for technology transfer in the *Fundació Bosch y Gimpera* explains:

> "Yes, surely [there is knowledge transfer; MO], since in these larger projects there are large and small firms, as well as universities. Therefore, through collaboration and subcontracting among each other technology transfer is realized. We are basically talking about the *CENIT* or *AVANZA* projects which are large projects with a duration of four years and in which many actors collaborate. In the allocation of these funds the government also implements its priorities in terms of innovation in certain sectors, as could be electro-mobility in automotive. But there is also a lot of collaboration in projects in the field of defense industry, in the development of new armament systems, or in aerospace. Also in these cases the state and the respective ministries are the initiators and carry large parts of the funding." (Interview CAT 1: 6; translation: MO)

With regard to the concrete forms these larger state induced innovation projects would take, the interview partners emphasized their formalized and contract based, but temporal and only project-related character. In most cases, the form of consortia would be chosen, but these would neither be institutionally stabilized, as e.g. through a network management organization, nor lead to the creation of stable network relations among the participating actors subsequently to these projects (Interviews CAT 1: 6; CAT 2: 12).

Scientific Parks and Technological Centers

Furthermore, also publicly founded *scientific parks*, mostly connected to the region's universities, and public-private *technology centers* were reported to be central instruments of knowledge transfer in Catalonia (Interviews CAT 1: 6f.; CAT 2: 2ff.). Currently (2008), there are 25 scientific parks located in Catalonia, and 20 technological centers (OECD 2010c: 114ff.). The added value of these entities for the actors in the Catalan knowledge transfer system would be, above all, physical proximity and continuous incentives for stable cooperation structures, as the interview partner from the *Fundació Bosch y Gimpera,* which itself is part of the managing board of the *University of Barcelona's* technology park, illustrates:

"Fundamentally, what they [technological centers and scientific parks; MO] provide are spaces to share installations. E.g., here in the Barcelona scientific park there is the complete pharma-ceutical sector represented, with departments collaborating with researchers in installations they share. This way, the departments of the firms can get many innovative impulses. If they would stay outside, they would not know what is going on. Being here, they share knowledge and staff, etc. The firms participate intentionally in the parks and centers to share know-how and knowledge. That is the main function of scientific parks, to basically work as mechanisms of technology transfer and of bringing firms, technological platforms and research groups together. These are stable structures, which were fostered ten years ago and we are already constructing the second phase of the scientific park." (Interview CAT 1: 6f.; translation: MO)

Technology Transfer Offices

Finally, referring to the role of the *technology transfer offices* at the Catalan scientific institutions in facilitating knowledge transfer between the scientific world and the business sector, the interviewed experts draw a differentiated conclusion (Interviews CAT 1: 8f.; CAT 2: 17f.; CAT 3: 11). On the one hand, in Catalonia, as in the rest of Spain, it would be obligatory for all researchers at public universities and research centers to pass through these transfer offices (*OTRIS—Oficinas de Transferencia de Resultados de Investigación*), since in Spain the results of scientific research are the property of the respective institutions. The OTRIS are, thus, in charge of all legal and contractual transactions connected to cooperation with the business sector and the transfer and commercialization of IPRs. Therefore, as an administrative unit they would play a central role in knowledge transfer processes in the region. On the other hand, the interview partners criticized that they would not be active in bringing firms and researchers together and in initiating cooperation. Moreover, as the director of the technology transfer department in the *Fundació Bosch y Gimpera* adds, the OTRIS should be more proactive in commercializing the scientific institutions' IPRs:

"This refers to the university-industry-relations. What the OTRIS are doing is to bring into order the administration and the contracts. From here on, since we are already doing this for 25 years now, we have to choose for more proactive ways, not just in administrating, but in the three fields I mentioned before: Promoting cooperative projects between firms and the scientific world, finding firms to buy licenses and patents, and finding and assessing researchers to create technology based businesses. This will be the role of the OTRIS in the future. It is also true that many OTRIS remained in being purely and encrusted administrative units. Therefore, they were not proactive enough." (Interview CAT 1: 8f.; translation: MO)

Especially the lack of cooperative attitude of the Catalan firms, and the pronounced mental barrier between firms and the scientific world in Catalonia, often preventing that firms would address the OTRIS to come into contact with the scientific institutes, would make a more proactive orientation of the OTRIS indispensible, as one interview partner from the *ACC1Ó* agency adds (Interview CAT 2: 17f.). In addition, what would be necessary in this connection would be the OTRIS to exercise a "switchboard function", acting as an interlocutor bringing researchers and firms into contact. Presently, this function would still be exercised by the *ACC1Ó* agency: First, through the *TECNIO* program and the corresponding labeling of the transfer oriented Catalan research institutes, centers and groups, making them visible to firms searching for scientific cooperation partners. And second, through the establishment of a *technology assessor network*, consisting of experts acting as intermediary actors in searching for contacts and establishing projects between research institutes, suppliers and firms (Interview CAT 2: 18).

State Induced Mechanisms in Catalan Automotive

These results for Catalonia as a whole were basically also confirmed when focusing on *automotive* (Interview CAT 3: 8f.). Public programs and measures would be decisive in initiating, promoting and financing knowledge transfer activities in the Catalan automotive sector. While the majority of the smaller firms in the sector would profit more from the funding provided by the Catalan Government, especially the two large OEMs in the region SEAT and NISSAN would also access funding from national and European programs. In particular, collaborations in larger projects initiated by the *CENIT* program, but also other programs offered by the *CDTI*, would be important instruments of knowledge transfer in the sector. Also technological centers and scientific parks would be important in the sectoral knowledge transfer activities, especially the sector specific and *TECNIO*-labeled center *CETEMMSA* in Mataró and the *Barcelona Technological Park* in Manresa (Interview CAT 3: 8f.). According to one interviewed expert for the Catalan automotive sector, the OTRIS would not be used intensively in the automotive firms, above all because for them the trade of IPRs would not play an important role (Interview CAT 3: 11). It can be assumed that one explanation for this is the fact that most R&D-activities in the sector are conducted outside of Catalonia.

State Induced Mechanisms in Catalan Biotechnology

Finally, also in the Catalan biotech sector, the state induced mechanisms of knowledge transfer were ascribed a high importance. Both, national and regional, but also EU-level policy measures and support programs would be important drivers of knowledge transfer processes in the region's biotechnology sector. The programs with the strongest impact in the sector would be, again, the *CENIT* program of the *CDTI* at the national, and the *NUCLIS* program at the Catalan level, as well as funding from the EU's *7th Framework Program* (Interview CAT 4: 3ff.). In addition, especially the state's activities in providing seed-capital have been highlighted in this connection, not least to foster start-up and spin-off activities in the very young and still emerging Catalan biotech sector. Nevertheless, one interviewed expert for the sector also highlighted that in later phases of the business life cycle the role of public funding decreases and mechanisms of auto-financing through the market would become more important. However, due to the overall immatureness of the sector this would only apply to a minority of firms (Interview CAT 4: 6).

Larger state induced innovation projects would also play a role for knowledge transfer in the Catalan biotech sector. Programs to create innovation consortia around concrete innovation projects and topics, like the *CENIT* and *NUCLIS* programs, through which the state also imposes some its own priorities to the market actors, would exercise a tractor effect and efficiently generate knowledge transfer among the participating actors (Interview CAT 4: 6). The significance of scientific parks and technological centers has been evaluated as extraordinarily high and vital for the young Catalan biotech sector. One representative of the *BioCat* cluster organization explains this as follows:

"They are vital actors. Without technological centers and scientific parks there would be no BioCluster. They are the interface between basic research and highly innovative technological platforms and research projects allowing small firms to access breaking point technology, and everything connected to assessment and incubation. Therefore, without parks and centers nothing would exist. It is the triple helix, it is indispensible that you have firms, as well as parks, centers, universities, hospitals and organizing bodies. If one of these actors fails, there is no cluster." (Interview CAT 4: 6f.; translation: MO)

In contrast to the region as a whole and to automotive, for the Catalan biotech sector also the OTRIS were reported to be very important in knowledge transfer processes between the academic and the business world. Since the OTRIS would be responsible of all contraction activities in terms of patenting, licensing, joint ventures and spin-offs, and since these would be core knowledge transfer activities in the young and emerging Catalan biotech sector, these transfer offices would be central and indispensible. However, also the expert for the biotech sector in Catalonia pointed to differences in the proactive attitude of the single OTRIS. One positive example in this connection would be the *Fundació Bosch y Gimpera* at the *University of Barcelona*, which developed a proactive strategy in promoting and assessing knowledge transfer activities in the Catalan biotech sector (Interview CAT 4: 9). Furthermore, it might be assumed that the strong ties between most of the biotech firms and their former academic institutes, as well as the specific academic socialization of large parts of their staff, lowers the mental barriers in contacting and interacting with the OTRIS.

6.2.2.5 Market Based Mechanisms of Knowledge Transfer

Market based mechanisms of knowledge transfer have been defined as mechanisms through which knowledge is "bought" and "sold" via the market. For capturing this market dimension of knowledge transfer, the interviewed experts have been asked for the role of external consultancies, mergers, acquisitions and hostile takeovers as well as the buying, selling and licensing of patents in knowledge transfer processes. The assumption derived from the literature on VoC and NIS is that external consultancies and mergers, acquisitions and hostile takeovers should only play a medium range role in the regional innovation systems of the predominantly state action driven MMEs like Spain, while the trade of IPRs has been assumed to not play a role at all.

External Consultancies

Looking at Catalonia as a whole, these results have only partly been confirmed by the interviewed experts. Consistently, the interview partners did not agree with ascribing *external consultancies* a medium range importance for knowledge transfer processes in the region, but evaluated them as only marginally important (Interviews CAT 1: 10; CAT 2: 19f.; CAT 3: 12). The main reasons for this would be a bad image of external consultancies in general, cost related factors for the predominantly small and micro firms in the region, and, again, the highly individualistic business culture in Catalonia, as the director of the technology networks department at the *ACC1Ó* agency accentuates:

> "Many people think that this has no value. Sometimes they think that the added value and the price are not corresponding. And the third factor is individualism. Just like the firms do not cooperate among each other, they do also not contract the services of a consultancy, since they believe to know how to do everything by themselves. Some of the support programs we talked about today must be translated into subcontracting external consultancies, which would then come and introduce new methodologies and elaborate a business plan. That means, nobody knows to do it all, but recognizing this is more difficult in the Latin Mediterranean culture, and in particular in Catalonia, than in it is in the Anglo-Saxon culture. Just like they do not subcontract external consultancies, they do also not subcontract research centers and university research groups. […] People here are more closed towards acquiring external knowledge and that is a problem, since competing in global markets requires acquiring knowledge worldwide." (Interview CAT 2: 19f.; translation: MO)

Therefore, even in this point, the Catalan Government would be active in addressing these market insufficiencies, through providing funding for firms engaging external consultancies in initial and early stages of cooperation with the scientific world within the framework of the *TECNIO* program (Interview CAT 2: 2ff.).

Mergers, Acquisitions and Hostile Takeovers

Mergers, acquisitions and hostile takeovers, on the contrary, would indeed have the predicted medium range importance as mechanisms of knowledge transfer. Although it would be nothing exceptional that companies would fusion or buy other companies, the intensity of these processes would be strongly dependent on the sector. Sectors like the pharmaceutical industry would be the most active ones in this regard. However, in many other industrial sectors this would be more difficult because of both the business structure with a vast majority of small and micro firms lacking the necessary resources, and the industrial structure basically consisting of production plants and with very few larger and/or multinational firms, which could be active in this regard (Interview CAT 2: 20).

Buying, Selling and Licensing of Patents

Finally, the *trade of licenses and patents* has been evaluated as almost irrelevant mechanism of knowledge transfer in Catalonia (Interviews CAT 1: 11; CAT 2: 21f.). Both, demand and supply dependent reasons would play a role in this, as the interview partner from the *ACC1Ó* agency summarizes, who again also refers to sectoral differences in this connection:

> "This only happens very few times, because in Catalonia patents are neither bought nor sold. I already told you that first of all they are not generated. In the public sector researchers create many papers but only a few patents. And there are also only a few patents because they cannot be sold. If they could be sold and if they could make money with them, they would do some. But they cannot be sold." (Interview CAT 2: 21f.; translation: MO)

The only sector that would be an exception in this would be the pharmaceutical sector. More likely would be technology transfer through the acquisition of machinery from abroad. Since most of the Catalan firms would not conduct own R&D-activities, introducing new technologies via this channel would constitute a usual mechanism of knowledge transfer, as the interview partner explains:

> "The majority of the people acquire technology buying equipment, not patents. The typical activity of the Catalan firm is to go to international trade fairs and if it wants to acquire any type of productive activity, what it does is to buy machinery. But it does not buy a patent, only if this machine would maybe come through a royalty, etc." (Interview CAT 2: 22; translation: MO)

Market Based Mechanisms in Automotive

For the field of *automotive*, these results were broadly confirmed by the interviewed expert for the sector. Contracting external consultancies to facilitate knowledge transfer would only occur in very few cases. Since most firms in the sector would lack an innovation-oriented business philosophy and culture, they would not subcontract external consultancies in topics related to knowledge and technology transfer (Interview CAT 3: 12). Mergers, acquisitions and hostile takeovers, on the contrary, would have some significance as mechanisms of knowledge transfer, but only to a very limited extent. The sector's mature and consolidated character and its business structure with predominantly small firms lacking the resources to acquire other companies would make the use of these instruments less frequent and rather exceptional (Interview CAT 3: 13). As in Catalonia as a whole, also in Catalan automotive buying, selling and licensing of patents has been reported to not play a role at all (Interview CAT 3: 14). It can be assumed that also the reasons might be similar and that, above all, most new technologies might be imported and transferred to Catalonia by the OEMs from their home countries abroad rather than being developed there using acquired IPRs.

Market Based Mechanisms in Biotechnology

A similar picture arises when looking at *biotechnology*: External consultancies would (still) not play a role in the very young Catalan biotech industry, above all because of a lack on the supply side. In particular, consultancy firms specialized in sector relevant topics like business development would still miss (Interview CAT 4: 10). Moreover, and even less than in the region as a whole and in automotive, mergers, acquisitions and hostile takeovers would not play a role at all yet. The explanation is, again, that the majority of the young firms in the sector still has not reached phases in their business life cycle in which these mechanisms would be relevant (Interview CAT 4: 10). Finally, also for considerable commercialization activities of IPRs most the firms in the sector would still be too young. In addition, also for the biotech sector the low patenting activities at Catalan research institutes and universities have been lamented in this connection (Interview CAT 4: 11).

6.2.2.6 Community Based Mechanisms of Knowledge Transfer

To address the community dimension of knowledge transfer processes, the interviewed experts have been asked for the importance of three mechanisms of knowledge transfer: Informal contacts, regional platforms and additional (regional) institutions, which allow for establishing trust-based relationships and build on a common regional identity, culture and tradition. Building on the theoretical discussion on national systems of production and innovation, it has been assumed that these community based mechanisms should be of a medium range importance within the mixed institutional framework of the Spanish national market economy, which combines elements of both market and non-market coordination of economic activities. This has been assumed to be complemented by strong state action compensating for cases of coordination failure, and less by strong community action.

Informal Contacts, Regional Platforms and other (Regional) Institutions

In general, the interviewed experts broadly confirmed these assumptions (Interviews CAT 1: 9; CAT 2: 18f.). *Informal contacts* were reported to be an increasingly important channel of knowledge transfer in Catalonia, despite the general individualistic attitude in the region, affecting all kinds of interaction, formal and informal. In Catalonia, *regional platforms* would especially be provided by the regional government through its *ACC1Ó* agency and, although to a lower degree and with a more local focus, by the chambers of industry and commerce. Other *(regional) institutions* have not been mentioned in this connection by the interview partners. As the director of the technology networks department at the *ACC1Ó* agency remarks, the goal of the creation of these regional platforms would be to overcome the communication and cooperation barriers among Catalan economic actors through offering them opportunities for low-level contacts, which could then turn into more stable interaction:

> "Some days ago we organized a conference named *InterACC1Ó*, where the actors can inscribe themselves and say that they would like to contact with people from certain sectors or centers, etc. Our role in this is to set up agendas and to offer a contacting platform. We also did *speed-dating*, which are meetings of five minutes to interact with each other. In other words: We try to arrange activities provoking continuous contacts with other firms and centers, etc." (Interview CAT 2: 18; translation: MO)

If the strong, common and historically grown Catalan *regional identity* is a supporting factor in establishing these informal contacts, cooperations and, thus, knowledge transfer processes, could not be judged clearly by one of the interviewed experts. The rest of the interview partners did not want to respond to this question.

Community Based Mechanisms in Catalan Automotive and Biotechnology

The estimations on the Catalan *automotive* sector concerning these community based mechanisms of knowledge transfer almost parallel those for the region as a whole. The only difference is that in this sector the role of chambers and business associations in setting up regional platforms has been evaluated as slightly more important (Interview CAT 3: 6f.). In *biotechnology*, on the other hand, informal contacts would generally be problematic because of IPRs protection. Nevertheless, in the very young biotech sector in Catalonia many informal contacts would exist, above all, among the proprietors of the many small firms (Interview CAT 4: 3). Regional platforms would be important, but not decisive mechanisms of knowledge transfer, and would mostly be offered by the sectoral business associations like *ASEBIO* on the national and *CataloniaBio* on the regional level, as well as by the *ACC1Ó* agency (Interview CAT 4: 4). Also the regional cluster organization *BioCat* would function as a regional platform in this regard. However, the number of concrete occasions to meet on these platforms would still be limited, especially because of the small size of the sector. Finally, also for the biotech sector the role of common Catalan *regional identity* as a supporting factor in establishing informal contacts, cooperations and knowledge transfer has not been evaluated clearly. Nevertheless, the majority of cooperation partners in the sector would be other Catalan actors, but there would also be other national and European cooperation partners (Interview CAT 4: 10).

6.2.3 Summary

In sum, the results of this qualitative case study on Catalonia provide both a broad correspondence of the findings with the assumptions derived from the literature on VoC and NIS for the knowledge transfer system in Spain's MME, and a few deviations from these assumptions, revealing at the same time specific characteristics and weaknesses of the Catalan knowledge transfer system. One of these weaknesses identified by all interview partners is strongly manifested on the network dimension of knowledge transfer processes: A highly individualistic business culture in the region and a low tradition and propensity of Catalan firms to cooperate with other actors of any kind resulted to be one of the most characteristic attributes of the Catalan knowledge transfer system. Apart from some knowledge transfer activities through informal networks, and a slightly higher relevance of vertical supply chain networks in certain sectors like automotive, all other forms of networks (horizontal business networks, institutionally stabilized networks) were reported to not play an important role in the region. In addition, also cooperations in R&D- and innovation related fields (inter-firm R&D-projects, cooperations with scientific institutions and universities) are only rudimentally developed in Catalonia, not least because of the atomistic business structure in the region, the low technology levels most firms are operating in and their lacking tradition in conducting own R&D- and innovation activities.

Another important deviation from the theoretical assumptions appeared with regard to the association based mechanisms of knowledge transfer. Both, business-, employers- and trade associations, as well as chambers of industry and commerce would not play any significant role for innovation processes in Catalonia. The decisive explanatory factors would be that most business associations would be mainly nationally oriented and would rather have a focus on interest representation than on innovation and knowledge transfer; a lack of coordination among the different local chambers on the regional level, as well as their inefficient internal organization structures and a lack of tradition in taking responsibility for the region as a whole and especially for innovation and knowledge transfer among their members.

A far reaching correspondence with the theoretical assumptions has been found concerning the organization based mechanisms of knowledge transfer. These have indeed been evaluated as medium range important mechanisms for the Catalan knowledge transfer system, showing considerable personnel transfer, headhunting and business start-up activities in the region. A lack of entrepreneurial mentality, like in the German regions, has not been detected in this connection. However, limits to these mechanisms would exist in form of sectoral specialisms, the technology level of the respective firms, firm size and venture-capital provision in later phases of the business life cycle. Overall, also in this dimension of knowledge transfer mechanisms the state has a central facilitating function: Through financing the employment of young researchers in firms, through initiating and funding innovation projects and R&D-consortia, as well as the labeling and promotion of transfer oriented academic institutes, the state contributes intensively to the creation of incentives for increased transfer and recruiting activities. In addition, through providing seed- and venture capital for business start-ups and academic spin-offs, it also enhances entrepreneurial activities in the region.

In general, the central dimension of knowledge transfer mechanisms in Catalonia clearly is, as assumed, the state. Regional and national policy measures in the fields of innovation and knowledge transfer, (larger) state induced innovation projects, as well as scientific parks and technological centers have consistently been reported as vital mechanisms of knowledge transfer in the region. In this regard, the state actively compensates for market failures and shortcomings in strategic coordination among the economic actors in almost all innovation and knowledge related fields, as well as in all analytical dimensions developed and investigated within this study. These system inherent failures are mainly rooted in Catalonia's atomistic business structure, its individualistic business culture, strong mental barriers between the business sector and the scientific world, the low technology level of many firms in the region, a tradition of importing technology rather than conducting own R&D- and innovation efforts, and corporatist actors lacking guiding and facilitating competences in the fields of innovation and knowledge transfer. Therefore, it is the state (national and regional, but also the EU-level) which is the all decisive player in the Catalan multilevel governance system of innovation and knowledge transfer, basically structuring and promoting knowledge transfer activities in the region top-down, allocating the decisive amounts of funding to it and creating the necessary structures and incentives.

In this, the considerable influence of the national level in terms concrete programs and amounts of funding must be highlighted, since it is to be expected that this might work as a homogenizing factor among the Spanish regions. On the other hand, the specific regional programs and measures initiated by the Catalan Government can be expected to work in the opposite direction. The regional policy's recent shift of focus towards a stronger labeling and promotion of transfer oriented scientific centers and institutes (*TECNIO*), the stronger promotion of network creation (*NUCLIS*) and the stronger centralization and coordination of the Catalan innovation and knowledge transfer policy within the regional development agency *ACC1Ó* can be interpreted as major differentiating features in this regard. However, like in the two investigated German regions, also in Catalonia the technology transfer offices at universities and research institutes (OTRIS) were evaluated as less significant than assumed. Especially their mostly administrative focus on legal and contractual issues has been criticized as too narrow. The interview partners emphasized the necessity to reform and modernize the OTRIS to be more proactive in initiating cooperation and knowledge transfer between the firms and the scientific world, and, therefore, to address the mental and cooperation barriers in the region described before.

Furthermore, many of the already discussed differentiating features of the Catalan knowledge transfer system also explain the overall low to medium range importance of the market based mechanisms of knowledge transfer. Despite the efforts of the Catalan Government to promote the use of external consultancies in the fields of innovation and knowledge transfer, they were reported to not play a role, because of the generally low propensity to cooperate with other actors, mental barriers and because of the low demand for these services generated by the lowly innovative firms. Moreover, the atomistic business structure in the region with many small and micro firms impedes higher numbers of mergers, acquisitions and hostile takeovers. Historically grown structural barriers within the Catalan research and higher education system disincentive researchers to generate patents, and the low technology level of Catalan firms and their low own innovation efforts are major barriers to a higher demand for patents and licenses.

Finally, the community dimension indeed resulted to exercise the assumed medium range importance for knowledge transfer activities in Catalonia. In this, occasions for informal contacts and regional platforms are, again, above all provided by the regional government through its *ACC1Ó* agency. Surprising is the fact that none of the interview partners neither wanted nor could report on the role of the common Catalan cultural identity in facilitating informal contacts and knowledge transfer. On the on hand, this might be explained with the highly controversial debate on the Catalan national identity and the Catalan separatist movement, which the interview partners did not want to touch, not even with regard to knowledge transfer. On the other hand, and taking into account the results from the other regional cases in this study, the role of strong and common regional identities appears to be clearly overemphasized in the literature. At least in the field of knowledge transfer structural factors, spatial proximity and market dynamics seem to be much more relevant than regional identities.

Table 7: Qualitative Relevance of Knowledge Transfer Mechanisms in Catalonia

MECHANISMS	ESP*	Cat	Auto	Bio
NETWORK BASED MECHANISMS				
business networks	O	–	–	–
informal networks	O	O	O	O
institutionally stabilized networks	O	–	O	O
customers and suppliers	O	O	O	–
cooperative inter-firm R&D-projects	–	–	–	O
cooperative R&D-projects with scientific institutions and universities	–	–	–	O
ASSOCIATION BASED MECHANISMS				
employers-, business-, and trade associations	O	–	O	–
chambers of industry and commerce	O	–	–	–
binding and common technical standards	O	???	???	???
ORGANIZATION BASED MECHANISMS				
transfer of technical and scientific staff	O	O	O	O
purposeful headhunting	O	O	O	O
business start-ups	O	O	–	+
STATE INDUCED MECHANISMS				
national government measures	+	+	+	+
regional government measures	+	+	+	+
(larger) state-induced innovation projects	+	+	+	+
scientific parks	+	+	+	+
technological centers	+	+	+	+
technology transfer agencies/companies	+	O	–	+
MARKET BASED MECHANISMS				
external consultancies	O	–	–	–
mergers and acquisitions, hostile takeovers	O	O	O	–
buying, selling and licensing of patents	–	–	–	–
COMMUNITY BASED MECHANISMS				
informal contacts	O	O	O	O
regional platforms	O	O	O	O
additional (regional) institutions	O	???	???	O

+ *Reported (*Hypothesized) as crucial mechanism of knowledge transfer*

O *Reported (*Hypothesized) as important but not decisive mechanism*

– *Reported (*Hypothesized) as not important at all*

??? Not reported by the interview partners

[dark grey box] = *Important deviation from hypotheses*

[light grey box] = *Modest deviation from hypotheses*

[white box] = *No deviation from hypotheses*

Overall, and at first sight surprisingly, also when looking at the two sectors of automotive and biotechnology no larger deviations from these results for the region as a whole have been detected. Focusing on automotive first, it has been shown that knowledge transfer processes in this sector are strongly determined by the following structural conditions: First, a lack of own OEMs in Catalonia, making the region rather a production plant for vehicles than a complete value added chain in the field of automotive. This contributes to the only limited and incomplete development of vertical supply chain networks and cluster-building. Second, and as a consequence, only marginal sectoral R&D- and innovation activities are conducted in Catalonia, with a tendency to import innovations from abroad rather than to develop them in the region, making R&D-cooperations among firms and with universities or scientific institutes rare, and the trade of patents an unusual and unnecessary activity. Third, a vast majority of small and micro firms shapes the sector's business structure, with only a few medium sized and larger firms, characterized by the same individualistic business culture identified for Catalonia as a whole, often impeding the creation of cooperation networks with any kind of other actors, including external consultancies and OTRIS. Fourth, the automotive sector in Catalonia is a mature and consolidated sector, giving little room for business start-ups or academic spin-offs. Fifth, the size of the sector and its resulting importance for the Catalan economy indeed makes the sector an important object of interest representation through business associations and chambers, but lets it receive less of their attention in the fields of innovation and knowledge transfer. Therefore, in general, also in the Catalan automotive sector the intervening state is the decisive actor in the field of knowledge transfer, with regional and national (and EU) policy measures, larger state induced innovation projects, scientific parks and technological centers as central channels of knowledge transfer.

Furthermore, also biotechnology in Catalonia shows only little deviations from the results for the region as a whole, although because of different structural reasons. The most important feature of Catalan biotechnology and one of its main differences in comparison to automotive is its young and emerging character, with most firms not being older than ten years. This explains why networks of all kind are still not established and consolidated, and business associations and chambers, as well as the central cluster management organization *BioCat* have still not found their role within the sector. In addition, most firms still have not reached phases of product development in which the trade of patents would play a role, although this can usually be seen as a vital mechanism in the sector. The sector's unconsolidated character also still leaves considerable room for business start-ups and academic spin-offs, which are some of its central mechanisms of knowledge transfer.

However, one major differentiating aspect in comparison to the region as a whole and to automotive is the higher affinity between the biotech firms and the academic world, as well as their more intensive R&D-orientation: Cooperative inter-firm R&D-projects as well as cooperations between firms and universities and research institutes, but also the use of the OTRIS are considerably stronger developed in biotechnology. In this regard, the strong ties many spun-out firms in the sector still maintain to their former institutes ease informal cooperation, even with a certain preference for other Catalan actors, but also the transfer of academic staff and technical personnel. Nevertheless, and surprisingly, the market based mechanisms, which could be expected to have a high significance in biotechnology, have been reported to be even less important than in automotive: While there would not exist sufficient external consultancies specialized on the biotech sector yet, mergers and acquisitions and hostile takeovers would not be realized because of the early stages of the business life cycle most firms would still be in.

Finally, also in the biotechnology sector public measures, programs and innovation projects, as well as public facilities such as scientific parks and technological centers are vital mechanisms of knowledge transfer. In this connection, the state has a considerable share in the emerging process of biotechnology in Catalonia, not least because of its highly supportive role in providing especially seed-, but also some venture-capital to the young firms in this highly capital-driven sector. All these results are summarized in Table 7.

6.3 The Case of the Madrid Community

6.3.1 Locating the Madrid Community in the Spanish Context

6.3.1.1 Selected Facts on the Madrid Community's Economic Structure

The second Spanish region investigated in this study is the Madrid Metropolitan Region, which will be, following many other studies on this region (OECD 2007b: 30ff.) defined as being covered by the territory of the Madrid Community (Comunidad de Madrid). The Madrid Community is one of the 17 Spanish Autonomous Regions, is located in the geographical center of the Iberian Peninsula and contains the Spanish capital city of Madrid, which is the clear and dominant center of the Madrid Metropolitan Region. The region's specific function as the national capital region since the 16th century strongly shaped its social and economic development. While for centuries Madrid has been the clear center of the political and social life in Spain, it has only started to become one of the country's economic centers in the beginning 20th century, especially because of the development of a resilient railway infrastructure in Spain. Before, its geographical isolation from Spain's main trade routes, but especially from the coasts and any navigable waterways, impeded stronger commercial activities in the region. Madrid's economic tradition has been, therefore, the one of a consumer market for luxury goods and of a reference market for the Castilian agriculture. Particularly after the Spanish Civil War, the Madrid region experienced a phase of late and catching up industrialization, above all in industrial sectors related to urban consumer goods, like chemicals, pharmaceuticals, electrical engineering, high-precision mechanics, or metallurgy. This process of late industrialization (compared to other Spanish Autonomous Regions like the Basque Country or Catalonia) was strongly favored by the concentration of national and international investments in and around the Spanish capital, and by Madrid being the seat of the Spanish Government, which indirectly provoked the location of a high number of headquarters of national and multinational firms in the region (OECD 2007b: 30ff.; COTEC 2004; COTEC 2003; Buesa 2004).

In the present, the Metropolitan Region of Madrid is one of Spain's most dynamic economic regions and an important hub in the global economy. In the decades after the end of the Franco-Regime the region developed into a powerful industrial pole within Spain, is the country's center of highly qualified services and became a nodal point in the national and international trade, service, communication and transportation infrastructure. Above all, a large supply of labor provided both by immigrants (especially Spanish-speaking natives from Latin America) and young educated fixed-term workers from other parts of Spain and Europe, and the presence of first-class transportation facilities, such as the Barajas Airport, the AVE-trains, or the dense highway network in Spain, enabled Madrid to mitigate the challenge of being a region in the European periphery (OECD 2007b: 33ff.; COTEC 2004; COTEC 2003; Buesa 2004).

Today, the Madrid Community is Spain's third largest Autonomous Region in terms of inhabitants after Andalusia and Catalonia, with about 6.4 million inhabitants in 2009, representing 13.7% of the country's total population (INE 2010b: 8). After Catalonia, the Community of Madrid is Spain's second strongest economic region contributing 17.7% (193 billion Euros) to the Spanish GDP in 2008. The Madrid Metropolitan Region is also one of Spain's main industrial poles, whose industrial sector's GDP contributed about 12% to the GDP of the Spanish industry in 2007 and where about 10% of all Spanish industrial firms are located (see also Figures 30 and 31; INE 2010: 39). Nevertheless, with regard to the national average of 17.3%, Madrid's industry has a comparatively low share in the region's economic structure, covering only 13.4% of the region's GDP (Catalonia: 18.9%). In addition, and as to expect for a metropolitan agglomeration area, the primary sector has a marginal share of only 0.1% in the region's GDP, while the construction sector with a share of 10.1% almost equals the national average of 11.6% (Figure 26; INEa 2010: 39; ICO 2009: 5; IESTADIS 2011a).

However, the most distinctive feature of Madrid's economic structure is its high degree of tertiarization, with services representing more than three quarters (76.5%) of the regional GDP, ranking high above the national average of 68.4% (Figure 26). In this, Madrid's service sector shows the typical structure of western metropolitan agglomeration areas and national capital city regions, with a comparatively elevated share of highly qualified services. Measured as a share of Madrid's service sector's GDP, the most important sectors are trade services (17.6%), business services (15.6%), transport and communication services (14.4%), housing and real estate services (12.6%), financial services (12.0%), and public administration (7.5%)[43] (IESTADIS 2011a; ICO 2009: 5ff.; INE 2010a: 39).

43 Own calculations based on data from IESTADIS 2011a, data for 2008

Pointing to Madrid's industrial structure, two sectors stand out: Energy and mining, with a share of 24.7% in the regional industry's GDP, and paper and printing (15.8%) are the two main industrial foci in the Madrid Metropolitan area. This is complemented by a diversified set of industrial sectors, of which the chemical industry (9.0%), electrical engineering (9.0%), automotive (7.5%), metalworking (6.8%), food processing and luxury foods (6.1%), and mechanical engineering (5.8%) are the most important[44] (Figure 36). The technology level of Madrid's manufacturing sectors is clearly more technology-intensive than in the rest of Spain, but ranks about average when compared to the OECD regions in general. Only 1.3% of the total employment in the Madrid Community is in high-technology manufacturing sectors, but 3.7% in medium-high technology sectors. 6.4% of the region's total workforce is employed in medium-low and low technology manufacturing sectors, which predominate clearly the region's manufacturing activities (OECD 2007: 75ff.).

Fig. 36: Madrid's Industrial Structure

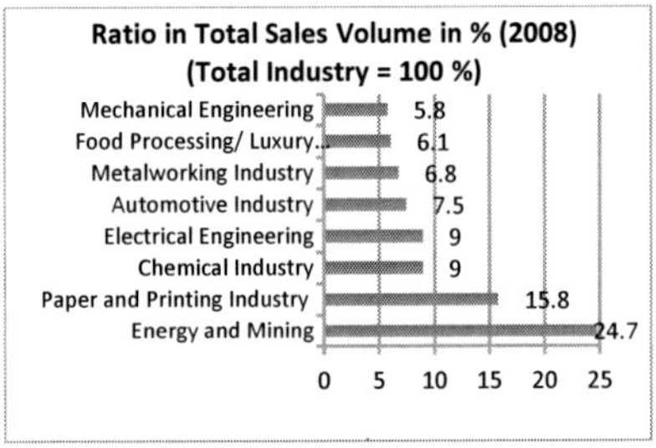

Source: Own Calculations and Illustration based on Data from IESTADIS (2011a)

Fig. 37: R&D-Intensity in Selected Spanish Regions

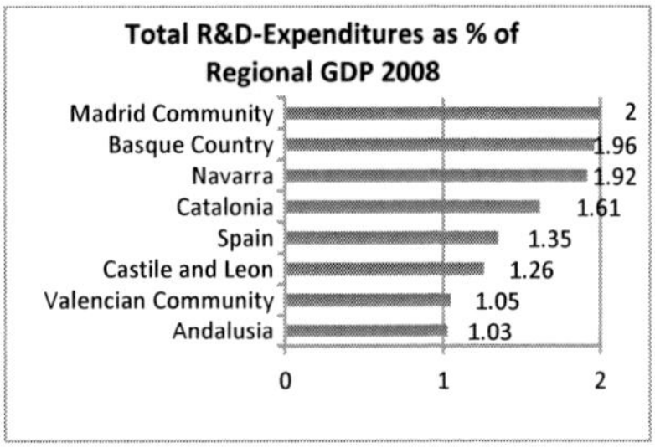

Source: Own Illustration based on COTEC (2010): 25

Like in Catalonia, also the Madrid Community's business structure almost equals the one on the national level, especially in being strongly dominated by (mostly family-owned) micro firms with nine or less employees, which represent 93.9% of the total number of firms in the region. Within the group of micro-firms those firms stand out that have no person employed: They represent 59.1% of the micro firms in the Madrid Metropolitan Region and 55.5% of its total number of firms. Small firms with ten to 49 employees represent 4.8% of all Madrilenian firms, medium sized firms

44 Own calculations based on data from IESTADIS 2011a, data for 2008

from 50 to 199 employees 0.9%. Only the value for large firms with more than 200 employees (0.4%) almost doubles the value for Spain (0.2%)[45] (Figure 38), underlining again the region's significance as a seat for large Spanish firms and the Spanish centrals of multinational firms. In this, firms with two or less employees represented 20.8% of Madrid's workforce, the micro-firms up to ten employees employed 11.7% of the region's total workforce, while small (11–49 employees), medium (50–249 employees) and large firms (with more than 250 employees) employed 20.6%, 21.0%, and 25.9% of Madrid's total workforce, respectively[46] (Figure 39; IESTADIS 2011b).

Fig. 38: Madrid's Business Structure I

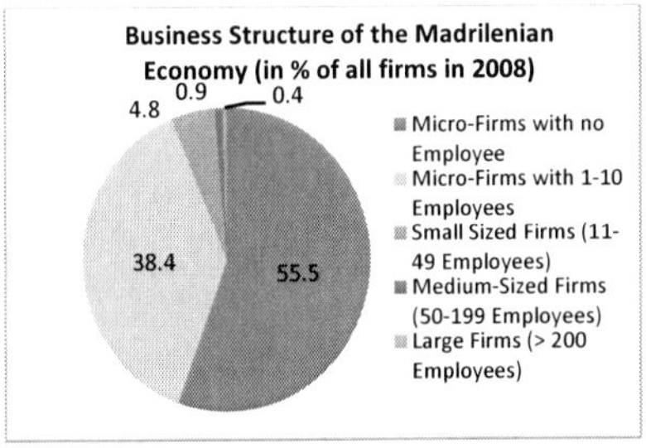

Source: Own Calculations and Illustration based on Data from IESTADIS (2011b)

Fig. 39: Madrid's Business Structure II

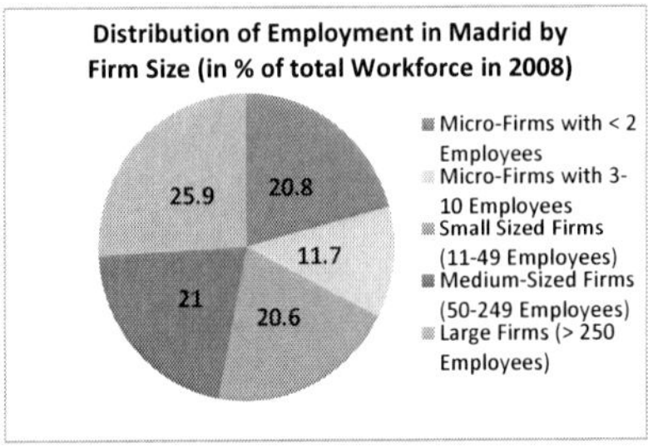

Source: Own Calculations and Illustration based on Data from IESTADIS (2011b)

Another important structural indicator describing a region's knowledge base is its R&D-activities. In 2008, Madrid had total R&D-expenditures of about 3.9 billion Euros. This represents 26.5% of Spain's total expenditures in R&D and is the highest contribution among the Spanish regions, clearly above those of Catalonia (22.4%), Andalusia (10.5%) and the Basque Country (9.2%) (Figure 33; COTEC 2010: 26). With that, the Community of Madrid is also Spain's top performing region in terms of R&D-intensity (total R&D-expenditures as a percentage of the regional GDP), which is of about 2.0%, and ranks high above the national average (1.35%) and the value for Catalonia (1.61%), but also above those for the Basque Country (1.96%) and Navarra (1.94%) (Figure 37; COTEC 2010: 25). 57.7% of the total Madrilenian R&D-expenditures were covered by the

45 Own calculations based on data from Idescat 2011, data for 2008

46 Own calculations based on data from IESTADIS 2011b, data for 2008

business sector and 42.3% by the public sector[47], giving the business sector's R&D-activities in Madrid a stronger position than those at the national level (54.9%) (COTEC 2010: 246), but a weaker one than in Catalonia (61.1%)[48].

A sectoral perspective on the business R&D-activities in the Madrid Community reveals that 34.8% of the R&D-expenditures of the Madrilenian business sector are realized by industry, 4.5% by construction and 60.6% by the service sector. 35.5% of the region's total business R&D-expenditures are covered by firms with less than 250 employees, while the large firms with more than 250 employees cover 64.4%. Given the region's business structure with a vast majority of SMEs and micro firms, this points to an enormous concentration of R&D-activities within the large firms. When only focusing on industry, the predominance of the large firms in terms of R&D-activities becomes even more pointed: Only 18.9% of the total R&D-expenditures of industry are covered by the SMEs, and 81.1% by the large firms. In terms of employment of R&D-personnel in the Madrilenian business sector, industry covers 27.6% of the total R&D-personnel, construction 6.8% and the service sector 65.4%. Differentiated by firm size, 48.7% of the business-R&D-personnel is employed in SMEs, and 51.3% in the large firms. This relation is again stronger differentiated in the industrial sector: The SMEs employ about 32.1% of the total R&D-personnel in industry, while the large firms cover 68.0%. In this, the most R&D-intensive industrial sectors in the Madrid region are automotive with a share of 31.6% of the region's industrial R&D-expenditures and 31.1% of its industrial R&D-personnel, followed by chemicals (25.0% and 42.6%) and electrical engineering (20.5% and 12.8%)[49].

6.3.1.2 Automotive—The Center of Madrid's Industrial R&D

A specific focus on the Madrilenian automotive industry[50] demonstrates that this sector has less weight for the regional industry than its Catalan counterpart: With a share of 7.5% in the regional industry's GDP it is only the fifth important industrial sector in the region (Catalonia: 10.9%). How-

47 Own calculations based on data from IESTADIS 2011c, data for 2008

48 Own calculations based on data from Idescat 2011, data for 2008

49 Own calculations based on data from IESTADIS 2011c, data for 2008

50 Because of data availability issues, this paragraph only analyzes data for the vehicle manufacturing subsector. Data for 2008

ever, with that, Madrid's automotive sector is the third most important pole of Spanish automotive, after Catalonia and the Basque Country/Navarra (Madrid Network 2011). The sector consists of 548 firms in the field of vehicle construction, representing 1.3% of all firms in Madrid's industry. These firms cover 8.3% of the total industrial employment in the region, of which 11.5% is employed in small firms (1–49 employees), 15.5% in medium sized firms (50–199 employees) and 72.9% in large firms with more than 200 employees, revealing a comparatively large overall firm size in Madrilenian automotive[51] (IESTADIS 2011b). In addition, Madrid also hosts a considerable number of supplier firms to these vehicle manufacturers, and is the third most important location of the automotive supply industry in Spain, with a share of 13.1% of all Spanish automotive supply firms, ranking only behind Catalonia (36.4%) and the Basque Country/Navarra (24.7%) (Madrid Network 2011).

However, in terms of its R&D-activities, automotive[52] is the region's most research intensive industrial sector. It has a share of 31.6% in the region's business R&D-expenditures of the industrial sector, of which only 2.2% are covered by firms with less than 250 employees, and 97.8% by large firms with more than 250 employees, demonstrating that R&D-activities in this sector are almost exclusively concentrated within the large firms. This relation is further underlined when looking at the R&D-personnel employed: Overall, automotive employs 31.1% of the R&D-personnel in Madrid's industry, of which only 2.9% are employed in firms with less than 250 employees, but 97.1% in firms with more than 250 employees[53] (IESTADIS 2011b). This concentration of R&D-activities in the large firms is not least explained by the fact that many of the larger firms in the sector have their main representations located in Madrid, where they also conduct considerable parts of their R&D-activities. Nevertheless, it should be taken into account that like the rest of Spain, also the Community of Madrid lacks of own OEMs in automotive, and is rather a production location for many internationally renowned OEMs. The most important of these production facilities in terms of sales volume and employment are those of *PSA Peugeot-Citroën España*, *IVECO España* and *John*

51 Own calculations based on data from IESTADIS 2011b, data for 2008

52 Because of data availability issues, this paragraph only analyzes data for the vehicle manufacturing subsector. Data for 2008

53 Own calculations based on data from IESTADIS 2011b, data for 2008

Deere, as well as the larger suppliers *Robert Bosch*, *Valeo*, *Faurecia*, and *Castellón S.A.* (Madrid Network 2011).

6.3.1.3 Madrid—The Center of Spanish Biotechnology

Like the Spanish biotechnology industry in general and like its Catalan counterpart, also Madrilenian biotechnology is a young sector with a majority of firms created in the last fifteen years, actively promoted by the regional and national governments. For 2008, the *Genoma España Report 2009* identifies 162 firms in biotechnology related fields in the Madrid Community, of which 60 are dedicated biotech firms (Genoma España 2009: 45). This represents 21.8% of all dedicated biotech firms in Spain and is the largest contingent among the Spanish regions, leaving behind Andalusia (20.4%) and Catalonia (17.1%). When looking at the broader category of firms in biotech related fields, the regional distribution is different, especially because of the strong role of chemical and pharmaceutical firms in Catalonia: 24.2% of the biotech related firms are located in Madrid, a value ranking behind Catalonia with 25.1%, but above Andalusia with 12.6%[54] (Genoma España 2009: 45). All biotech related firms in Madrid together created a total sales volume of about 214.7 million Euros, representing 30.4% of the Spanish biotech industry's total turnover, and about 1.2%[55] of the total sales volume of the Madrilenian industry. These firms employed around 1.800 persons, representing 42.5% of the total workforce in biotechnology in Spain[56] (Genoma España 2009: 45; ASEBIO 2010: 34ff.; BioCat 2009: 21ff.; Madrid Emprende 2007: 5ff.).

6.3.2 Structures and Mechanisms of Knowledge Transfer in the Madrid Community: Six Governance Dimensions

6.3.2.1 Networks as Mechanisms of Knowledge Transfer

Building on the theoretical assumptions developed in section 3.3 and on the discussion of quantitative indicators describing the Spanish (section

54 Own calculations based on Genoma España (2009): 45, data for 2008

55 Own calculations and estimations based on Biocat (2009), ASEBIO (2009), and Genoma España(2009), data for 2008

56 Own calculations and estimations based on Biocat (2009), ASEBIO (2009), and Genoma España(2009), data for 2008

6.1) and the Madrilenian economy in the previous section, in the following sections the results of the empirical qualitative case study of Madrid's knowledge transfer system will be detailed. The first dimension of knowledge transfer mechanisms that has been investigated in this regard is the network dimension of knowledge transfer, for which the theoretical assumptions have mostly been confirmed when looking at the Madrid Community as a whole.

Horizontal and Vertical Networks

Horizontal business networks, might they be *informal* or *institutionally stabilized*, and *vertical supply chain networks* between firms and their customers and suppliers have been reported to play a medium range role in the Madrilenian knowledge transfer system. In this, like in Catalonia, also the interviewed experts for the Madrid region identified a strongly pointed aversion of firms to cooperate with other business actors (Interviews MAD 1: 11f.; MAD 2: 3; MAD 3: 4; MAD 4: 4ff.; MAD 5: 9). The reasons mentioned by the experts are manifold and quite similar to those given by the Catalan experts: First, the comparatively low technology level of the majority of Madrilenian firms would make cooperations with other firms in the fields of innovation and knowledge transfer less necessary. Second, a lack of tradition of the firms in the region in conducting own R&D- and innovation activities, and a tendency to rather import innovation and technology from abroad would make innovation efforts through cooperation less likely. Third, the atomistic business structure in the region, with a vast majority of small and micro firms, mostly lacking the necessary financial and human resources to innovate and cooperate with other economic actors. Fourth, an individualistic business culture, making inter-firm cooperation and network-building generally unlikely, as the director of the department of innovation and competitiveness at the *Madrid Chamber of Industry and Commerce* explains:

> "There is only very few inter-firm cooperation in Madrid. [...] There is no tradition to cooperate like in other territories. The Madrilenian entrepreneur is different in this sense. Maybe in the service sector there is slightly more cooperation, but very few in the industrial firms. The Madrilenian entrepreneur has been a solitarian gentleman which had his factory and scarcely collaborated. There is no collaborative culture like it possibly exists in Germany or in the UK." (Interview MAD 4: 4; translation: MO)

Despite this generally limited propensity of Madrilenian firms to cooperate, the director of the innovation department at the *Madrid Confederation of Employers and Industries (CEIM)* argues that inter-firm cooperation would still have a slightly higher significance for Madrilenian firms in comparison to other Spanish regions, since the share of larger firms would be higher, leading to a concentration of R&D- and innovation activities in the Madrid Community, what would then also generate more cooperation in these fields. In this regard, he also points to a higher degree of formalization in these cooperations in the Madrid region, and explains:

> "This is a question of business culture. And inter-firm cooperation is relatively low. [...] In general, these cooperations are stronger formalized in Madrid. I.e. that those firms that conduct R&D- and innovation activities in the Madrid Community have a generally higher cooperation culture than in other regions. Why? Because they are closer to a high number of agents providing them information about technology transfer. [...] And this stronger cooperation culture indeed leads to a major formalization of cooperation. If there is cooperation, since it is not excessively elevated." (Interview MAD 3: 4; translation: MO)

Like in Catalonia, also in the Madrid Community all interview partners consistently reported that the state would exercise a decisive role in compensating these communicative and interactive market shortcomings in the fields of innovation and knowledge transfer. One major emphasis in the regional innovation policy in this regard would be on cluster-building and networks (Interviews MAD 1: 1ff.; MAD 2: 7ff.; MAD 3: 11f.; MAD 4: 10f.; MAD 5: 5ff.), constituting an important difference in comparison to Catalonia, where the focus is more on supporting single and short term, but less on long term and institutionally stabilized cooperation projects. The decisive institution in this context would be *Madrid Network*, which is a public-private partnership initiated by the *IMADE* (Madrid Institute for Regional Development) as a regional development body of the *Madrilenian Ministry of Economy and Finance*, *CEIM*, and the *Madrid Chamber of Industry and Commerce*. Created in 2007, *Madrid Network* is one of the central coordinating, managing and assessing bodies in the Madrilenian innovation system, responsible for all regional clusters and scientific parks in the region. In this, as a central cluster management organization it would be a focal point for already existing regional clusters, and as a regional development body it would be actively promoting the identification and creation of new regional cluster structures in distinct sectors, including large and small firms, universities, research centers and the region's scientific parks. As the director of clusters at *Madrid Network* remarks, it is one decisive goal of *Madrid Network* and the regional innovation policy to diffuse an innovation culture to the regional actors:

"Madrid Network is a tool basically building upon the regional *Plan of Technological Innovation* of the Madrid Community in which the conviction is formulated that the best way to approximate the small, medium sized and micro-firms in the Madrid Community are the concept and the culture of innovation. The conviction is that the creation of a network of scientific parks and sectoral clusters in the Madrid Community is an excellent tool to produce this approximation and to facilitate technology transfer. […] Basically it is about bringing science, administration and firms to the same platforms so that knowledge can circulate and technology is transferred." (Interview MAD 1: 2; translation: MO)

Therefore, *Madrid Network* would address almost all dimensions of networking and cooperation of the region's economic actors: Through centrally coordinating the different regional clusters it would bring together firms from different sectors in horizontal interaction in joint events, conferences, and (mostly publicly funded) cooperation projects. Second, the twelve sectoral clusters under the roof of *Madrid Network* in the fields of aerospace, automotive, logistics, biotechnology, health, ICT security, ICT audiovisual, graphical arts and communication, renewable energies, financial services, tourism and Spanish language would have a strong focus on sustaining vertical cooperation among firms and other actors along the respective value added chains, especially in the traditional industrial sectors. Third, participation in *Madrid Network* and its clusters would be formalized with fixed association agreements, building on a membership statute, and cooperation in concrete projects initiated, promoted, assessed or (partly) funded by *Madrid Network* would mostly be contractually fixed. Cooperation in the single clusters would be institutionally stabilized, initialized, supported and assessed by the respective cluster management departments at *Madrid Network*. Finally, also the informal dimension would be addressed by *Madrid Network*, through bringing the actors together in information talks, conferences, events and other platform activities (Interviews MAD 1: 2ff.; MAD 2: 8; MAD 3: 11ff.; MAD 4: 4; MAD 5: 5ff.).

However, also inter-firm cooperation, networking and cluster-building within the framework of *Madrid Network* would predominantly be initiated top-down, and only scarcely bottom-up by the firms themselves, as both the directors of clusters and for technology at *Madrid Network* admit:

"Clusters can be promoted by the firms, or, like in this case, by the administration. In Madrid we have sectors which have organized themselves within clusters and are not part of *Madrid Network*. […] There are clusters emerging spontaneously, and others, like almost all of our clusters, are the result of an analysis identifying twelve strategic sectors in the Madrid Community. There was a relevant activity within these sectors and, therefore, we tried to structure them as clusters. We saw the problem that in Madrid we had a very high activity in all these sectors, but there was not much coordination, since there is a high concentration of large firms which sometimes appear to be auto-sufficient and which probably do not need any cluster structures. But one of the important points for a really functioning cluster is that both, small and large firms benefit from the triple helix relation." (Interview MAD 1: 11f.; translation: MO)

Inter-Firm R&D-Cooperations

Cooperative inter-firm R&D-projects were reported to be slightly more important in the Madrid Community than assumed for the Spanish case (Interviews MAD 1: 2ff.; MAD 2: 8ff.; MAD 3: 14; MAD 4: 2ff.), and also as reported for Catalonia. Even if also for this region the generally low R&D- and innovation activities and the little cooperative attitude of the business sector have been emphasized as major barriers to cooperations in R&D in the small, medium and micro firms, the specifically high concentration of large firms and their respective R&D-departments and facilities in the region would make inter-firm R&D-activities, especially among the large firms, and among the large firms and their customers and suppliers, more significant than in other Spanish regions. As the director of the innovation department at *CEIM* points out, these inter-firm R&D-projects would generally base on fixed cooperation contracts, securing confidentiality and IP-protection:

"Formalized R&D-cooperations are mostly established with the suppliers, there are fixed contracts, clearly defined collaboration contracts. There is not so much cooperation as we would wish, but there is cooperation and it is perfectly established and formalized. My experience is that the large firms apply for projects funded by the European or the national level only for topics that are not strategic for their activity, but which have a stronger focus on general knowledge generation. The strategic topics are financed and developed by the large firms themselves with their closest suppliers. This would then be completely formalized and with strong agreements on confidentiality and IP-protection." (Interview MAD 3: 14; translation: MO)

R&D-Cooperations between Business and Academia

Also *cooperative R&D-projects between firms and universities or other scientific institutions* have been evaluated as having a medium range importance as mechanisms of knowledge transfer in the Madrid Community (Interviews MAD 1: 2ff; MAD 2: 9ff.; MAD 3: 4; MAD 4: 15; MAD 5: 7ff.). Even if also in Madrid strong mental and institutional barriers between the business sector and the academic world would exist, and despite the generally low cooperation culture in the business sector affecting cooperation with the scientific world, the high concentration of large firms, firm headquarters and business R&D-departments in the region would make these cooperations more frequent than in other Spanish regions. For the rest of the SMEs and micro firms, cooperation with the scientific world would be very unlikely, and would only be realized on the basis of public funding, as the director of the department of innovation and competitiveness at the *Madrid Chamber* illustrates:

> "University-industry relations are very rare and are only produced if there is public funding, otherwise it is not done. In a natural way these relations are not produced. [...] To make yourself an idea, you can look at a collaboration indicator of Madrilenian or Spanish firms with firms in their environment: We can see this in the structure of the project consortia presenting themselves to the *7th Framework Program*. In many cases Spanish and Madrilenian public research centers apply collaborating with foreign firms and not with Spanish ones. The Madrilenian firm still strongly distrusts the university." (Interview MAD 4: 15; translation: MO)

The reasons for these barriers between the two worlds would be both cultural and institutional. One factor would be that the decisive ministries of both the regional and national government, which could be the central actors in breaking open these barriers, would be predominantly equipped with former academic staff and professors. These would often lack insight into and experience with the business world (Interview MAD 4: 15). Another reason is provided by the head of the technology transfer department at the *Madrilenian Ministry of Economy and Finance*, who identifies the highly traditional orientation of some important universities in the Madrid region, as e.g. the *University of Alcalá de Henares* or the *Complutense University*, as one reason for the persistence of these mental barriers. However, the interview partner also highlights a more open attitude towards the business sector in the younger Madrilenian universities, like the *Politéchnica* or the *Carlos III University*, and expects an incremental change of mentality with a new generation of researchers (Interview MAD 5: 7).

To overcome these cultural, mental, structural and institutional barriers between the firms and the world of science, the Madrid Community developed the *madri+d* system[57], which is one of the most differentiating elements of the Madrilenian innovation system and one of the region's most striking features in terms of knowledge transfer between science and economy. The *madri+d* system is a professional network having as members large parts of the region's public and private research institutions and universities, as well as business associations and intermediary actors like the OTRIS, covering the essential aspects of interaction between the knowledge producing sector and the business world, with the aim to improve knowledge transfer in the Madrid region. The *madri+d* network has been founded in 1997 as public-private cooperation between the *madri+d Foundation* of the *Madrilenian Ministry of Education*, the *CEIM* and *APD* (Association for the Progress of Management) as major partners. *Madri+d* is a central coordination body in Madrid's knowledge transfer system, with its six thematic offices representing its main fields of action: Creation of technology based firms, commercialization of scientific results and technology transfer, participation in the European research area, diffusing and improving scientific culture in society, promotion of young researchers, and coordination of scientific infrastructures (madri+d 2008).

All interview partners consistently evaluated the *madri+d* system as a positive, decisive, and efficient element in the Madrid Community's knowledge transfer system (Interviews MAD 1: 15; MAD 2: 1ff.; MAD 3: 4ff.; MAD 6: 20). This is further underlined by the fact that the website of *madri+d* is the world's most visited Spanish language website in science and technology, with more than 60 million accesses per year. With that, the *madri+d* system is the central nodal point of most networking activities between science and business, making visible and accessible a large share of the regional knowledge producing and knowledge demanding actors. In its structure and mission it is, therefore, comparable to the *Innovationsnetzwerk Niedersachsen*, which exercises similar functions, although on a lower scale, for the MR HBGW. As the director of the innovation department at *CEIM* accentuates, from a business point of view, the *madri+d* system has a central function in connecting both worlds, although he also points to the fact that recently the Madrilenian innovation policy puts a stronger focus on cluster-building, what would also be actively supported by *CEIM*:

57 The term "madri+d" is a wordplay using the Spanish language equivalent of R&D: I+D.

"The firms in Madrid have access to a higher number of intermediary agents providing them information on technology transfer. I think in this regard the *madri+d* system has had an absolutely relevant role in the sense that it achieved that the scientific world, universities and research centers, cooperates to provide a common offer to the business world, especially to the innovative SMEs and small firms. [...] The *madri+d* system has been a benchmark for many regions, of course in Spain, but also in the rest of Europe. Therefore, it also won two awards of the European Commission, for the most innovative regional policy in the field of promoting technology transfer. What the system *madri+d* did was to put into contact and to coordinate the academic and the business world and to create mechanisms and services to enhance their interaction. This worked very well for twelve years now, but recently there is a certain tendency to develop cluster policies." (Interview MAD 3: 4, 11; translation: MO)

Also the interviewed expert from the *Madrid Chamber* supports this shift towards a more cluster oriented policy. He criticizes that from the point of view of many SMEs in the region represented by the chamber the *madri+d* system recently developed too much into an academic research network, and lost some of its focus on the business sector, as he explains:

"*Madri+d* has been created to close the gap between the university and the firm. What happens is that *madri+d*, or its foundation, is strongly connected to the Ministry of Education. This fact strongly limits its freedom of action, since it is committed to the interests of the universities, and a bit less to the interests of the firms. Therefore, we still have a considerable gap between knowledge generation and its transfer to the productive sphere. [...] Now there are high expectations concerning the network of parks and clusters [Madrid Network; MO], [...] since it does create an environment in which the entrepreneur thinks he will be well represented and will be able to defend his interests." (Interview MAD 4: 3, 13; translation: MO)

Network Based Mechanisms in Madrilenian Automotive

Focusing on the Madrilenian *automotive* sector, the results for the region as a whole are mostly confirmed, except a higher significance of vertical cooperations along the value added chain (Interview MAD 6: 3ff.). Also for this sector a generally low cooperative business culture, an atomized business structure and a low tradition in innovation and R&D have been mentioned as major reasons for cooperation barriers in horizontal networks, formal and informal cooperations, as well as in inter-firm R&D-projects. In this, automotive is addressed by the regional policy as one of the strategic clusters of *Madrid Network*, from where the *Madrid Cluster Automoción* is coordinated. Although cooperation within the cluster would be more vital as it is impulsed by the cluster management, only a small part of the firms would be members and it would be hard to convince them to join the cluster, as the managing director of the *Cluster Automoción* remarks:

> "There has been a necessity to opt for an integrative system for the firms of the Madrilenian automotive sector. The sector here is highly atomized, composed of many SMEs, more smaller than medium firms […], and many micro-firms. Therefore, it is very difficult to promote the integration and the grouping of these firms, since all of them have their business rhythms, preferences and different languages. […] Not all of them are members of the cluster. We have seven founding members among which are the three manufacturers in the region *PSA Peugeot-Citroën España*, *IVECO España* from the *FIAT Group* and *John Deere,* as well as a number of SMEs. From there onwards we keep growing, we are 21 members now, of the 300 which are our goal in principle. […] The problem of the atomization of the firms makes it difficult to gain the confidence with regard to the utility of the cluster, especially in the recent economic crisis. […] However, once the crisis is over, we expect more firms to join the cluster." (Interview MAD 6: 2f.; translation: MO)

Within the cluster, cooperation would be formalized with fixed contracts and also institutionally stabilized through the cluster management. Also outside the cluster, informal cooperation and knowledge transfer would be rare, and especially vertical cooperations would always be contractually underpinned (Interview MAD 6: 4f.). These vertical cooperations between firms and their customers and suppliers would be more important in automotive than in other sectors (Interviews MAD 1: 6f.; MAD 4: 6; MAD 5: 9; MAD 6: 5ff.). Very intensive vertical cooperations would exist in the aerospace branch, which in general is part of the automotive sector, but within *Madrid Network* constitutes a separate cluster. In this particular branch vertical cooperation would be more intense than in the terrestrial vehicle construction branch, since in the former the value added chain would be more complete than in the latter (Interviews MAD 1: 7; MAD 5: 9), which is characterized by the already discussed absence of own OEMs and rather consists of production plants of foreign MNEs. This high dependence on strategic decisions from abroad would also lead to a recent destabilization of customer-supplier networks in the region, since suppliers could easily be exchanged and many supplied products and modules would be available worldwide and would not have to be necessarily ordered from regional firms (Interview MAD 6: 5f.).

Inter-firm R&D-cooperations would exist, as well as cooperations between firms and academic institutions, especially in the larger firms in the sector, and they would be essential channels of knowledge transfer in the limited number of cases they would be realized. Especially cooperations with foreign firms and research institutes in European cooperation projects, as for example within the *Seventh Framework Program*, would be particularly fruitful for the regional firms, which would sometimes have a lower technology level than their European partners. However, also these cooperations would predominantly be initiated by the different policy levels and would mostly depend on public funding, since also in automotive strong mental barriers between the business and the academic world would exist. In addition, also in this sector the *madri+d* system would be used to promote cooperation with the scientific world and to foster knowledge transfer (Interview MAD 6: 5ff.).

Network Based Mechanisms in Madrilenian Biotechnology

A slightly higher importance of the network based mechanisms of knowledge transfer has been reported for Madrilenian *biotechnology* (Interview MAD 7: 1ff.). Horizontal and vertical cooperations would indeed take place and especially within the framework of the *Madrid Biocluster*, which is also part of *Madrid Network*. Moreover, firms in the sector would have a less regional but rather international perspective and would only to a limited extent cooperate with regional partners. Any kind of cooperation in the sector would be highly formalized and contract based, since any cooperation in this highly research driven sector would automatically include massive flows of knowledge. Within the cluster, cooperation would actively be initiated and promoted by the cluster management department, and, therefore, be institutionally stabilized. Cooperative inter-firm R&D-projects and cooperations of firms with academic institutes would frequently be realized and would constitute essential channels of knowledge transfer in the sector. In this, apart from cooperation projects initiated and funded by the regional and national governments, and, above all, by the *CDTI*, especially the European policy level would be decisive in fostering cooperative R&D-activities in the Madrilenian biotechnology sector, as the managing director of *Madrid Biocluster* explains:

"Especially important in this case is the international policy developed by the EU. Europe has an enormous quantity of programs and projects to make the biotech sector participate at the European level. From the *7th Framework Program*, the *Regional Competitiveness Program*, to the *Regions of Knowledge Program* the EU has a multitude of programs in which biotech firms can participate. For these programs and projects, as a minimum, actors from three different countries or more must participate. All that not only enables the construction of research projects, might they be business or scientific projects, but also fosters technology transfer in a very important way. That is so to say the principal impulse giver to the development of the industrial biotech sector and biotechnological research." (Interview MAD 7: 11; translation: MO)

6.3.2.2 Association Based Mechanisms of Knowledge Transfer

For the association based mechanisms of knowledge transfer (namely employers-, business-, and trade associations; chambers of industry and commerce; binding and common technical standards), the assumed medium range importance has not only been confirmed by the interviewed experts when looking at the Madrid Community as a whole, but even evaluated as higher than assumed. This constitutes a major difference in comparison to the region of Catalonia, where these mechanisms have been reported to only play a very limited role.

Employers-, Business-, and Trade Associations

All interview partners consistently estimated the role of employers-, business-, and trade associations and chambers of industry and commerce in knowledge transfer processes as active, important and positive (Interviews MAD 1: 7f.; MAD 2: 4f.; MAD 3: 1ff.; MAD 4: 1ff.; MAD 5: 11f.). The decisive organization in this regard would be the *Madrid Confederation of Employers and Industries* (CEIM), which is the roof organization of Madrid's business associations and, therefore, directly represents and coordinates the vast majority of these organizations and indirectly about 95% of the Madrilenian firms (Interview MAD 1: 1ff.; CEIM 2009). This centralized structure of business associative activities has been evaluated as a distinctive feature of the Madrid region in comparison to other Spanish regions, where especially regionalized tranches of national associations would exist, and, on the other hand, atomized local and sectoral business associations with only little coordination on the regional level (Interviews MAD 5: 11; MAD 8: 9). As a multi-sectoral "meta-association" with a clearly defined focus on the Madrid region as a whole it would be the central player and interlocutor representing the business side in the Madrilenian innovation and knowledge transfer policy. In this, and differentiating *CEIM* from many other Spanish business associations, it would have recognized early the importance of actively promoting innovation and knowledge transfer on the regional level, as the director of the innovation department at *CEIM* explains his function, also pointing to the specific challenges to coordinate a multi-sectoral confederation of business associations:

"The main function of the department for innovation and new technologies is to bring into contact the innovation, research and technological development policies of the Madrid Community with the business world. Obviously there is a problem to differentiate. A sectoral association has a clear idea of which aspects of R&D and innovation it has to develop, since there is a direct demand by their members. In a multi-sectoral confederation like *CEIM* the definition of adequate innovation services is more complex. In any case, because of its position in the cupola of the business sector in the Madrid Community, *CEIM* has, therefore, a mission to connect the necessities of the collectivity of the firms to the innovation policy. Let us say that we are interlocutors." (Interview MAD 3: 2; translation: MO)

CEIM would be a major cooperation partner in and initiator of many central Madrilenian innovation policies, strongly supporting the interaction of academia and business through the *madri+d* system, and the recent shift towards cluster policies through *Madrid Network* (Interviews MAD 1: 7f.; MAD 2: 4f.; MAD 3: 4ff., 11f.; CEIM 2009: 98ff.). Apart from this engagement, CEIM would also offer a broad variety of conferences, information talks and formation services, which have been evaluated as positively and frequently used be the interviewed experts. Moreover, CEIM would also connect the regional firms to the *Enterprise Europe Network* in terms of innovation and knowledge transfer. Finally, also the technological expertise brought in by CEIM in these policy mechanisms would be appreciated by the regional actors, as the director of the *Technological Entrepreneurs Office* at *madri+d* illustrates representatively for other interview partners:

"We are lucky to have CEIM which normally eases our work when we have to include the firms into what we are doing. If we have e.g. a program for private investment to entrepreneurs, through CEIM we have facilities to diffuse the message to the possible investors from the business world. If we have e.g. a topic of technology transfer in the audiovisual sector, they can also diffuse the message to the firms in the sector. We know more about the technology that is generated in Spain and in other countries. Sometimes we organize events in Madrid to attract technology from other places and to offer them in Madrid. But the in depth knowledge of technology of the firms is in the business associations." (Interview MAD 2: 4; translation: MO)

Chambers of Industry and Commerce

Similarly positive the interview partners estimated the role of the *Madrid Chamber of Commerce and Industry*. Also the Madrid Chamber would be an active player in the Madrilenian knowledge transfer system, and a major initiator and cooperation partner in the different regional innovation policies (Interviews MAD 1: 7f.; MAD 2: 4f.; MAD 3: 1ff.; MAD 4: 1ff.; MAD 5: 11f.). This active role in innovation and knowledge transfer exercised by the Madrid Chamber strongly differentiates it from its Catalan counterparts, which have been reported to only marginally engage in these fields, to lack necessary human and organizational resources for it and to have a more local rather than a consistent regional perspective. The Madrid Chamber, on the other hand, would have been among the first Spanish chambers to recognize the importance of and to engage in these fields (Interview MAD 4: 1ff.). Furthermore, the centralized structure with only one chamber of commerce in the Madrid Community (represented by several local offices in the distinct communalities) would make it easier to develop a coherent perspective on the region as a whole and to be a visible, powerful and efficient cooperation partner for the regional innovation policy (Interview MAD 5: 11f.). Also the concentration of all innovation and technology related activities of the Madrid Chamber within one of its organizational units, the department of innovation and competitiveness, has been highlighted in this connection (Interview MAD 4: 2). In addition, the Madrid Chamber would have a long tradition in being part of the regional economic policy and, therefore, taking responsibility for the Madrid region as a whole. In this regard, and from the point of view of the regional government, the head of the technology transfer department at the *Madrilenian Ministry of Economy and Finance* points to the role of the Madrid Chamber as "diffusion instrument" for the programs of his ministry and to the close links between the ministry and the chamber:

> "What resulted to be an extraordinarily interesting and very useful instrument is the Madrid Chamber of Commerce. It is a very large and well organized chamber, with which we have contracts signed including their help in enhancing R&D and innovation activities in the firms through their knowledge of and contacts to the firms. [W]e have seen that doing this via the Madrid Chamber is much more efficient." (Interview MAD 5: 11; translation: MO)

Although, as already indicated before, the Madrid Chamber's engagement in the *madri+d* system is lower than the one of CEIM (not least because *madri+d* is an initiative of the Ministry of Education), especially its intensive participation in *Madrid Network* and the respective cluster policies has been highlighted by the interview partners as very important and fruitful (Interviews MAD 1: 7f.; MAD 2: 5; MAD 4: 3; MAD 5: 11f.). Moreover, also the classic instruments like conferences, information talks, assessment services and, above all, professional formation services have been listed by the interview partners as further important and positive contributions of the chamber to the knowledge transfer system of the Madrid Community. Also the relation between the Madrid Chamber and the associations represented by CEIM would be, apart from some overlapping in the provided services in past, positive and cooperative. As the director of the innovation department at CEIM accentuates, there would be a clear division of labor which is not least rooted in the larger financial resources the Madrid Chamber could manage:

"There are differences between the two institutions. CEIM is a private entity and affiliation to CEIM is voluntary and the associations pay a voluntary contribution. Therefore, our disposable financial resources to offer services are very limited. The chamber of commerce is an entity of public law, depends on or is protected by the Ministry of Economy and Finance and through the law it has an enormous budget. And a certain part of this has to be invested in activities fostering technological innovation in general. Therefore, the type of services we offer is different. The chamber offers services directly, i.e. through subcontracted consultants, offers services for preparing projects, for the participation in European projects, informs about the incorporation of ICT, etc. [...] We do almost the same, but on a much lower scale, since we depend on funding through the different administrations. One could say we work parallel to each other, but they have much more resources than we have. [...] Actually I preside two joint commissions with the chamber on innovation and new technologies and entrepreneurs, so that there is an excellent relation." (Interview MAD 3: 8f.; translation: MO)

Associations and Chambers in Madrilenian Automotive

Differentiated by sectors, a similar picture arises for Madrid's *automotive* industry. Both, employers-, business-, and trade associations, like the business association of the Madrilenian metalworking industry and CEIM, and the Madrid Chamber, would play an active and important role for the knowledge transfer activities in the sector (Interview MAD 6: 7f.). In this, the contribution of the associations would be more indirectly through their engagement in the *madri+d* system and, above all, in the *Cluster Automoción* and *Madrid Network*. As the managing director of the *Cluster Automoción* at *Madrid Network* explains, a clear division of labor would exist between the business associations and the cluster, whereby the former would outsource most knowledge transfer, innovation and R&D-related activities to the latter. This would also have been one major motive for the associations to initiate and promote this cluster policy. The chamber, on the other hand, would also directly offer services in the field of innovation and knowledge transfer, as has already been described before (Interview MAD 6: 7f.).

Associations and Chambers in Madrilenian Biotechnology

For the Madrilenian *biotechnology* sector, the role of chambers and associations has been evaluated as less significant than in traditional industrial sectors, especially because of its strong international orientation and its more research-driven character (Interview MAD 7: 6f.). However, and differentiating the Madrid region from Catalonia, the Madrid Chamber and business associations like *BioMadrid* and *CEIM* would have already recognized the importance of the biotech industry and would increasingly offer specialized services to the biotech firms in the region, like the participation in and organization of trade fairs, congresses, international seminars, assessment in financial questions and in start-up activities (Interview MAD 7: 7). However, in biotechnology, associations and the Madrid Chamber would not be members of *Madrid Biocluster* and a clear division between both would exist, as the managing director of *Madrid Biocluster* accentuates:

> "I perfectly distinguish between the roles of the business associations and the cluster: A business association is a place of interest representation, of administration, of fiscal novelties etc., while a cluster is a place to do business. In other words: If we compare this whole universe with a firm, the business association would be the 'back office' and the cluster the 'front office'." (Interview MAD 7: 6; translation: MO)

Common and Binding Technical Standards

Finally, the interview partners only scarcely reported on the role of *common and binding standards* as association based mechanisms of knowledge transfer in the Madrid region. However, independently of the sector, they particularly stressed the role of chambers and associations in developing, negotiating, regulating and adapting standards for professional and vocational training courses as important channels of knowledge transfer (Interviews MAD 1: 8; MAD 2: 5; MAD 3: 6ff.; MAD 4: 2ff.; MAD 5: 13).

6.3.2.3 Organization Based Mechanisms of Knowledge Transfer

With regard to the organization based mechanisms of knowledge transfer (transfer of technical and scientific staff, purposeful headhunting, business start-ups) the interviewed experts for the Madrid region basically confirmed the assumed medium range importance for knowledge transfer processes when looking at the region as a whole (Interviews MAD 1: 12ff.; MAD 2: 11; MAD 3: 15; MAD 4: 16ff.; MAD 5: 20f.).

Transfer of Technical and Scientific Staff/ Purposeful Headhunting

The *transfer of technical and scientific staff* and *purposeful headhunting* activities would indeed be used as mechanisms of knowledge transfer, although, like in Catalonia, it would be strictly limited to the larger firms with own R&D- and innovation activities, firms and sectors with a higher technology level and firms that already employ academically trained staff. For all other firms, financial and human resources related barriers, as well as mental and administrative barriers would impede intensive fluctuation of highly qualified personnel to achieve knowledge transfer. This is summarized by the director of the innovation department at *CEIM*, who, overall, does not recognize intensive competition and fluctuation on qualified staff among Madrilenian firms:

"First of all there is an important mental barrier, but there is also an administrative barrier. Researchers form universities and research centers are doing their curriculum and are evaluated depending on their scientific publications. [...] Therefore, the relations to the firms inevitably remain second priority. [...] I think that the new *Science Law* [Ley de la Sciencia] intents to resolve this, but, of course, we do absolutely not have the mobility between the scientific and the business world which exists in other countries. Much less than in the USA. [...] My impression is that first of all firms conducting R&D have qualified personnel who they take care of. Then there are not so many firms with a strong, potent and well equipped crew of R&D-personnel. Therefore, I think that the problem of competition on qualified staff does not exist." (Interview MAD 3: 15; translation: MO)

Like in Catalonia, also in the Madrid region the state would addresses this field with several programs and measures, of which the most prominent would be the national *Torres Quevedo* program promoting the incorporation of doctoral students and post-doc researchers in the firms (Interview MAD 2: 11; MAD 4: 16; MAD 8: 14). Like in Catalonia, also the experts for the Madrid region evaluated this measure as not successful from a firm's perspective, as an exemplary statement of the director of the department of innovation and competitiveness of the *Madrid Chamber* illustrates:

"There is a national program, the *Torres Quevedo* program, [...] which not only in the Madrid Community, but also on the national level has a very low demand from the firm's side. In this point I would not locate the problem at the universities, but at the firm's side. The firms lack the necessary culture and the vision to understand that the incorporation of a person with such a qualification could bring them into a better position. There has not been a general fluctuation, although there have been efforts in this direction. But this mechanism has not been very successful." (Interview MAD 4: 16; translation: MO)

More positive have been evaluated measures on the regional level, especially those through the *madri+d* system. On the one hand, the *Office for Young Researchers* would offer a broad variety of mechanisms and services to researchers interested in changing to the business world, especially through specific training and formation services. On the other hand, the website of *madri+d* would be the region's most important job-portal for scientific personnel, bringing together demand and supply for highly qualified staff and accelerating transfer activities (Interviews MAD 1: 12ff.; MAD 2: 11).

Business Start-Ups

As assumed, *business start-ups* and academic *spin-offs* would be medium important mechanisms of knowledge transfer, like in Catalonia (Interviews MAD 1: 13f.; MAD 2: 12f.; MAD 3: 16; MAD 4: 17f.; MAD 5: 21ff.). In technology related sectors these start-up activities would especially occur around the region's scientific parks. However, despite extensive support of these activities, the absolute number of newly created firms in the Madrid region would still not be satisfying and their success rates would be low. An insufficient entrepreneurial mentality, especially among the region's researchers, a lack of business and management skills in many academic spin-offs and an insufficient provision of seed- and venture-capital have been the main reasons mentioned by the interview partners. Many public measures and programs would address these business start-up activities, of which the most positively evaluated are the ones by the *Technological Entrepreneurs Office* at *madri+d*, which are summarized by its director as follows:

"In our section of *madri+d* we provide support from the business plan to specialized services, like mentoring services for entrepreneurs, support services in IP-related topics, we co-finance reports on IP- and technology-assessment, financial rounds for start-ups with venture capital, we have a network of private investors, we organize two large forums with twenty projects each, and, in addition, many more projects for entrepreneurs circulate in our business angel and private investors network. We have more than 80 investors inscribed in our network. Moreover, we provide much information on national programs etc., internationalization services and network creation." (Interview MAD 2: 12f.; translation: MO)

However, other support measures in this field, like the services provided by the region's scientific parks and business incubators, would not bring the pretended returns in terms of a cost-benefit-relation. This would also be true for the respective services provided by the *Madrid Chamber*, as one of its interviewed representatives openly admits (Interview MAD 4: 17f.). With regard to the availability of seed- and venture capital for start-up activities, the interview partners stated that the state would provide sufficient seed-capital for the early phases of business development. What would be missing are private investors, venture-capital firms and professional business angels in the seed-phase and especially in the later phases of business development, where venture-capital would only insufficiently be provided (Interviews MAD 1: 18; MAD 2: 16f.; MAD 3: 19f.; MAD 4: 21f.; MAD 5: 23). Especially a lack of professionalization in the fields of innovation and new technologies of the already existing business angels and venture-capital suppliers has been added in this connection as a major barrier to more successful knowledge transfer activities via business start-ups. Moreover, the majority of the venture capital suppliers, and also of the banks, would lack a tradition in investing in innovative technologies and academic spin-offs, and would rather invest in traditional fields like construction (Interview MAD 3: 19f.). Finally, and turning around the perspective, the representative of the *Madrid Chamber* also points to a lack of adequate start-up projects, so that often funds provided by the regional government or by the *Madrid Chamber* itself would remain unaccessed (Interview MAD 4: 21f.).

Organization Based Mechanisms in Madrilenian Automotive

For the Madrilenian *automotive* sector the reported picture is slightly different. While the transfer of technical and scientific staff as well as graduates from the scientific sector to the automotive firms would be very low, especially because of strongly pointed mental barriers between the two worlds and a lack of demand from the firm side, the transfer of staff between firms would be frequent. Because of the difficulties to find adequate personnel coming directly from the universities or scientific institutes, firms would rely more on recruiting personnel which has already been trained in other firms. Purposeful headhunting activities would, therefore, absolutely be usual in the sector. What would still be missing are specific policy measures addressing the specific needs of automotive in this regard to intensify transfer from the academic world and graduates. Also the *Cluster Automoción* would not yet be active in this direction (Interview MAD 6: 16ff.). Like Catalonia, also the Madrilenian automotive industry would not be characterized by vital start-up activities. The consolidated and closed character of the sector would impede these activities, and especially successful academic spin-offs, which would be very rare. In addition, a low entrepreneurial mentality among graduates in automotive related fields would reinforce this situation. Also in this point a lack of specific policy support has been remarked, since the *Cluster Automoción* would have an exclusive focus on already existing firms and less on business creation (Interview MAD 6: 18f.).

Organization Based Mechanisms in Madrilenian Biotechnology

Also in the Madrilenian *biotechnology* sector, the transfer of technical and scientific staff would only play a subordinated role, although because of different reasons. The majority of the staff in the mostly research driven biotech firms would stem from universities and research institutes, but there would not be strong fluctuation between the two worlds. On the one hand, most biotech firms would remain small in size during their whole business life cycle, and, on the other hand, even in this sector a mental barrier would exist making it unlikely that researchers in the field would be motivated to transfer to industry. Even less important would be transfers between firms and purposeful headhunting activities, since the firm's main recruiting pool would be the scientific world. To overcome certain shortcomings of qualified staff in the sector the *Madrid Biocluster* would be active in attracting qualified persons to the Madrid region, apart from the respective activities conducted through the scientific parks, as its managing director remarks (Interview MAD 7: 11ff.). Business start-up-, and especially academic spin-off-activities, on the contrary, would be very intensive and would be decisive mechanisms of knowledge transfer. Madrilenian biotechnology would still be an emerging and dynamically growing sector, with an annual growth rate in number of firms of more than 20%, as the interview partner states. He accentuates, that a comparatively high entrepreneurial mentality would exist in the sector, making it one of the most dynamic sectors in the region (Interview MAD 7: 14). However, also for biotechnology the provision of venture capital would be one of the major bottlenecks for business foundations, complemented by an insufficient professionalization of venture-capital suppliers, business angels and banks to the specific needs of the sector. Also in this regard *Madrid Biocluster* would be a major interlocutor between the capital providers and the firms (Interview MAD 7: 19ff.).

6.3.2.4 State Induced Mechanisms of Knowledge Transfer

Like in Catalonia, and like assumed for the Spanish regions, also for the Madrid Community the interviewed experts ascribed most of the state induced mechanisms of knowledge transfer (regional and national government measures, (larger) state-induced innovation projects, scientific parks) a decisive role in the regional knowledge transfer system, except technological centers and technology transfer offices.

National and Regional Government Measures

In this, also the Madrilenian innovation and knowledge transfer system has been described as clearly public action driven. *National and regional government measures* would basically shape and vitalize knowledge transfer activities in the region (Interviews MAD 1: 8ff.; MAD 2: 6ff.; MAD 3: 9ff.; MAD 4: 12ff.; MAD 5: 2ff.; MAD 8: 2ff.). While national policy measures, especially those by the *CDTI*, like the *CENIT* programs, would be major sources of funding and central incentives for cooperative R&D- and innovation projects as well as for other knowledge transfer channels, the regional level in the Madrid Community would especially be relevant in creating network structures and institutional support for the economic actors, not least within the framework of the *PRICIT* (Regional Scientific Research and Technological Innovation Plan). As the most important regional policy measures the interviewed experts consistently listed the activities of the *Madrilenian Economic Development Agency* (IMADE), the *madri+d* system as a central regional platform connecting science with business, and *Madrid Network* with its recent shift towards a more cluster oriented regional innovation policy. This policy focus on networks and clusters is a major differentiating feature of the Madrilenian knowledge transfer system when compared to Catalonia, where the central role of the *ACC1Ó*-agency has been reported to dominate the regional policy landscape and where a more single project oriented strategy is persecuted.

Another distinctive factor would be Madrid's specific position among the Spanish autonomous regions as the country's national capital city (Interviews MAD 1: 2ff.; MAD 3: 10f.; MAD 5: 2ff.). As the director of the innovation department at *CEIM* highlights, this would make the Madrid Community a preferred target for national government measures, not least because of the spatial proximity between regional actors and national administrative bodies on the region's territory:

> "In any case, the Madrid Community is different again. It is different again, because the *Spanish National Research Council* is in Madrid and about 50 research centers are here in Madrid. In a certain way the relation to a national center is much more fluent on the regional level. *CIEMAT*, the *National Research Center for Energy, Environment and Technology*, depending on the Ministry of Industry, is here in Madrid. Therefore, interaction with the Madrilenian firms is much easier. *CDTI*, as financing entity of business innovation is also in Madrid. I want to say that Madrid, because of its political position, has some very different characteristics in terms of location of research centers and firms compared to other autonomous regions." (Interview MAD 3: 10f.; translation: MO)

Another striking difference would be the active and intensive participation of the corporatist actors in the Madrilenian innovation policy (Interviews MAD 1: 7ff.; MAD 2: 4f.; MAD 3: 1ff.; MAD 4: 2ff.; MAD 5: 11f.). Especially *CEIM* and the *Madrid Chamber* would be usual and important partners for the regional government in initiating, promoting and implementing the respective policy measures. However, the interview partners also point out that the close links that these actors maintain to the regional ministries, and the dependence of their activities on public funding, would mostly leave the protagonism to the state, so that in certain fields these corporatist actors appear to be more implementing instruments for regional policy measures than independent impulse givers, as it would be the case e.g. in the German region of Baden-Württemberg.

(Larger) State Induced Innovation Projects

Furthermore, also *(larger) state induced innovation projects* would be important channels of knowledge transfer in the Madrid region (Interviews MAD 3:

12ff.; MAD 4: 10ff.; MAD 5: 3ff.; MAD 8: 2ff.). Especially through specific cluster projects (aerospace, automotive, biotechnology) on the regional, the *CENIT* projects funded by the *CDTI*, and the *Spanish Technological Platforms* on the national, as well as their counterparts on the EU-level (e.g. European Technological Platforms), the state would foster innovation and knowledge transfer in specific technological fields, above all in the defense industry and the aerospace sector, which would be essential for the Madrid region. In these projects, larger and small firms, as well as scientific institutions would cooperate horizontally and vertically on a contract based and project specific basis and would mutually exchange knowledge.

Scientific Parks and Technological Centers

Scientific parks, of which six are located within the Madrid Community, would be essential mechanisms of knowledge transfer (Interviews MAD 1: 11; MAD 2: 8; MAD 3: 13f.; MAD 4: 14; MAD 5: 19f.; MAD 8: 13). These parks would offer physical space for firms, universities and research institutes to interact in spatial proximity, what has been evaluated as very positively by all interviewees. Especially their role as business incubators for technology based start-ups and spin-offs has been highlighted in this connection. In addition, also the recent efforts to network the region's scientific parks, to centrally coordinate them and to bring them into contact with the regional clusters under the roof of *Madrid Network* have been acknowledged positively by the experts. However, these scientific parks would mostly be connected to the region's universities and, therefore, have a too strong focus on basic research, reflecting Madrid's excellent research landscape, but would often not fit to the region's business sector's needs. In this connection, the absence of *technological centers* with a stronger focus on applied research and development activities has been remarked and criticized by some of the experts. As the director of the department of innovation and competitiveness at the *Madrid Chamber* points out, this would reflect the regional policy's historical exclusive focus on scientific excellence, and its negligence of transfer activities to the firms:

"There are some very singular characteristics of the Madrid Community, which are probably different from what we can find in Catalonia or […] in the Basque Country. And this refers to the fact that we have a large number of public universities and we lack of technological centers. Policy always pretended to broaden the field of action of the universities, forgetting the role technological centers could play. In this regard, also the firms have not been too active in dynamizing the creation of these technology transfer oriented infrastructures. […] We have a focus on education and basic research, but a low impact to the productive sector and a scarce contribution to the firms' competitiveness. This has been a traditional question." (Interview MAD 4: 2f.; translation: MO)

Technology Transfer Offices

Finally, similar to the picture drawn by the Catalan experts, and those drawn by the experts in the German regions, the interviewed experts evaluated the *OTRIS* as less significant than assumed (Interviews MAD 1: 15f.; MAD 2: 13; MAD 3: 17f.; MAD 4: 3f.; MAD 5: 9f.; MAD 8: 15f.). On the one hand, like in the rest of Spain, it would be obligatory for all researchers at public universities and research centers to pass through these transfer offices, since in Spain the results of scientific research are the property of the respective institutions. The OTRIS would, thus, be in charge of all legal and contractual transactions connected to cooperation with the business sector and the transfer and commercialization of IPRs. Therefore, as an administrative unit they would play a central role for knowledge transfer processes in the region. On the other hand, also in Madrid the OTRIS have been evaluated as mostly inefficient, bureaucratic and not proactive enough. Moreover, they would often lack of knowledge on the needs of the business sector and of a clear valorization analysis of their portfolios. As the director of the innovation department at *CEIM* critically formulates, the OTRIS would be outdated institutions which are in a situation to be overtaken by newer players like the *madri+d* system:

"I think that the OTRIS are a relatively antique mechanism and that they are not valid anymore. The *madri+d* system has initiated a 'technological marketplace', which has more ease. The OTRIS are, at the end, entities of the university, and work, thus, with the same administrative slowness. […] They should be modernized. In fact, they should be the universities' commercial departments. But that is absolutely not the case!" (Interview MAD 3: 17; translation: MO)

State Induced Mechanisms in Madrilenian Automotive

For the *automotive* sector the results almost equal those for the region as a whole. National and regional government measures, especially those within the frameworks of the *National Plan for Scientific Research, Development and Technological Innovation* and the *National Integral Plan for the Automotive Industry*, as well as the regional *Business Innovation Plan* by IMADE, would be important impulses and mechanisms of knowledge transfer in the sector (Interview MAD 6: 9f.). Moreover, also the *Cluster Automoción* under the umbrella of *Madrid Network* would be an important instrument of knowledge transfer in this connection. Larger state induced innovation projects, however, would only play a medium range role for knowledge transfer in Madrilenian automotive (Interview MAD 6: 12). Programs to create innovation consortia around concrete innovation projects and topics, like the *CENIT* program, through which the state also imposes some of its own priorities to the market actors, as it would be the case with the development of electric mobility concepts or in certain subsectors of aerospace, would indeed exercise a tractor effect and efficiently generate knowledge transfer among the participating actors, in the limited number of cases in which Madrilenian automotive firms would participate in them (Interviews MAD 5: 3ff.; MAD 6: 12). Scientific parks would also for automotive be important facilities to exchange knowledge with the scientific world in close spatial proximity. Although the only technological center in the region (INSIA) would be related to automotive, it would be too small and much too specialized (in bus security) to exercise any remarkable effects on the overall knowledge transfer activities in the sector (Interview MAD 6: 13). Finally, like in Catalonia, also for the Madrilenian automotive industry the OTRIS have been reported to not play any role at all, not only because of strong mental barriers to interact with these facilities, but also because of the fact that the trade of IPRs would not be relevant for the firms in the sector (Interview MAD 6: 20ff.).

State Induced Mechanisms in Madrilenian Biotechnology

Also in the Madrilenian *biotech sector*, regional and national government measures would be important initiators and drivers in almost all sector-relevant knowledge transfer activities (Interview MAD 7: 8ff.). The very young, dynamically growing and atomized character of the sector would be a major explanation of this strong dependence on public policy. Again, the respective national and regional policy plans, but especially the national *CENIT* projects by CDTI, several EU-policy measures and the regional cluster policy with *Madrid Biocluster* under the roof of *Madrid Network* have been highlighted by the interviewed expert for the sector. Moreover, (larger) state induced innovation projects would also play a role for knowledge transfer in Madrid's biotech sector. Programs to create innovation consortia around concrete innovation projects and topics, like the CENIT program, through which the state also imposes some its own priorities to the market actors, would exercise a pulling effect in the sector and efficiently enhance knowledge transfer among the participating actors (Interview MAD 7: 8f.). Furthermore, and like in Catalonia, also scientific parks have been evaluated as potent motors of knowledge transfer (Interview MAD 7: 10). In five of the region's six scientific parks biotechnology would play a major role, and the sixth would even have an exclusive focus on the sector, what has been highlighted as very positive for the sector's knowledge transfer activities, since physical proximity of firms, clients, suppliers, service providers, hospitals, research centers and universities would be vital for its innovation performance. Technology centers in the sector would not exist. Finally, the OTRIS would indeed be used intensively in the sector, not least because IP-related topics would be crucial in the field. However, also the expert for biotechnology criticized the OTRIS' financial and human resource related underequipment, as well as their too passive interpretation of their role (Interview MAD 7: 8ff.).

6.3.2.5 Market Based Mechanisms of Knowledge Transfer

As concerns the market dimension of knowledge transfer mechanisms (external consultancies; mergers, acquisitions, hostile takeovers; buying, selling and licensing of patents), like in Catalonia, the interviewees only partly confirmed the theoretical assumptions when looking at the Madrid Community as a whole.

External Consultancies

Detailing the results for the single mechanisms, the use of *external consultancy services* for knowledge transfer purposes has been reported to be very rare (Interview MAD 2: 15; MAD 3: 18; MAD 4: 19ff.; MAD 5: 24; MAD 8: 16). A lack of demand for these services generated by the firms, a lack of need to use these services in the predominantly not innovating regional firms, an individualistic business culture impeding cooperation with external consultants and a lack of tradition would be the major reasons for this. As the director of the innovation department at *CEIM* remarks, this low use of external expertise not only in the region, but also in Spain in general, would be one major disadvantage in comparison to other European countries and would be a decisive open task for the future:

> "There are very few [external consultancies; MO]. I absolutely agree that there should be more firms helping the researcher in defining his technology, in making a market study, protecting intellectual property rights, finding adequate clients. In general, this should be a function of the OTRIS [...]. But, logically, consultancies are much more flexible, more active, and, above all, they are making money with that and are, therefore, more efficient. These firms do not exist because there is no tradition of consultancy, neither in Spain, nor in Madrid. I think that this is an important deficit we have in comparison to Germany, France, the UK and the USA. There is no tradition of good consultants in technology, and, on the other hand, there is a too strong administrative character of the universities not permitting an easy way of transferring technology. [...] Evidently, until now the Spanish firms could compete well in acquiring machinery and processes with incorporated technology. But this does not work anymore, since everybody can buy technology. Technology is a commodity. [...] I think Spanish firms should generate more early demand for technology." (Interview MAD 3: 18; translation: MO)

An exception would be some of the larger firms in the region, which would more likely use external consultancy services, or, in a very limited number of cases, also maintain own internal consultancy departments. To compensate this low use of external expertise in knowledge transfer processes, public policy institutions like *IMADE*, the different offices of *madri+d* and *Madrid Network* would also offer consultancy services, or partly contribute to the funding when external consultants are contracted. However, also this public supply would not be used intensively by the regional firms (Interviews MAD 2: 15; MAD 4: 20f.; MAD 5: 25).

Mergers, Acquisitions and Hostile Takeovers

Furthermore, also *mergers, acquisitions and hostile takeovers* would not play an important role in knowledge transfer (Interviews MAD 1: 17; MAD 2: 15f.; MAD 3: 18f.; MAD 4: 21; MAD 5: 25). First, most firms would not be listed on the stock exchange, making these mechanisms comparatively difficult to realize. Second, mergers, acquisitions and hostile takeovers would, if they occur, mostly not be motivated by acquiring knowledge and technology, but by accessing new markets or increasing production capacity. Third, the interview partner from *CEIM* adds a general lack of "growing culture" in both the regional and the Spanish firms, which would remain small over their whole business life cycle:

"From this department I am promoting projects which aim to increase the average size of Spanish firms. We have to consider that in Spain and Madrid we have less than half of the employees employed in large firms than in Germany, France or the UK, and the double of employees in micro-firms with less than 10 employees. Therefore, the average dimension of the Spanish firms is very small and their capability to compete is low. […] Essentially, this is a cultural question. Here, a small firm is born and it remains small for a life time. There is no tradition of growing. I always mention the examples of Bosch, Daimler and Siemens, which were persons who founded companies which are now multinationals. This would be the right mechanism." (Interview MAD 3: 19; translation: MO)

Buying, Selling and Licensing of Patents

Finally, also when asked for the importance of trading IPRs as mechanism of knowledge transfer, the experts broadly confirmed the results obtained for Catalonia: *Buying, selling and licensing of patents* would not play any significant role in the Madrid Community's knowledge transfer system (Interviews MAD 1: 18f.; MAD 2: 17; MAD 3: 19; MAD 4: 22f., 18f.; MAD 8: 17f.). Above all, a lack of demand generated by the firms, as well as insufficient incentives and tradition of patenting in the scientific world would explain the huge gap between strong research activities and extraordinarily low patenting and licensing activities in the region.

Market Based Mechanisms in Madrilenian Automotive

Divided by sectors, in the Madrilenian *automotive* industry external consultancies have been reported to be frequently used, which is contrasting clearly the results from Catalonia. Coming from a situation without any consultancy activity in the sector, especially a recently emerging branch of sector specific consultancies would be actively used by the automotive firms, especially in questions of accessing public funding for innovation and knowledge transfer activities (Interview MAD 6: 21). On the other hand, mergers, acquisitions and hostile takeovers would also in automotive not play an important role as mechanisms of knowledge transfer. If anything, these would be conducted by foreign MNEs to secure their strategic position, but only in a limited number of cases to acquire know-how (Interview MAD 6: 21f.). Like in the region as a whole, and because of the same reasons, also in Madrilenian automotive the trade of IPR would almost be irrelevant (Interview MAD 6: 24).

Market Based Mechanisms in Madrilenian Biotechnology

In *biotechnology*, external consultancies would only marginally be used. An insufficient supply of consultancies specialized in the biotechnology field and a still low demand generated by the very small and only emerging sector would be the main reasons for that (Interview MAD 7: 18). Also mergers, acquisitions and hostile takeovers, as well as the trade of IPR would still not play an important role in Madrid's biotechnology sector, since most firms would still operate in very early phases of their business life cycles, in which they would still not be attractive candidates for being acquired or would have produced any patentable results. In addition, most firms in the sector would not be listed on the stock exchange (Interview MAD 7: 19ff.).

6.3.2.6 Community Based Mechanisms of Knowledge Transfer

With regard to the group of community based knowledge transfer mechanisms (informal contacts, regional platforms, additional (regional) institutions), which allow for trust-based relationships to be established and base on a common regional identity, culture and tradition, the interviewed experts broadly confirmed the theoretical assumptions (Interviews MAD 1: 4ff.; MAD 2: 14; MAD 3: 17; MAD 4: 19; MAD 5: 24).

Informal Contacts, Regional Platforms and other (Regional) Institutions

Informal contacts would be one mechanism among others in the Madrid Community, without being specifically relevant, and would be affected by the generally individualistic attitude in the region, being barrier to all kinds of interaction, formal and informal. Also *regional platforms* would be of medium range relevance for knowledge transfer activities. Those platforms would be particularly provided by the regional government through the *IMADE*, the *madri+d* system and increasingly by *Madrid Network*, as well as by corporatist actors like the *Madrid Chamber* or *CEIM* and would be actively used, although with the limitations discussed above. Other *(regional) institutions* relevant for knowledge transfer processes in the region have not been mentioned in this connection by the interview partners.

According to the interviewed experts, a strong, common and historically grown Madrilenian *regional identity* would not exist, and would, therefore, not be a supporting factor in establishing these informal contacts, cooperations and, thus, knowledge transfer processes (Interviews MAD 2: 14; MAD 3: 17; MAD 4: 19; MAD 5: 24). On the one hand, as the national capital city area, Madrid would have a more open and international, rather than regional self understanding in comparison to other Spanish regions, as a representative of *Madrid Network* points out:

> "The fact of always being the capital of Spain provoked that there is not such a localist mentality than in other regions […]. Therefore, actors here always had an outward oriented perspective, towards Europe […]. But it is true, compared with other Autonomous Regions we do not have such a regional spirit they have." (Interview MAD 6: 14; translation: MO)

On the other hand, this lack of regional identity would be compensated by the density and proximity of institutions and actors in the region, as well as the region's compact character (Interviews MAD 2: 14; MAD 3: 17). What would be decisive for certain channels of knowledge transfer would be spatial proximity, rather than a common cultural identity. This would also be the position of the regional policy in this regard, as the director of the *Technological Entrepreneurs Office* at *madri+d* explains:

> "I think that regional identity is not so important for doing business. Identity is a cultural topic, and it is something completely different to do business or to find cooperation partners. If you need them to be in geographical proximity, then maybe regional identity could help, since interests conjugate or you could use consolidated regional forums for this kind of contacts. But if you are looking for partners from around the world, regional identity is less important. And it could even be an obstacle, since it limits your perspective. We state it very clear, we are a regional network, but with an international perspective. We are not only trying that firms from Madrid use services in Madrid, but also that what is done in Madrid is known in the world and is used outside of Madrid, and vice versa." (Interview MAD 2: 14; translation: MO)

Community Based Mechanisms in Madrilenian Automotive and Biotechnology

Differentiated by sectors, the results for the Madrilenian *automotive* industry fully parallel those for the region as a whole (Interview MAD 6: 4, 8, 14). Also for *biotechnology* informal regional contacts would have some significance, although they would be clearly less important than formalized interaction, essentially because of the strong fear of IP-losses. As most important regional platforms the *Madrid Biocluster*, the *BioMadrid* association and the *Spanish National Center of Biotechnology*, which is located in Madrid, have been mentioned. However, overall the sector would absolutely not have a regional, but rather a national and international orientation, so that these platforms would predominantly be used to access contacts on these higher levels. For the same reason, a strong regional identity would absolutely not play a role for knowledge transfer processes in the Madrilenian biotechnology sector (Interview MAD 7: 16f.).

6.3.3 Summary

Summarizing these results, also the qualitative case study on the regional knowledge transfer system of the Madrid Community revealed both broad correspondence with the theoretical assumptions derived from the literature on VoC and NIS for the Spanish MME, but also considerable deviations from some of these assumptions, as well as from the results obtained for the region of Catalonia, pointing to regional structural specialisms, distinct actors constellations and different regional policy approaches. Most of the dimensions of knowledge transfer investigated in this study on the Madrid Community are, therefore, affected by the region's atomistic business structure, the low technology level of the business population, a lack of tradition in conducting own R&D- and innovation activities and the tendency to rather import new technologies, a very traditional orientation of parts of the scientific world not acknowledging transfer activities as one of their key missions, as well as an individualistic business culture.

On the network dimension, these regional specificities of the Madrid Community are manifested in a stronger focus of the regional policy to address the individualistic business culture and the generally low cooperative attitude of the regional actors with network and cluster-building measures, of which the *madri+d* system and *Madrid Network* are the most striking examples. It is these two measures which stand in the center of the Madrilenian knowledge transfer system, supporting horizontal and vertical, as well as informal and institutionally stabilized cooperation, and which are some of its most distinctive characteristics in comparison to Catalonia. Not least through these measures, but also because of a comparatively high concentration of larger firms and business-R&D-activities in the region, especially cooperation in R&D resulted to be considerably higher than in Catalonia and as assumed for the Spanish case.

A second region specific finding is the comparatively strong role of the association based mechanisms of knowledge transfer. Especially business associations, represented by *CEIM*, and the *Madrid Chamber* resulted to be central players in the Madrilenian knowledge transfer system, decisively contributing to the regions overall innovation and knowledge transfer activities. In this, their centralized structure, their internal cohesion and coordination, their organizational and personal equipment, their early acknowledgement of innovation and knowledge transfer as key topics, as well as their tradition of developing a perspective on the Madrid region as a whole, taking responsibility for it and being an accepted partner in the initiation, development and implementation of regional innovation policy measures are differentiating them clearly from their Catalan counterparts.

As expected, and like in the Catalan case, the organization based mechanisms of knowledge transfer (personnel transfer, headhunting, business start-up activities) result to only play a medium range role in the Madrid Community. The use of these mechanisms is strictly limited to the larger firms with own R&D- and innovation activities, firms and sectors with a higher technology level and firms that already employ academically trained staff. For all other firms, financial and human resources related barriers, as well as mental and administrative barriers impede intensive fluctuation of highly qualified personnel to achieve knowledge transfer. In this connection, the *madri+d* system unfolded to be a key instrument and platform in bringing together demand and supply for highly qualified and scientific staff in the region. Entrepreneurial activities in technology based fields are mostly publicly supported, especially within the framework of the region's scientific parks and their business incubators. In this, even in the Madrid Community as Spain's financial industry center the state is also the central provider of seed-capital, pointing to underdeveloped markets for professionalized seed- and venture capital supply and business angel services for regional start-ups and spin-offs.

Overall, also the Madrilenian knowledge transfer system can be described as clearly public action driven. The different policy levels resulted to be decisive impulse givers, initiators, funders, assessors and coordinators in all investigated dimensions of knowledge transfer processes. Characteristically for the Madrid Community's innovation and knowledge transfer policy is a strong focus on network creation and cluster-building, as is manifested in the *madri+d* system and *Madrid Network*, as well as an intensive participation of the corporatist actors CEIM and, above all, the Madrid Chamber, in all phases of the policy process, but especially in its implementation. Another focus lies on the creation, maintenance and coordination of scientific parks as physical spaces of interaction between the business sector and the scientific world, especially in the field of basic research. In contrast to Catalonia, technological centers as locations for joint applied research and product development activities do not exist in Madrid and have been reported to neither be a priority for the regional policy, nor for the business sector, what has been seen critically by the experts. Furthermore, also in the Madrid region the OTRIS would not play their predicted central role as interfaces between business and science. Again, their too strong focus on administrative and contractual issues, their insufficient equipment with financial and human resources, as well as their too passive role interpretation have been listed as major shortcomings in the regional knowledge transfer system. This more proactive role is increasingly assumed be the *madri+d* system, which is perceived to have a more professional, network oriented and global perspective.

Surprisingly low for a metropolitan region resulted to be the market based mechanisms of knowledge transfer (external consultancies; mergers, acquisitions, hostile takeovers; buying, selling and licensing of patents). Many of the structural barriers to a more intensive use of these mechanisms characterizing the Spanish economy as a whole also apply for the Madrid Community. One of these barriers is an individualistic business culture generally impeding cooperation with other actors, including external consultants. In addition, the lowly innovative firms generate no demand for these services. Second, the atomistic business structure in the region with many family owned small and micro firms, which are mostly not listed at the stock exchange, impedes more intensive acquiring and taking over activities. Moreover, structural barriers within the Madrilenian research and higher education system disincentive researchers to generate patents, and the low technology level of the regional firms and their low own innovation efforts are major barriers to a higher demand for patents and licenses.

Finally, the assumed medium range importance of the community based mechanisms of knowledge transfer like informal contacts and regional platforms has been confirmed by the experts. The decisive enabling factor for these mechanisms to work efficiently would especially be the compact character of the region with a high density of regional actors allowing interaction in close geographical proximity. However, not significant in this regard would be a common and strong, historically grown regional and cultural identity, which would not exist in Madrid. As the national capital area the actors in the Madrid Community would have a stronger national and international, rather than a regional orientation.

From a sectoral perspective, automotive almost equals these results for the region as a whole. Despite a generally atomized business structure and an individualistic business culture in the sector, networks of all kind play a medium range important role as mechanisms of knowledge transfer, and, above all, vertical supply chain networks, which are especially strong developed in the aerospace branch. In the other subsectors, also the Madrilenian automotive industry lacks of own OEMs, so that the region could more correctly be described as a production plant than a complete automotive value added chain. However, cluster-building efforts through the *Cluster Automoción* under the roof of *Madrid Network*, and a concentration of headquarters and R&D-departments of automotive firms in the region partly compensate for these structural barriers, and also lead to more intensive inter-firm R&D-cooperations and joint R&D-projects with the scientific world than has been reported for Catalonia. These networking and cooperation activities are actively accompanied by business associations and the Madrid Chamber, much stronger than in the Catalan case. Strong mental barriers between the automotive sector and the scientific world, as well as a tradition to rather import technology than to develop it in house, not only impede strong transfers of scientific staff between the two worlds, but also contacts to the OTRIS and an intensive trade of IPR. Furthermore, as a mature and consolidated sector, Madrilenian automotive is characterized by very low start-up and spin-off activities. In sum, therefore, also in this sector the intervening state in collaboration with the corporatist actors is the key actor in most knowledge transfer activities, with regional, national and EU policy measures and scientific parks as central channels of knowledge transfer.

Table 8: Qualitative Relevance of Knowledge Transfer Mechanisms in the Madrid Community

MECHANISMS	ESP*	Mad	Auto	Bio
NETWORK BASED MECHANISMS				
business networks	O	O	O	O
informal networks	O	O	O	–
institutionally stabilized networks	O	O	O	O
customers and suppliers	O	O	+	O
cooperative inter-firm R&D-projects	–	O	O	+
cooperative R&D-projects with scientific institutions and universities	–	O	O	+
ASSOCIATION BASED MECHANISMS				
employers-, business-, and trade associations	O	+	+	O
chambers of industry and commerce	O	+	+	–
binding and common technical standards	O	O	O	O
ORGANIZATION BASED MECHANISMS				
transfer of technical and scientific staff	O	O	–	O
purposeful headhunting	O	O	O	–
business start-ups	O	O	–	+
STATE INDUCED MECHANISMS				
national government measures	+	+	+	+
regional government measures	+	+	+	+
(larger) state-induced innovation projects	+	+	O	+
scientific parks	+	+	+	+
technological centers	+	–	–	–
technology transfer agencies/companies	+	O	–	+
MARKET BASED MECHANISMS				
external consultancies	O	O	O	–
mergers and acquisitions, hostile takeovers	O	–	–	–
buying, selling and licensing of patents	–	–	–	–
COMMUNITY BASED MECHANISMS				
informal contacts	O	O	O	O
regional platforms	O	O	O	O
additional (regional) institutions	O	???	???	O

+ *Reported (*Hypothesized) as crucial mechanism of knowledge transfer*

O *Reported (*Hypothesized) as important but not decisive mechanism*

– *Reported (*Hypothesized) as not important at all*

??? Not reported by the interview partners

= Important deviation from hypotheses

= Modest deviation from hypotheses

= No deviation from hypotheses

Notedly different structures stand in the center of knowledge transfer activities in Madrid's biotech sector. Its very young and emerging character with a majority of small firms allows for dynamic start-up and spin-off activities, but does still not produce noteworthy merging, taking over and acquiring activities, as well as tradable IPRs. Like its Catalan counterpart, also Madrid's biotech sector maintains intensive knowledge transfer channels to academia, especially through cooperative R&D-projects, the transfer of staff, the OTRIS or collaboration in the physical proximity of scientific parks. Recent cluster-building efforts in the sector through *Madrid Network* and its *Biocluster* support these links. While most of the interaction in the sector is strictly formalized, informal flows of knowledge are clearly less significant because of the strict awareness of knowledge drains. The sector's general strong focus on the national and international arenas lower the role of the regional corporatist actors in facilitating knowledge transfer processes and regional platforms. Finally, also in biotechnology, through public measures, programs and innovation projects, as well as public facilities such as scientific parks and hospitals, the state is the key player in the sectoral knowledge transfer processes. Not least, the state exercises this role through the provision of seed- and venture-capital, being a vital basis for the emergence and the expansion of the Madrid's highly capital-driven biotech sector. Table 8 summarizes these results for the Madrid Community.

6.4 Lessons from the Spanish Case

Pointing to the two research questions guiding this study, the two qualitative case studies on the regions of Catalonia and the Madrid Community revealed both broad correspondence of the findings to some of the predicted structures and mechanisms of knowledge transfer for the Spanish case, as well as considerable variation from others. In this, both regions showed similar structural conditions strongly shaping and also limiting the concrete forms their respective knowledge transfer systems take, but also considerably different approaches in dealing with these conditions, basing on distinct historical traditions, actors constellations and policy strategies, making it difficult to talk of the Spanish case as a coherent and homogeneous innovation and knowledge transfer system.

The investigated Spanish regions join an atomistic business structure, with a vast majority of family owned small and micro firms, mostly lacking the financial and human resources for innovation and knowledge transfer, an individualistic business culture impeding cooperative action and network-building, a comparatively low technology level of the business sector, making innovation and knowledge transfer generally unlikely activities, and significant mental and structural barriers between the business sector and the scientific world. These structural conditions only allow for innovation and knowledge transfer activities on a low level, and make them occur only to a limited extent out of purely market, entrepreneurial or corporatist action. This leaves the protagonism in the regional knowledge transfer activities in both regions clearly to the different policy levels of the state, making it the central actor in initiating, funding, organizing and assessing these activities.

However, even if both investigated regional knowledge transfer systems can be described clearly as public action driven, the respective strategies chosen by regional innovation policy, as well as distinct regional actors' constellations and their degree of inclusion into the regional innovation policy and knowledge transfer processes constitute important drivers of regional incoherence and heterogeneity within the Spanish innovation and knowledge transfer system. In this, the central role of the different policy levels works in two opposite directions: While the strong position of the national Spanish government in terms of funding larger policy measures has a pointed homogenizing effect across the Spanish regions, the considerable degrees of autonomy of the Spanish regions allows them to follow markedly heterogeneous regional policy strategies in the fields of innovation and knowledge transfer.

While Catalonia resulted to persecute a highly centralized innovation and knowledge transfer policy, concentrating and coordinating major shares of it in and through the regional development agency *ACC1Ó*, the Madrid Community has a more pluralistic policy approach, including business associations and the *Madrid Chamber* in developing and implementing the regional policy measures. In this, the focus of the Catalan innovation and knowledge transfer policy is strongly oriented towards initiating single technology oriented cooperation projects and networks (NUCLIS program) and towards making transfer oriented regional scientific institutes visible through labeling and supporting them, as well as to professionalize their interaction with the business sector within the framework of the

TECNIO program. Given the (even for the Spanish case) extraordinarily pointed individualistic attitude predominating the Catalan business sector and the strong mental barriers between the business sector and the scientific world, these policy measures can be interpreted as first steps in transmitting an innovation and collaboration culture to the regional actors and to motivate them to mutually transfer knowledge.

Contrasting that, the focus of the Madrid Community's innovation and knowledge transfer policy lies more on cluster- and network-building, and, therefore, on establishing long term, rather than project oriented, cooperation structures, especially through the *madri+d* system as central interface, technological marketplace and assessing body for connecting firms and scientific actors, and *Madrid Network* as cluster management and coordination facility for the regional clusters and scientific parks. Although it might be questioned if the Catalan policy approach is far reaching enough to not only punctually, but markedly and sustainably improve the overall knowledge transfer activities in the region, and if the Madrilenian cluster and network policy strategy might not be too optimistic and advanced for the similarly individualistic business sector in the Madrid Community, these distinct policy approaches lead to a considerably different use of structures and mechanisms of knowledge transfer in the two regions.

Moreover, apart from distinct policy approaches, the two regions also pointedly differ in the role association based mechanisms play in the respective regional knowledge transfer systems. While business associations and chambers are almost marginalized in the Catalan knowledge transfer system, they are active, influential and well accepted actors in its Madrilenian counterpart. Like in the investigated German regions, also for Madrid and Catalonia associations and chambers decisively differ in their structures, self understandings and traditions, but especially in their degree of centralization, their internal cohesion and coordination, their organizational and personal equipment, their early acknowledgement of innovation and knowledge transfer as key topics, as well as their tradition of developing coherent regional perspectives and taking responsibility for the region they represent. While the Catalan associations and chambers were reported to rank low in all these dimensions, their Madrilenian counterparts *CEIM* and *Madrid Chamber* resulted to be active partners in the initiation, development and implementation of regional innovation policy measures. This allows for describing the Madrilenian innovation system as more pluralist and inclusive, while in the Catalan case the state cannot rely on strong corporatist partners in the fields of innovation and knowledge transfer, giving public action the exclusive protagonism in this regard and creating more centralized policy structures.

Fig. 40: Regional and Sectoral Variation from Spain's Mediterranean Model: Catalonia

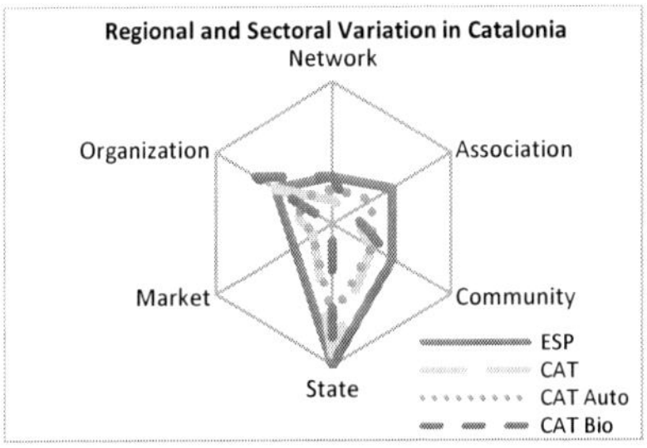

Source: Own Illustration

Fig. 41: Regional and Sectoral Variation from Spain's Mediterranean Model: The Madrid Community

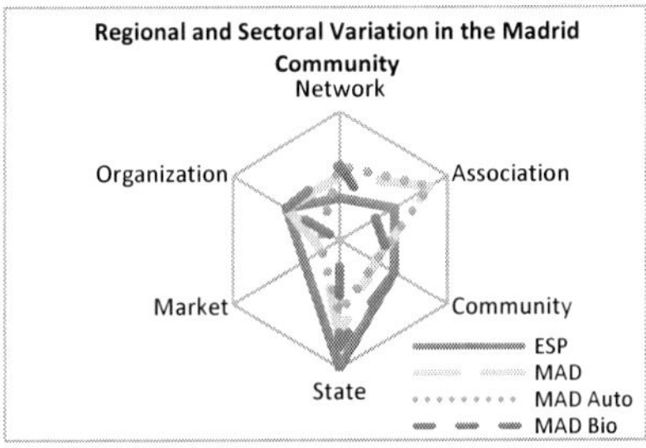

Source: Own Illustration

Further differences between the two regions can be traced back to their distinct features as historically grown industrial regions and metropolitan regions. For the Madrid Community, its role as national capital city and Spain's major urban agglomeration area is combined with a privileged equipment with business headquarters, larger firms, business R&D-departments, public administration bodies and research facilities, giving it certain structural advantages in comparison to Catalonia. What resulted to be the most differentiating point in this connection is the compact character of the Madrid region, its density in terms of actors and, above all, intermediary actors, allowing for and easing interaction in close physical proximity. A slightly higher propensity of cooperative inter-firm R&D-projects and cooperations between firms and scientific institutions in the Madrid Community might not least be explained by these specific structural conditions. However, other assumed differentiating metropolitan factors have not appeared in the analysis: Neither is the Madrid Community an outstanding entrepreneurial breeding ground for business start-ups and spin-offs, nor a more fluent job market for scientific and technical staff. One major reason for this might be that Madrid's rich metropolitan university and research landscape is mainly oriented towards basic research and, above all, higher education, and less on applied research and technology development. As a consequence, it not fully finds expression in elevated innovation and transfer activities, which is, not least, reflected in the absence of any technological centers in the region. In this regard, it very much resembles its German counterpart MR HBGW. Especially surprising is the fact that not even the market based mechanisms of knowledge transfer resulted to be more important in the Madrid region, although being the country's clear center of the business and financial service sector. A general tendency towards more liberal and market based interaction models diffusing through a stronger connection to global networks in the metropolitan regions has, therefore, not been found in the Spanish case (see also Figures 40 and 41).

In addition, the qualitative case studies on the Spanish regions also revealed certain sectoral differences with regard to the structures and mechanisms of knowledge transfer as explaining factors for regional variation within national systems (Figures 40 and 41). The structures and mechanisms of knowledge transfer applied in the automotive industry of both regions resulted to be basically shaped by its specific structural conditions, describing it as mature and consolidated sector, as lacking of any larger own OEMs, as being more production plants for foreign multinationals, as only marginally conducting R&D- and innovation activities, as rather importing technology from abroad than developing itself, as having a comparatively low technology level, an atomized business structure, an individualistic business culture, as well as strong mental barriers towards the scientific world. Therefore, overall, the automotive sectors in both regions strongly resemble their respective regional environments in terms of the structures and mechanisms of knowledge transfer applied, also paralleling the previously described regional specificities. While this could have been expected for the case of Catalonia as a historically grown industrial region, constituting a "natural" environment for a traditional industrial sector like automotive, the far reaching match of the automotive sector with the structures and mechanisms of knowledge transfer of the Madrid Community as a metropolitan region requires further explanation.

Although the automotive sectors in both regions are characterized by a more intensive use of vertical supply chain networks and a stronger involvement of corporatist actors in knowledge transfer processes, which can be traced back clearly to sectoral and functional specificities, both factors are considerably stronger developed in the Madrid Community, reflecting the region's pointed focus on cluster- and network-building-policy, as well as its specific actor's constellation in terms of strong and engaged associations and chambers. Paradoxically, in these two points the metropolitan region of Madrid resembles more the classical ideal type of an industrial region than Catalonia. However, it could be argued in this connection that in strongly public action driven regional innovation and knowledge transfer systems like Madrid and Catalonia, the respective policy strategies also mark off strongly to the sectors. In addition, also well organized and engaged associations and chambers will encounter a fruitful breeding ground in the automotive sector's modus of institutionally stabilized and incremental innovation in close interaction of firms, customers and suppliers. Therefore, in the Spanish case, regional and sectoral factors mutually reinforce each other, leading, on the one hand, to slightly incoherent structures of knowledge transfer between the two sectors of the two investigated regions. On the other hand, specific metropolitan effects on the structures and mechanisms of knowledge transfer in the Madrilenian automotive sector have not been found.

These ambivalent results also apply to the Spanish biotech sector, although because of different reasons. The first insight which has to be taken into account in this connection is the sector's extraordinarily young and emerging character, with the biotech firms in Catalonia being even younger than their Madrilenian counterparts. This has several implications: First, in their emerging phase the biotech sectors in both regions still depend strongly on public funding and support in all discussed dimensions of knowledge transfer activities. Second, especially in the Catalan case, but also in the Madrid Community most structures and channels of knowledge transfer in the sector are not fully developed yet, including specialized job markets, business associations, links to the chambers, specialized seed- and venture-capital providers and business angels, specialized external consultancies, marketable IPRs, or regional platforms. Therefore, many expected "typical" structures and mechanisms of knowledge transfer of a high-tech and radically innovating sector within the environment of a MME can (still) not be observed in the investigated regions.

Table 9: Qualitative Relevance of Knowledge Transfer Mechanisms in two Spanish Regions and Sectors: Catalonia and the Madrid Community

MECHANISMS	ESP*	Cat	Aut	Bio	Mad	Aut	Bio
NETWORK BASED MECHANISMS							
business networks	O	–	–	–	O	O	O
informal networks	O	O	O	O	O	O	–
institutionally stabilized networks	O	–	O	O	O	O	O
customers and suppliers	O	O	O	–	O	+	O
cooperative inter-firm R&D-projects	–	–	–	O	O	O	+
cooperative R&D-projects with scientific institutions and universities	–	–	–	O	O	O	+
ASSOCIATION BASED MECHANISMS							
employers-, business-, and trade associations	O	–	O	–	+	+	O
chambers of industry and commerce	O	–	–	–	+	+	–
binding and common technical standards	O	???	???	???	O	O	O
ORGANIZATION BASED MECHANISMS							
transfer of technical and scientific staff	O	O	O	O	O	–	O
purposeful headhunting	O	O	O	O	O	O	–
business start-ups	O	O	–	+	O	–	+
STATE INDUCED MECHANISMS							
national government measures	+	+	+	+	+	+	+
regional government measures	+	+	+	+	+	+	+
(larger) state-induced innovation projects	+	+	+	+	+	O	+
scientific parks	+	+	+	+	+	+	+
technological centers	+	+	+	+	–	–	–
technology transfer agencies/companies	+	O	–	+	O	–	+
MARKET BASED MECHANISMS							
external consultancies	O	–	–	–	O	O	–
mergers and acquisitions, hostile takeovers	O	O	O	–	–	–	–
buying, selling and licensing of patents	–	–	–	–	–	–	–
COMMUNITY BASED MECHANISMS							
informal contacts	O	O	O	O	O	O	O
regional platforms	O	O	O	O	O	O	O
additional (regional) institutions	O	???	???	O	???	???	O

+ *Reported (*Hypothesized) as crucial mechanism of knowledge transfer*
O *Reported (*Hypothesized) as important but not decisive mechanism*
– *Reported (*Hypothesized) as not important at all*
??? Not reported by the interview partners
[dark shading] = *Important deviation from hypotheses*
[light shading] = *Modest deviation from hypotheses*
[no shading] = *No deviation from hypotheses*

Nevertheless, what resulted to be biotechnology specific features in terms of knowledge transfer across the Spanish regions is the sector's markedly higher affinity to R&D-activities and to the scientific world, especially manifested in more intensive R&D-cooperations, an active use of scientific parks and the OTRIS, as well as frequent transfers of staff from research institutes to the firms, reflecting the specific needs of a strongly research driven sector. On the other hand, and despite the similar structural conditions shaping the sector in both regions, like for automotive, the analysis also revealed region specific factors influencing knowledge transfer processes in both regional biotech sectors. The stronger policy focus on institutionally stabilized networks and vertical cluster structures, as well as the more active role of business associations in the Madrid region are also displayed in its biotech sector, while the Catalan biotech sector is clearly shaped by the absence of these policy approaches and actor's constellations. However, also these points might be traced back to the strong dependence on public action of a sector *in statu nascendi*, on which the respectively persecuted policy strategies and historically grown actor's constellations still mark off strongly.

In sum, these results for the Spanish case reveal encompassing insights into the knowledge transfer system of a MME. Like for the German case, also Spain cannot be regarded as coherently structured national innovation and knowledge transfer model, but rather as a regionally variegated set of knowledge transfer governance systems. Regional variation of the two investigated regions is, however, considerably lower than in Germany, which can, above all, be explained with the similar structural barriers as well as with a strong role of central government. In addition to the explanatory factors for regional variation already discussed for the German case and for the analysis of Spain so far, this last aspect refers to the degree of federalism and regional policy autonomy in national models, which are not sufficiently taken into account in the approaches of VoC and NIS up to now. A more varieties-centered research agenda on innovation and knowledge transfer should stronger focus on these and the other driving factors of regional variation in national systems, as will be discussed in detail in the concluding section 8. The results on Spain are summarized in Table 9.

7. UK—Knowledge Transfer Systems in a Liberal Market Economy

7.1 Portraying the UK—A Service Dominated Economy

In section 3.2, the UK has been illustrated as a LME according to the VoC-approach (Hall/Soskice 2001b). Building on this discussion, the UK has been selected as an example of a LME in Europe in the case study at hand. This section will give a brief overview on some fundamental quantitative empirical indicators to describe the UK's current economic structure and its knowledge generating capacity.

With 61.4 million inhabitants in 2008 the UK is the third largest country in the EU (ONS 2011a), and also Europe's third largest economy, with a GDP of 1.82 trillion Euros, representing 14.5% of the EU's GDP (Eurostat 2010: 98). As in the other EU-countries, also the UK's economic structure is mainly dominated by the service sector, which is, with a share of 75.4% in the total GVA, considerably stronger developed than in Germany (69.2%) and Spain (68.4%). Especially the high share of business and financial services of 32.2% in the total GVA differentiates the UK from the other two investigated countries, which rank markedly lower with 29.4% and 22.6%, respectively. By contrast, the UK's primary sector is comparatively small with a share of 0.8% in the total GVA, being less than one third of the value for Spain (2.8%), but almost equaling the value for Germany (0.9%). The industrial sector in the UK (without construction), on the other hand, is characterized by a strong decline in the last decade (1998–2008) from 23.4% to 17.6%, and has now a share in the total GVA comparable to Spain, which declined in the same period from 21.8% to 17.3%. Germany, by contrast, even expanded the share of its industry in the total GVA, from 25.3% to 25.6%, and has a pointedly stronger industrial focus than the UK (Figure 6; Eurostat 2010: 100).

The UK's business structure is, similar to Spain's, strongly dominated by small and especially micro-firms. In 2008, 95.7% of all firms in the UK were micro-firms with less than ten or no employees. Small firms with ten to 49 employees represented 3.6% of all firms in the UK, medium sized firms from 50 to 249 employees 0.8% and large firms with more than 250 employees only 0.1%. Within the group of micro-firms those firms stand out that have no person employed: They represent about 77.2% of all micro-firms, and 73.9% of the total number of firms in the country. In this, the micro-firms only employed 33.1% of the country's total workforce in the private sector, while small, medium and large firms employed 14.4%, 11.7%, and 40.8%, respectively, pointing to a markedly high average size of the UK's large firms[58] (Figures 42 and 43; BIS 2010a: 2ff.).

Fig. 42: Business Structure UK I

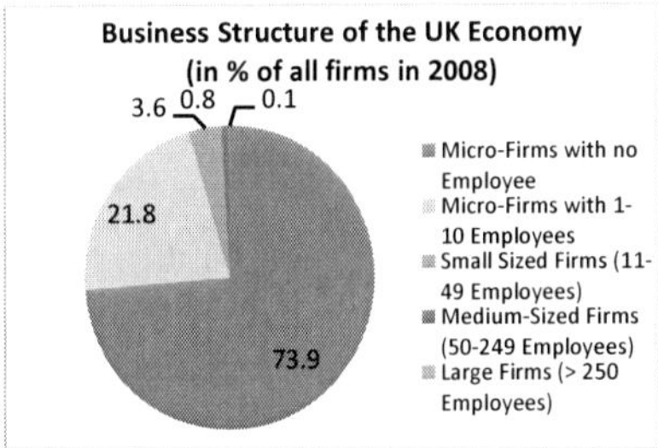

Source: Own Calculations and Illustration based on BIS (2010a): 2ff.

Fig. 43: Business Structure UK II

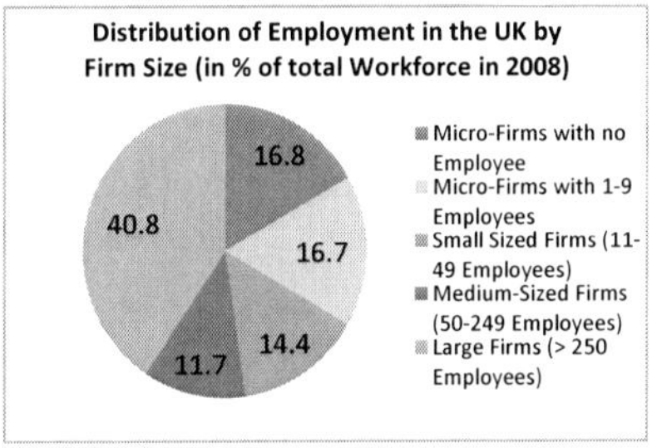

Source: Own Calculations and Illustration based on BIS (2010a): 2ff.

The sectoral structure of UK manufacturing is mainly shaped by high and medium-high technology sectors. In this, and similar to its Spanish counterpart, also UK manufacturing is characterized by a high sectoral diversity and, therefore, by a low degree of specialization, with no manufacturing sector accounting for much more than 14% of manufacturing's GVA. The most important manufacturing sector is food processing and luxury foods, with a share of 14.6% in the total GVA of manufacturing, followed by chemicals, pharmaceuticals and bioscience (12.6%), publishing and printing (10.3%), automotive (10.2%), metalworking (8.7%), mechanical engineering (8.2%), rubber and plastic (4.6%), other manufacturing and recycling (4.6%), and medical and precision equipment (4.0%)[59] (Figure 45; BIS 2010b: 2).

---

58 Own calculations based on BIS (2010a): 2ff.; data for 2008

59 Data in this paragraph refer to 2009

Fig. 44: Sectoral Structure of Selected UK-Regions in International Comparison

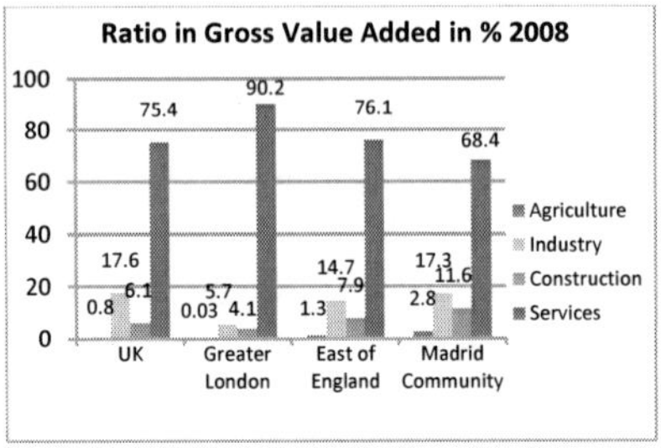

Source: Own Calculations & Illustration based on Data from ONS (2011b); IESTADIS (2011)

Fig. 45: Industrial Structure of the UK

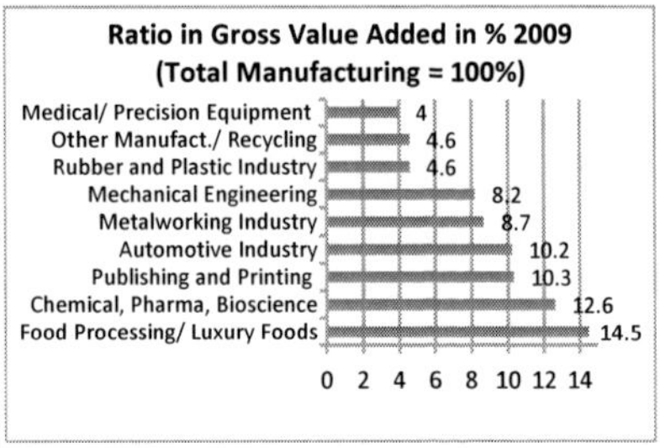

Source: Own Illustration based on BIS (2011b):

To describe the UK's knowledge generating capacity, some structural indicators of the country's R&D-activities will briefly be detailed. In 2008, 152,000 persons were employed full time in the field of R&D in the UK's business sector, representing 0.96% of the country's total workforce in the business sector of 15.9 million[60] (ONS 2011c). The UK's total R&D-expenditures were about 26.6 billion Pounds in the same year, representing 1.79% of the UK's GDP (GERD for Germany: 2.68%; Spain: 1.35%; EU: 1.92%; see Figure 46), and of which 45.7% were covered by the business sector, 36.7% by the public sector plus other non-commercial public and private research institutions, and 17.6% by funding from abroad. 9.3% of these expenditures went into the defense industry (ONS 2011d: 11f.; COTEC 2010: 246). When differentiating again in this context between smaller companies (< 99 employees), medium sized firms (100–399 employees) and large firms (> 400 employees), the small firms had a share of 9.7% in the business sector's total R&D-expenditures, and employed 17.1% of its R&D-personnel, while the medium sized firms covered 15.3% and 17.8%, and the large firms 58.2% of the expenditures and 53.3% of the R&D-personnel, respectively[61]. Given the UK's overall business structure, this points to a strong concentration of the country's R&D-activities within the large firms. A sectoral perspective reveals that, furthermore, most of the UK's business-R&D-activities are strongly concentrated in only a few key industries: Measured by their share in total R&D-expenditures and R&D-employment of manufacturing, chemicals (including phar-

60 Own calculations based on ONS (2011c), data for 2008

61 Own calculations based on ONS (2011e), data for 2008

maceuticals and bioscience) is the UK's most R&D-intensive sector (42.2% and 31.8%), followed by automotive (26.9% and 29.0%, of which 54.2% and 55.2% are covered by the aerospace subsector, respectively), electrical engineering (11.3% and 13.1%) and mechanical engineering (7.7% and 11.2%)[62] (Figure 47; ONS 2011e).

Fig. 46: R&D-Intensity of the UK in International Comparison

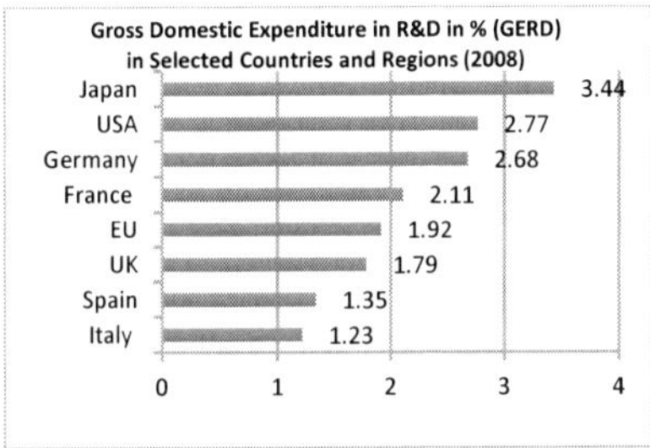

Source: Own Illustration based on Data from Eurostat (2011a)

Fig. 47: Sectoral R&D-Performance in UK Manufacturing

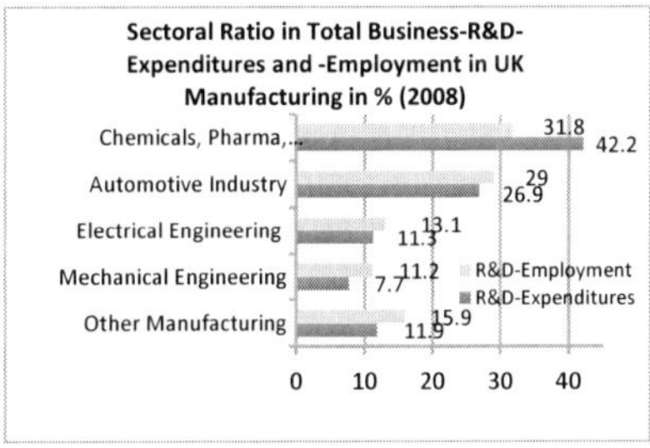

Source: Own Calculations and Illustration based on ONS (2011e)

One of the two sectors focused on by this study is automotive, which is the UK's fourth most important industrial sector. In 2007, the sector's total GVA was of 10.2 billion Pounds, representing a share of 6.4% in the total GVA of the UK's manufacturing sector. In the same year, automotive employed about 194,000 persons, representing about 5.9% of the total workforce in UK manufacturing. In this, the sector consisted of 3,300 firms, representing about 1.1% of all manufacturing firms in the UK (Shaw 2011: 61). 55.2% of these firms are micro-firms with less than four employees, 32.4% are small firms with between five and 49 employees. Medium sized firms with between fifty and 249 employees represent 8.9% of the sectoral business population, and 3.6% of the firms in UK automotive are large firms with more than 250 employees[63] (ONS 2009: 92f.). This demonstrates that in the UK's automotive sector medium sized and large firms have a considerably higher weight than in the UK economy as a whole. When looking at the persons employed in the distinct firm-size classes, this is further underlined: While the micro, small and medium sized firms only employ 6.4%, 7.3% and 12.3% of all employees in the sector,

62 Own calculations based on ONS (2011e), data for 2008

63 Own calculations based on ONS (2009): 92f., data for 2008

respectively, the large firms employ 73.7%. This pointedly dominant position of the large firms in the sector is also illustrated by their share in the sector's total sales volume of 87.2%, to which the medium sized firms only contribute a share of 7.7%, the small firms with 3.5%, and the micro firms with 1.7%[64] (BIS 2010a)[65].

In this, and like its Spanish counterpart, also the UK's automotive sector lacks of large indigenous OEMs. The country lost most of its OEMs within the past 25 years, which partly disappeared from the market or are now owned by foreign multinationals. Apart from some smaller niche product manufacturers, therefore, the UK must be predominantly regarded as an assembly location for many of the large automotive MNEs and suppliers operating worldwide. This implies that important strategic decisions in the sector are in most cases not taken in the UK, but in the OEMs' headquarters abroad, and will, moreover, strongly depend on foreign investments (Holweg et al. 2009: 45). The most important of these production facilities in terms of sales volume are those of Nissan, Toyota, Honda, BMW (MINI), Land Rover, GM (Vauxhall), Ford, and Jaguar. These volume production facilities are supplemented by a broad variety of high value and luxury vehicles manufacturers serving niche markets, like the Formula 1, and which are characteristic for the country's automotive sector: Among them are Aston Martin, BMW (Rolls-Royce), Caterham, Connaught, Gibbs, Morgan, Proton (Lotus), TVR, VW (Bentley), McLaren and Mercedes-Benz. The most important suppliers operating in the UK include Aisin, Bosch, Calsonic, Dana, Delphi, Denso, GKN, Johnson Matthey, Pilkington, Siemens VDO, TRW and Visteon (Holweg et al. 2009: 22ff.; BIS 2011). The regional distribution of business activities in automotive in terms of number of firms located shows main concentrations within the UK regions of North West, West Midlands, South East, Greater London, East of England and Scotland (Holweg 2010 et al.: 9).

Although on a pointedly lower level than Spanish automotive, the UK's automotive sector can absolutely be described as a predominantly supply oriented industrial sector, in which the dedicated vehicle producing firms

64 Own calculations based on BIS (2010a), data for 2008. Data include automotive retail, service and maintenance firms.

65 Data used in this paragraph of BIS (2010a) and ONS (2009) rely on differently large sectoral business populations, and are, therefore, not completely comparable. BIS (2010a) data include automotive retail, service and maintenance firms. ONS (2009) data only include automotive manufacturing firms.

only represent about 29% of all the firms (but 58% of all employees) in the sector, while the suppliers to these production plants represent about 71% of the firms (but only 42% of the employees) [66] (Holweg 2010 et al.: 6). Also in the UK's automotive sector assembly plants and supplying firms (components, materials, services) are interconnected intensively with each other as customers and suppliers along the value added chain. Around 70% of the sector's total production value stems from supplied advanced performances, reflecting its strong intra- and inter-sectoral division of labor. This especially implies supplies of suppliers among each other and supplies between suppliers and the assembly plants of the larger OEMs located in the UK. However, and markedly stronger than in the cases of Spain and Germany, considerable shares of these supplied advanced performances are imported from outside to the UK (Holweg 2010 et al.: 6ff.).

Focusing on the knowledge generating and R&D-activities of UK automotive, it becomes obvious that it is a R&D-intensive sector within the UK economy: In 2008, around 19.9% of the total R&D-expenditures of the UK business sector and about 26.9% of those of UK manufacturing have been realized by automotive, representing the second largest share in the total R&D-expenditures among the country's industries, right behind the chemical industry (including pharmaceuticals and bioscience). In addition, 19.1% of the R&D-employment in the UK business sector, and 27.1% of the R&D-employment in the UK manufacturing sector is in automotive. Particular attention should receive the fact that a large share of these R&D-activities is covered by the aerospace subsector, accounting for 54.2% of the R&D-expenditures and 55.2% of the R&D-employment in automotive[67].

In this, also UK automotive shows a strong concentration of the sectoral R&D-activities within the large firms: When differentiating again between smaller (< 99 employees), medium sized (100–399 employees) and large firms (> 400 employees), the small firms had a share of 0.9% in the total R&D-expenditures of automotive, and employed 3.4% of its R&D-personnel, while the medium sized firms covered 6.6% and 6.9%, and the large firms 92.5% of the expenditures and 89.7% of the R&D-personnel, respectively[68]. However, when looking at the sector's R&D-intensity of 1.9% (share of R&D-expenditures in total sales volume in

66 Own calculations based on Holweg et al. (2010): 6, and Shaw (2011): 61

67 Own calculations based on ONS (2011e), data for 2008

68 Own calculations based on ONS (2011e), data for 2008

2006), UK automotive still ranks behind the average R&D-intensity for the country's manufacturing sector of 2.4%, and is characterized by a pointedly downward trend over the last decade. In addition, and although its R&D-intensity is clearly higher than the one of Spain (0.9%), UK automotive is markedly less R&D-intensive than its main competitors like e.g. France (4.8%), Germany (4.4%), Japan (4.3%), and USA (3.4%) (Figure 21; Legler et al. 2009: 94; Holweg et al. 2010: 45). As Holweg et al. (2010) state, one major reason for this comparatively low R&D-intensity has to be seen in the fact that R&D-activities are likely to be conducted in the home country of the OEMs, which lack in the UK (Holweg et al. 2010: 45).

The UK automotive sector can, therefore, be regarded as a traditional and mature industrial sector showing some trends of decline and dissolution, and losing its central position in the UK's industrial landscape in the last decades. Biotechnology in the UK, on the other hand, is, in contrast to most of its European peers, an already consolidated, but still dynamically growing sector increasingly shaping the country's economic portfolio. In 2010, the UK biotech sector consisted of 997 firms active in the fields of medical and industrial biotechnology, generating a turnover of about 5.8 billion Pounds[69] (BIS 2010c), representing about 0.2% of the UK business sector's total turnover.[70] These firms employed around 37.780 persons, representing 1.5% of the total employment in UK manufacturing. About 75% of these firms are specialized service providers such as consultants, material and equipment suppliers and drug development services, while about 25% are mainly active in developing, manufacturing or selling therapeutic products[71]. This diversified structure of UK biotechnology indicates a well developed supply chain with a significant number of companies able to outsource non-core activities, something that differentiates the sector from its continental counterparts (BIS 2010c: 28).

The UK biotech sector is dominated by SMEs, with more than 97% of the firms having less than 250 employees. The sector's business structure shows 49% of micro-firms with no up to four employees, 35% of small firms with five to 49 employees, 13% of medium sized firms with between 50 and 249, and 3% of large firms with more than 250 employees[72] (BIS 2010c: 33ff.). While the share of large firms more than doubles the Ger-

69 Own calculations based on BIS (2010c), data for 2010

70 Own estimation based on BIS (2010b): 2 and BIS (2010c)

71 Own calculations based on BIS (2010c), data for 2010

72 Own calculations based on BIS (2010c): 33, 44; data for 2010

man one (1.4%), it is only about one third of the Spanish case (8.5%), pointing to considerable growth dynamics in Spanish biotechnology, but also to a high number of larger Spanish chemical firms shifting their main dedication towards biotechnology. However, German biotech firms remain smaller in size than those in the UK, with a share of 87% having less than 50 employees, while this share is 84% in the UK and 79% in Spain, indicating a lower growth performance of the sector in Germany (OECD 2009b: 21; OECD 2011a). In this, the UK biotech sector is markedly more consolidated than its only emerging counterparts in Spain and Germany: Looking at the sector's business demography, 41% of the firms are mature firms older than ten years, and 45% are already established firms operating for between four and nine years. 12% of the firms are two to three years old, and only two percent are young and start-up companies being less than two years old[73] (BIS 2010c: 34ff.).

Concerning the geographical distribution of biotechnology in the UK, business activities are, in terms of location of the number of companies, turnover and employment mainly concentrated in the East of England, which accounts for 27% of total turnover, 24% of the employees and 24% of the total number of companies in the sector. The second major pole of biotechnology activities is the South East with 26% of the total turnover, 21% of the employees and 14% of the total number of companies in UK biotechnology.[74] Further concentrations of the sector can be found in Scotland, Greater London and the North West (BIS 2010c: 35ff.).

Like in Spain and Germany, also the UK biotech sector is highly R&D-intensive. Actual structural data for R&D-activities in the sector are not available at this point (BIS 2010c: 20). However, apart from basic research, in particular in the field medical biotechnology, the development of new products, and especially of new pharmaceuticals, requires long clinical test phases (phase I–III) of up to ten years or more, before a new pharmaceutical is approved for commercial distribution. In total, UK biotech firms had 235 drug candidates in phases of clinical trials, markedly more than Germany (141) and Spain (35), and representing a share of about 20% of all drug candidates in the European pipeline. Looking at the distribution of UK drug candidates in the three clinical phases, in 2009, 29% of the drug candidates were in phase I, 57% in phase II and only 13% in phase III[75].

73 Own calculations based on BIS (2010c): 34, 45; data for 2010

74 Own calculations based on BIS (2010c): 35ff., 47; data for 2010

75 Own calculations based on Ernst & Young (2010): 45, data for 2009

With that, and in a comparison to the mean values of all other European countries active in biotechnology, the UK biotech firms only have minor comparative strengths in drug candidates in phases II and III, but are slightly behind in drug candidates in phase I, pointing to the overall matureness of the sector with many businesses being in later phases of their business life cycles (Ernst & Young 2010: 42).

7.2 The East of England Case

7.2.1 Locating the East of England in the UK Context

7.2.1.1 Selected Facts on the East of England's Economic Structure

The East of England is one of the nine English Government Office Regions, which together with Scotland, Wales and Northern Ireland constitute the UK (ONS 2003). The East of England is located in the east of the British Islands, containing three sub-regions: East Anglia, Essex and Bedfordshire/Hertfordshire. It borders on the regions of South East, Greater London and the East Midlands, as well as on the North Sea to the east, giving the region around 240 kilometers of coast and estuary (Eurostat 2004a). Historically being strongly shaped by the agrarian sector, the region paralleled the early industrialization processes of the 19th century in other British regions only to a limited extent, and did so especially in sectors related to the mechanization and intensification of agriculture, like grain milling, brewing, textiles, shoe manufacturing, small scale ship-building, fishing and food processing. While other industrial regions in the UK had a stronger focus on heavy industries like iron and steel, coal, ship-building, railway engineering or heavy chemicals, which declined strongly from the 1960s on, the East of England region could at this time expand its share in newer, technology-dynamic and research intensive manufacturing sectors, like electrical and electronic engineering, scientific instruments, mechanical engineering, metal products and engine construction. This growth of newer industries, which first emerged in the region in the early 20th century, was particularly accelerated by early spin-off activities around Cambridge, as well as by manufacturing firm relocation processes to the East of England from the South East Region and Greater London, converting the

East of England to one of the UK's fastest growing manufacturing regions from the 1980s on (Keeble 1998: 7ff.; Eurostat 2004b).

Fig. 48: Sectoral Structure of the East of England in Comparison

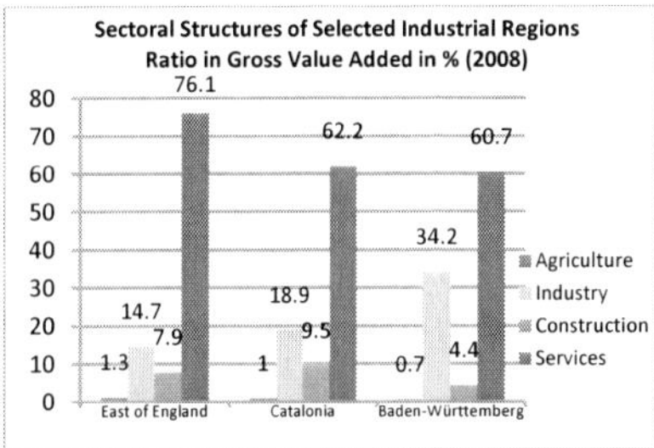

Source: Own Calculations and Illustration based on Data from ONS (2011b), IDESCAT (2011) and SLABW (2011)

Fig. 49: The East of England's Industrial Structure

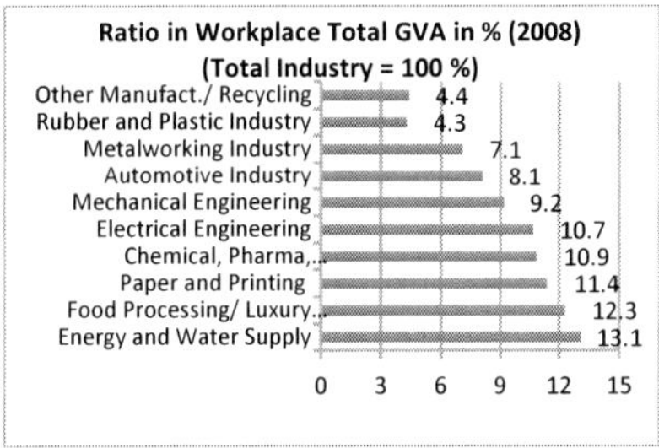

Source: Own Calculations and Illustration based on Data ONS (2011b)

Today, (2009) with a population of about 5.8 million the East of England is the UK's fourth largest region in terms of inhabitants, right behind Greater London, the South East and the North West Region, and represents 9.3% of the UK's total population. The East of England is also the UK's fourth strongest economic region, with a total GVA of about 107 billion Pounds, representing 8.7% of the UK's total GVA (ONS 2011b: 12), and is one of the country's leading regions in life sciences, engineering, advanced manufacturing, ICTs and media (Eurostat 2011b). The economic structure of the East of England almost equals the national average, only with the primary sector covering a significantly higher share of the regional total GVA (1.3%) than the UK average with 0.8%, still displaying the region's historic agricultural tradition. Apart from that, the region experienced an important tertiarization process, with industry (without construction) now contributing 14.7% to the region's total GVA (and 9.2% to the total GVA of the UK industry), a share being higher than the UK average of 14.0%, but being one of the lowest among the UK regions, with only Greater London (5.7%) and the South East (11.9%) having lower shares of industry. Services, however, account for 76.1% of the regional total GVA, slightly less than the UK average of 78.8%[76] (Figure 48; ONS 2011b: 20ff.).

76 Own calculations based on ONS (2011b): 20ff., data for 2008

The East of England's industrial structure is characterized by a high degree of diversity, with no industrial sector accounting for much more than 13% of the region's industry's total GVA. In this, the region's most important industries are energy and water supply with a share of 13.1% in the regional industry's total GVA, food processing and luxury foods (12.3%), paper and printing (11.4%), chemicals (10.9%), electrical engineering (10.7%), mechanical engineering (9.2%), automotive (8.1%), and metalworking (7.1%)[77] (Figure 49). When looking at the technology level of the East of England's manufacturing sectors, it can be stated that it is clearly more technology-intensive than the rest of UK, and also than the investigated German and Spanish regions. High-technology and medium-high technology predominate clearly the region's manufacturing activities (EEDA 2009).

Fig. 50: Business Structure East of England I

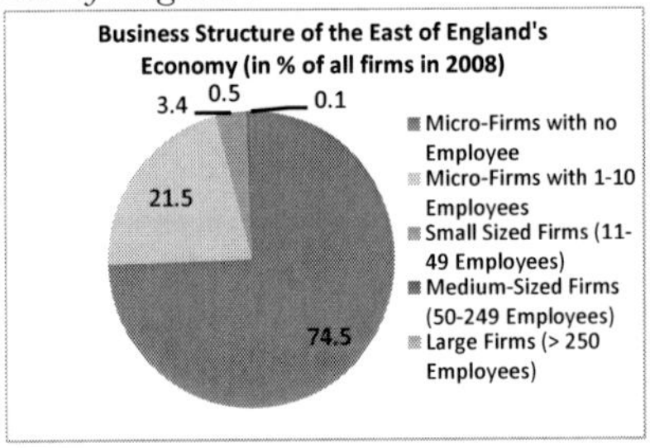

Source: Own Calculations and Illustration based on Data from BIS (2010a)

Fig. 51: Business Structure East of England II

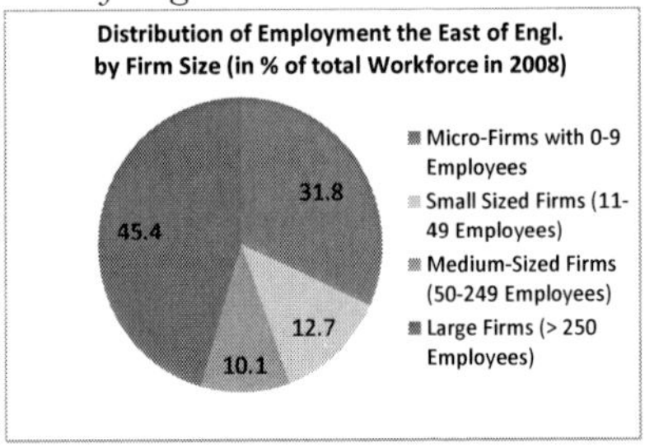

Source: Own Calculations and Illustration based on Data from BIS (2010a)

Also the East of England's business structure is similar to its national UK counterpart, especially in being strongly dominated by micro firms with less than nine employees, which represent 96.0% of the total number of firms in the region's private sector (UK: 95.7%). Within the group of micro-firms those firms stand out that have no person employed: They represent 77.6% of the micro firms in the region and 74.5% of the total number of firms in the East of England's private sector. Slightly lower is the region's share of small businesses (10–49 employees) and medium sized firms (50–249 employees) of 3.4% and 0.5%, when compared to the national average of 3.6% and 0.8%, respectively. Finally, also the share of

77 Own calculations based on data from ONS 2011b, data for 2008

larger firms with more than 250 employees of 0.1% almost reaches the average value for the UK[78]. In this, the micro firms employed 31.8% of all employees in the private sector, while the small, medium sized and large firms employed 12.7%, 10.1% and 45.4%, respectively. In sum, this points to an even higher average firm size of the large firms in the region, which represent a markedly higher share of employment than the national average of 40.8%[79] (Figures 50 and 51; BIS 2010a).

Fig. 52: R&D-Intensities in Selected UK Regions

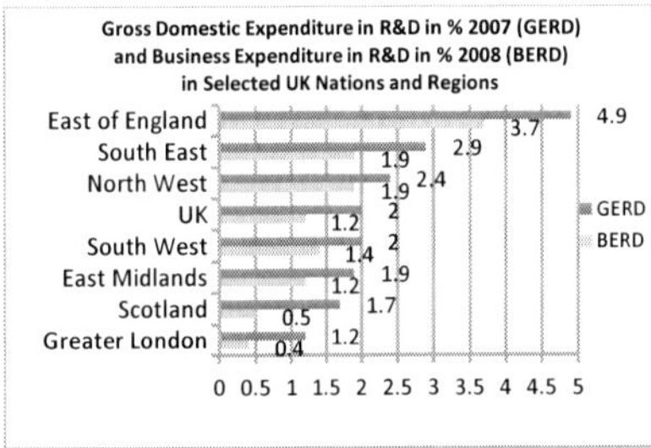

Source: Own Calculations and Illustration based on Data from ONS (2011f)

Fig. 53: Sectoral R&D-Performance in East of England Manufacturing

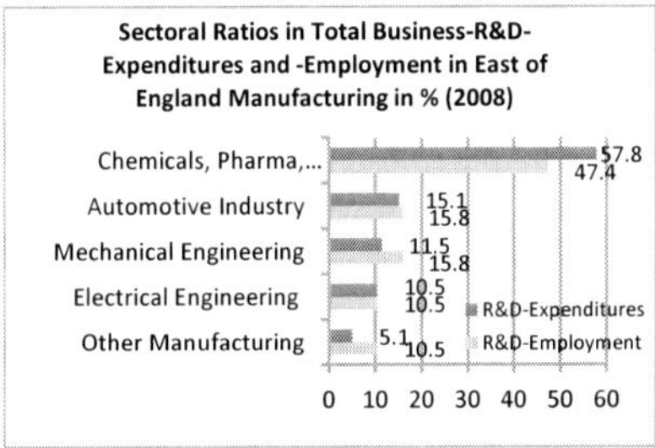

Source: Own Calculations and Illustration based on Data from ONS (2011e)

Another important structural indicator describing a region's knowledge generating capacity is its R&D-activities. The East of England's R&D-intensity (gross domestic expenditure on R&D as percentage of total workplace based GVA) is of about 4.9%, and is high above the national average (2.0%). It also leaves far behind all other UK regions, of which the South East (2.9%), the North West (2.4%) and the South West (2.0%) follow on the subsequent ranks[80] (Figure 52). These values identify the East of England as a center of R&D-activities in the UK, and also as one of its scientific poles of excellence. This is further underlined by the region's business sector R&D-efforts: In 2008, the East of England had total business R&D-expenditures of 4.1 billion Pounds, representing a share of 26% of the total business R&D-expenditures in the UK. This is by far the largest contribution to the total business R&D-expenditures among the UK regions, leaving behind the South East (21.6%), the North West

78 Own calculations based on data from BIS (2010a), data for 2008
79 Own calculations based on data from BIS (2010a), data for 2008
80 Own calculations based on data from ONS (2011f), data for 2007

(14.4%), the South West (8.6%), and Greater London (6.7%)[81] (ONS 2009: 18). The region's business sector's R&D-intensity (business R&D-expenditure as percentage of total workplace based GVA) of 3.7% more than triples the national average of 1.2%, and leaves behind again the South East, the North West (both 1.9%), and the South West (1.4%)[82] (Figure 52). This points to a remarkably R&D-intensive business sector in the East of England, and also accentuates again its high average technology level with a strong focus of high and medium-high technologies.

A sectoral perspective on the business R&D-activities in the East of England reveals, on the one hand, and compared to the national average, a markedly more R&D-oriented regional service sector: 36.7% of the total business R&D-expenditures and 31% of the total business R&D-employment in the region are covered by service firms, while the values for the UK average rank considerably lower with 24.7% and 28.3%. On the other hand, it shows that these activities are concentrated strongly in manufacturing, covering 62.6% and 65.5% of business R&D-expenditures and -employment, with the region's most important manufacturing sectors also being the most R&D-intensive ones: The chemical industry (including pharmaceuticals and biosciences), accounting for 57.8% of manufacturing R&D-expenditures and 47.4% of its R&D-personnel, automotive (15.1% and 15.8%), mechanical engineering (11.5% and 15.8%), and electrical engineering (10.5% and 10.5%)[83] (Figure 53).

7.2.1.2 *Automotive: Niche Sector Specialization in the East of England*

A specific focus on automotive shows that it is one of the East of England's most important manufacturing sectors, accounting for 10.2% of the total GVA in the region's manufacturing[84]. Furthermore, the region also had a significant share in UK automotive, with the automotive firms located in the region covering a share of 11.8%[85] of the sector's total sales volume (IMI 2007: 18). In 2008, there were around 14,500 people employed in the East of England's automotive sector, representing about 9% of the UK automotive workforce (SEMTA 2011: 29). In this, the sector

81 Own calculations based on data from ONS (2009): 18, data for 2008
82 Own calculations based on data from ONS 2011f, data for 2008
83 Own calculations based on data from ONS 2011e, data for 2008
84 Own calculations based on data from ONS 2011b, data for 2008
85 Own calculation based on data from IMI (2007): 18; data for 2007

consisted of around 595 companies, representing 11.5% of all UK automotive firms. The sector's business structure is characterized by a larger share of SMEs and large firms when compared to the region as a whole, as well as by a lower proportion of micro-firms. While the latter (0–9 employees) represent 76.5% of all firms in the sector (of which 82.4% have less than four employees), the small (10–49), medium sized (50–249) and large (>249) firms account for 15.1%, 5.9% and 2.5% of all firms in the East of England's automotive sector, respectively[86]. In this, automotive presents itself as highly research intensive, accounting for 15.1% of the East of England's manufacturing R&D-expenditures and 15.8% of its R&D-personnel, being with that the second highest contributor to the region's R&D-activities behind the chemical industry[87].

It is especially the automotive manufacturers for highly specialized niche markets that decisively coin automotive in the East of England. On the one hand, the region is an important motorsport manufacturing site, and is, as such, part of the cross-regional UK motorsport cluster "Motorport-Valley", reaching from central to southern and eastern Britain (MRA 2004: 12ff.; UKTI 2007: 1f.; EofE 2011a). Moreover, also some of the large MNEs in the sector like Ford, General Motors, Lotus, Nissan, TWI and Perkins are located in the region, maintaining both assembly facilities and R&D-centers (EofE 2011b). On the other hand, the region also has considerable expertise in the aerospace subsector, with more than 80 aerospace manufacturers supported by a diversified supply chain, including BAE Systems, Smiths Industries, GKN Westland, Raytheon, Rolls Royce, and Lockheed Martin (EofE 2011c).

7.2.1.3 The East of England: Europe's Powerhouse in Biotech

Switching to the East of England's biotech sector, it can be stated that it is one of the region's most significant and distinctive industries, both in national and international terms. The region hosts the largest and most successful life sciences cluster outside the United States, and includes the Greater Cambridge cluster, Norwich Research Park and Colworth Science Park (EofE 2011a). With 256 biotech companies, 100 medical technology companies and 370 service companies supporting the life sciences sector, the East of England is the most mature, complete and successful cluster of

86 Own calculations based on data from ONS 2011b, data for 2009

87 Own calculations based on data from ONS 2011e, data for 2008

biotechnology and life sciences companies in Europe (EEDA 2011). Among the most renowned biotech and biopharma companies operating in the East of England are Astra Zeneca, Eisai, Genzyme, GSK, Illumina, Napp, Pfizer, Takeda and Unilever, as well as the medical device manufacturers Baxter Healthcare, Bayer Diagnostics, Olympus KeyMed and Philips Research (EofE 2011d). Within the UK, the East of England region is the major pole of business activities in biotechnology, concentrating 27% of the total turnover, 24% of the employees and 24% of the total number of companies in the sector. With that, it leaves behind clearly the South East with 26% of the total turnover, 21% of the employees and 14% of the total number of companies in UK biotechnology[88] (BIS 2010c: 35ff.).

Because of its dominant position within the biotech sector in the UK, it appears to be plausible to assume that the business structure of the East of England's biotech sector is similar to the one of the UK as a whole, which is mainly dominated by SMEs with less than 250 employees (UK: 97%)[89] (see section 7.1; BIS 2010c: 33ff.). Furthermore, as one of the breeding grounds of biotech activities in the UK, it might be assumed that like its national counterpart, also the East of England's biotechnology is a mature and consolidated sector, with a business demography of about 41% of mature firms being older than ten years, 45% of already established firms operating for between four and nine years, 12% of firms being two to three years old, and about two percent are young and start-up companies being less than two years old[90] (BIS 2010c: 34ff.). In this, the sector's intensive growth dynamics and R&D-activities are supported by a singular scientific environment, including research institutes like the John Innes Centre, the Institute of Food Research, Babraham Institute, Rothamsted Research, the Genome Analysis Centre, Wellcome Trust Sanger Centre, EMBL, twelve Medical Research Council Institutes, as well as by the region's world class universities of Cambridge, East Anglia, Cranfield, Hertfordshire, Anglia Ruskin and Essex (EofE 2011d).

---

88 Own calculations based on BIS (2010c): 35ff., 47; data for 2010

89 Own estimations based on BIS (2010c): 33, 44; data for 2010

90 Own estimations based on BIS (2010c): 34, 45; data for 2010

7.2.2 Structures and Mechanisms of Knowledge Transfer in the East of England: Six Governance Dimensions

7.2.2.1 Networks as Mechanisms of Knowledge Transfer

In the following sections the results of the empirical qualitative case study of the East of England's knowledge transfer system will be analyzed, building on the theoretical assumptions developed in chapter 3.2 and on the discussion of quantitative indicators describing the UK's and the East of England's economies in the previous section. The first dimension of knowledge transfer mechanisms that has been investigated in this regard is the *network dimension*. In this, the theoretical assumptions derived from the literature on VoC and NIS were that in LMEs like the UK, where atomistic market coordination predominates, informal networks, institutionally stabilized networks like business clusters, vertical networks between firms and their customers and suppliers and cooperative R&D-projects with scientific institutions and universities should be of a medium range importance for knowledge transfer activities. Moreover, both horizontal business networks as well as cooperative inter-firm R&D-projects should not be important as knowledge transfer mechanisms.

Horizontal Business Networks

These assumptions have only partly been confirmed by the surveyed interviewees. Concerning *horizontal business networks* between competitors, the experts concordantly identified a slightly higher significance for knowledge transfer processes than assumed. These networks would have at least a medium range importance, although strongly dependent on the sector (Interviews EoE 1: 2ff.; EoE 2: 5f.; EoE 3: 2ff.; EoE 4: 13). One prime example would be life sciences with *ERBI* (*Eastern Region Biotech Initiative*), tracing back to an initiative and support of the *East of England Development Agency* (*EEDA*). In those sectors, interaction between the members would be formalized with fixed contracts, but would also offer opportunities for informal interaction and social networking. Especially *Cambridge Network* has been highlighted in this connection, a cross-sectoral network of high-tech-companies with a focus on the Greater Cambridge area, but also beyond. Furthermore, and being cross sectoral as well as cross regional, the *London Technology Network* would connect high-technology firms and universities in the South East, Greater London and the East of England with each other, basing on the initiative of the respective Regional Development Agencies (RDAs). This strong and active role of central government bodies like the RDAs in promoting business networking activities comes as a surprise for a LME environment like the UK and will to be discussed in detail below.

However, even if these cooperations would sometimes take the form of business consortia around concrete topics or projects, and even if these consortia would in some cases be fostered by public funding from the EU or national government, the majority of these networks would only to a limited extent be used for knowledge transfer in an innovation context, but, more likely, to leverage collective resources, information exchange and for connecting the actors to knowledge about the framework conditions of business activities in the region. Therefore, it could be stated that these horizontal networks in the East of England are rather used to improve the operative environment than for substantial collaborative innovation and knowledge transfer activities (Interviews EoE 1: 2ff.; EoE 2: 5ff.; EoE 3: 2ff.; EoE 4: 13).

Vertical Networks of Firms with their Customers and Suppliers

In addition, also *vertical networks of firms with their customers and suppliers* along the value added chain would be medium range important mechanisms of knowledge transfer (Interviews EoE 2: 5; EoE 3: 4f.; EoE 4: 14). These vertical business clusters would exist e.g. in industries like the wireless sector with *Cambridge Wireless*, mechanical engineering, automotive, or renewable energies with *East of England Energy*. However, the intensity of cooperation within these agglomerations of firms would again be strongly dependent on the sector, and would generally be lower than e.g. in comparable German clusters, as the CEO of *Cambridge Network* explains:

> "The wireless sector is historically strong here, […] we have a lot of device development and you have people who are doing other things in this field and that would be again formalized as network called *Cambridge Wireless*, which formalized that supply chain relationship. […] There is also a *printing cluster* here, which is very strong, so there are many companies that do inkjet printing, and there is also a cluster created around *organic semiconductors*. These are the two very strong examples that I would point to. […] Largely, these are clusters that exist because there were people that created a series of spin-offs, and these groups of spin-offs ended up being located at the same place, even though they are competing. But I do not

think that they collaborate very much, it is not a very strong collaboration. It is more that it happens to be a set of skills and then these people happen to be located here. So there is advanced manufacturing, but it is not like the German model of having excellence in machine tools which then lead to excellence in manufacturing products and so you have the whole set of second tire manufacturers who supply into automotive and are very dedicated. We do not have that kind of strong big industry focus here." (Interview EoE 3: 4f.)

All interviewed experts consistently highlighted that with regard to both horizontal and vertical networks, it would be the specific R&D-intensive character of the region and its industries, especially in the Greater Cambridge area, and the common reference point of Cambridge University and other research institutions, to which many firms would be connected to and from which many of them would have spun-out, which would be an important and differentiating driver of these network activities in comparison to other UK regions, as well as to other parts of the East of England (Interviews EoE 1: 2ff; EoE 2: 5ff.; EoE 3: 2ff.).

Cooperative Inter-Firm R&D-Projects

With regard to *cooperative inter-firm R&D-projects*, the interviewees confirmed the assumed low relevance and emphasized that these R&D-cooperations would only scarcely occur (Interview EoE 1: 6; EoE 2: 8f.; EoE 3: 7f.; EoE 4: 15). Apart from IP-protection issues impeding these cooperations, especially competitive standard-setting races in the predominantly high-technology sectors in the region would make competitors unlikely partners in business R&D-efforts.

Cooperative R&D-Projects with Universities and other Scientific Institutions

Contrariwise, *cooperative R&D-projects with universities and other scientific institutions* would be absolutely central channels of knowledge transfer for large parts of the firms located in the Eastern Region, and would be more common than in other UK and European regions. These cooperations would take the whole range of possible forms, starting from informal exchange of ideas and results to formalized material transfer up to contract research, joint research groups, personnel transfer and consultancy (Interviews EoE 1: 6f.; EoE 2: 9f.; EoE 3: 8ff.; EoE 4: 3f.). The reasons given for that are manifold: First, many national firms or MNCs would explicitly locate themselves in the region to enter into cooperation with Cambridge University and other excellent research units. Second, the high-technology and research-intensive character of the regional firms would require strong collaborative ties to universities and research institutes, as the CEO of *Cambridge Network* accentuates:

"In Cambridge it will probably be more common than elsewhere in the UK, and the reason is that there are many local companies that are doing research intensive work and for them the differentiator is that they are located close to the strongest research university in Europe. Thus, that we have more collaborations here, has to do with the nature of the university and with the nature of the local companies." (Interview EoE 3: 8)

Third, the interview partner also points to lower mental and cultural barriers between the academic and the business world, relieving cooperation in the region:

"There is also the case that within Cambridge there is a very well developed path between the university and local companies. Many of these local companies were founded by academics or are spin-offs that were funded by academics. So the cultural difference is small. There is a strong acceptance that there is a movement. […] In Cambridge you find less of that barrier, because of the history of successful actors. If you look at the key actors that are most cited in the region, they are very often former academics in terms of the high-tech industry. And within the university, there are many people who are still active in the university and also operate a spin-out. So it makes it much simpler to cooperate." (Interview EoE 3: 8ff.)

Fourth, the interview partner continues highlighting distinctive features of the higher education and research systems in Europe, and also points to specific university policies and self understandings, which would make Cambridge University much more an entrepreneurial university than other UK or European universities, and the East of England a much more collaborative region than other UK or European regions:

"You can contrast these kinds of networks and compare them to the German system where you have mainly civil servant academics, who are primarily doing research as a public good, and then the very powerful research institutes like the *Fraunhofer Institutes*. These civil servants in the universities will have very little links to large companies, because there is a very strong tradition of people going between industry and academia, especially in the engineering departments, but they do not necessarily have an interest in creating spin-outs or small companies themselves, because there is a strong civil service tradition of not creating private benefits from public money. The big research institutes will be doing a lot of interesting work, but they will do it as *contract research*. This does not create such a broader base or such a diverse set of actors as you have it here in Cambridge.

By contrasting Cambridge, the academics are not civil servants, they have very little job security, the way they will succeed is by being quite *entrepreneurial* themselves in finding ways to fund more research students, finding ways to fund more technical staff, and also secretaries. All these things can get funded if you get access to industry money and work with somebody from industry. So the academics start up being quite entrepreneurial. We do not have these very strong research institutes like the *Fraunhofer*. What we have instead are technology consultancies like *Cambridge Consultance*, the *Technology Partnership*, *PA* etc., and these organizations do the contract research, but they also create many spin-outs. Thus, the difference is in the kind of actors in the system. There is a much larger network around Cambridge, because it is a very large and diverse and mixed scale network compared to the world." (Interview EoE 3: 2f.)

Although the interview partners consistently constrain these aspects to the Greater Cambridge area, and remark that in other universities and subspaces in the East of England the "entrepreneurial university" model (see 3.2) is less developed, the majority of the East of England firms would maintain close links to Cambridge University and, not least through being member of *Cambridge Network*, also to other actors around it (Interviews EoE 1: 6f.; EoE 2: 9f.; EoE 3: 3). Altogether, therefore, markedly strong ties between academic and business sector can be identified as a characteristic feature of the East of England's knowledge transfer system.

Network Based Mechanisms in the East of England's Automotive Sector

A sectoral perspective on *automotive* reveals that horizontal cooperations between competitors are seen as less relevant mechanisms of knowledge transfer by the interviewed experts (Interviews EoE 3: 5; EoE 5: 2f.). Although these networks would exist, and especially in the sector's highly specialized niche markets like luxury vehicles, motorsport and aerospace, they would only lowly be institutionalized, more likely be used for general interaction and (informal) information flows, as well as for the arrangement of the framework conditions of business activities in the sector. One expert for the automotive sector in the East of England traces this back to highly competitive niche sectors, the strong research focus of the regional automotive firms and to the lack of own OEMs in the region:

"[A] general feature of UK automotive is that we lost almost all our indigenous large volume manufacturers within the last decades. The OEMs operating in the country are, therefore, foreign owned companies. However, many of these large firms operating in the Eastern Region only have R&D-facilities here, and no production plants. And these businesses appreciate the region's scientific excellence. But R&D-cooperations with other larger OEMs are rare. Where it is more likely to see networks is in the niche market sectors, like motorsport. Here we see more b-to-b cooperations, and these are in many cases grown for decades. Other important examples would be luxury cars and aerospace. [T]here certainly is knowledge transferred. The question is if the knowledge transferred in these networks really drives innovation forward. Especially the niche markets of aerospace and motorsport are highly competitive environments, and that is a major barrier to cooperation between competitors. Take the *Motorsport Valley*. This is a cross-regional cluster, so it is transcending the Eastern Region and connecting it to other sites in the UK. Here you got cooperation and network activities, but especially joint marketing efforts etc. leveraging the benefit of the whole cluster. This includes many of the most important motorsport manufacturers like McLaren, Lotus, Mercedes and Red Bull Racing. And you can be sure that cooperation among them is really limited. So I would generally see cooperation between these manufactures as less relevant." (Interview EoE 5: 2f.)

More important would, on the contrary, be vertical supply chain networks. These would be vital channels of knowledge transfer in the East of England's automotive sector and its different subsectors (Interview EoE 5: 3). However, two factors would limit these supply chain effects in the case of the East of England: On the on hand, the focus of many firms would be on conducting R&D-activities and connecting with Cambridge, Cranfield and other universities in the region, while only a few larger volume production facilities would exist. On the other hand, one interview partner points to a UK-wide trend of a supply chain hollowing out, which would also affect the East of England region:

"[W]e recently did a larger study on the automotive supply chain in the UK, including the Eastern Region, and what we found was that it is becoming increasingly difficult for the firms to find adequate UK suppliers. Like the UK lost most of the OEMs in the last decades, it is obviously now losing its suppliers base. The reasons are again mainly cost related, skills related, and they give room for foreign suppliers to dominate the UK market. The business leaders we surveyed are very preoccupied about that. It might become increasingly difficult in the future to maintain those close supply chain relationships, especially those in close physical proximity, what has been a strength of the industry in the past." (Interview EoE 5: 3)

As for the region as a whole, also for automotive the interview partners did not ascribe collaborative inter-firm R&D-projects a high significance for knowledge transfer (Interview EoE 5: 5). Although these cooperations would exist in a few cases, they would be strictly limited by the highly competitive business environments in the niche markets dominating the regional automotive landscape, in which smallest improvements like new designs or aero dynamical tools would make the difference and would strongly incentive the actors to protect their breaking point knowledge. However, some interview partners expect that there will be more intensive collaborative R&D-activities in the sector in the future around the low carbon topic, which would currently be strongly incentivized by the central government, the RDAs and EEDA in particular, with many projects already planned (Interview EoE 5: 5). Much more intensive would be, on the contrary, cooperation in R&D-activities with universities and research institutes, in all stages of automotive business processes. Many firms in the sector would have a strong focus on R&D, and some of them would have explicitly chosen the East of England as a location to operate in close proximity to its highly renowned universities and institutes. And again the interview partners emphasize that a high number of automotive firms in the East of England is specialized in high technology fields like aerospace and racing cars, where intensive cooperation with engineering institutes would be required (Interview EoE 5: 5f.).

Network Based Mechanisms in the East of England's Biotechnology Sector

In the East of England's *biotechnology* sector, on the other hand, horizontal networks have been reported to even play a central role (Interviews EoE 4: 13; EoE 2: 5f.; EoE 3: 3f.). The interviewees consistently named *ERBI* as the central and most important of these sector specific networks, and *Cambridge Network* as cross-sectoral network also including large parts of the regional biotech firms. However, the character of interaction within these networks should be noted, especially when comparing them to horizontal networks in Germany and to networks in traditional sectors like automotive: As one interview partner from *Cambridge Enterprise* illustrates, cooperation in these networks is mostly related to accessing a common pool of expertise, skilled labor, training services, lab space, and marketing activities, while cooperation in core competences like joint drug development activities would be unusual (Interview EoE 4: 13). Rather, business-to-business cooperation in biotechnology would especially consist of interaction between the biotech firms and highly specialized service providers. And it would, above all, be the access to these service providers which would be a particular strength of networks like *ERBI* or *Cambridge Network*. As a consequence, these formalized membership networks would also have a large "social network" component, making the informal dimension of interaction an important feature of knowledge transfer processes within these networks, apart from using formal membership services (Interview EoE 4: 13f.).

Two points should be added in this regard: First, the interview partners also highlighted the role of EEDA and, therefore, indirectly of central government, in establishing and initially funding *ERBI*. Although this network is now self-sustaining, this momentum of public intervention in biotechnology network-building should be noted. Second, after conducting the interviews for the East of England case, *ERBI* merged with the *London Biotechnology Network* into the cross-regional *One Nucleus* life sciences network in April 2010, expressing with that the development of a UK life sciences super-cluster. As in the case of the automotive *Motorsport Valley*, networks creating trans-regional innovation and knowledge transfer contexts seem to be characteristic for the East of England region and obviously also for the UK, what might not least be traced back to a comparatively lower historical, economic, political and socio-cultural contextualization within the UK regions, as will be discussed in detail below.

Vertical supply chain networks in a classical sense would not play any role in the East of England's biotechnology sector (Interview EoE 4: 14). If anything, subcontracted service providers could be interpreted as suppliers in this context, as well as equipment suppliers, which both, however, would not be involved in the core business activities of the biotech firms and would only marginally contribute to knowledge transfer processes in the sector, as the technology manager for the biotechnology sector at *Cambridge Enterprise* explains:

> "[...] I have never heard a CEO of a local biotech company emphasizing this as important. I think in the biotech sector it might not play an important role. What I guess is that [for; MO] the bigger pharmaceutical companies, when they enter into a collaboration or do a license or invest in them, then this sort of relationship will become important. It depends on who your technology partners are, and it will play a role with the equipment suppliers. What they would also want to achieve is that their patent agents are good and professional." (Interview EoE 4: 14)

Much stronger would, thus, be the supply chain relation between the biotech firms as suppliers of new drug candidates and the pharmaceutical firms as their customers, in which often the whole biotech company would be acquired by the latter, also under mediation of *ERBI* as a cluster organization (see also Interview EoE 2: 5), as the CEO of *Cambridge Network* highlights:

> "[T]he most clear examples of that are particularly in biotech, where there is a very strong relationship between drug discovery biotech and pharmaceutical companies, because the drug discovery biotechs are looking to be acquired by the pharmaceutical companies, so that those are trying to sell them candidates. That is a supply chain type and a sector that is strong and there is an organization for the Eastern region biotech industry called *ERBI* which tries to link all those people together. They run *partner events* which are basically about people working out how to exit or how to find potential partners." (Interview EoE 3: 4)

Furthermore, cooperative inter-firm R&D-projects would be, like in automotive, very rare. Since R&D-activities would be the most sensitive and core parts of their business activities, the firms in the sector would be highly aware of avoiding IP-contamination and would explicitly manage competition through being pointedly restrictive in conducting joint R&D-projects (Interview EoE 4: 15). R&D-related cooperations with universities and research institutes, on the contrary, would be, like in all other investigated biotech regions, vital channels of knowledge transfer in the East of England's biotech sector. In this, all possible forms of interaction would be used intensively, starting from informal exchange of ideas and results to formalized material transfer up to contract research, joint research groups, personnel transfer and consultancy. What would be a specifically differentiating characteristic of the East of England in this regard would be the close links between the academic community and the regional biotech firms, especially in the Greater Cambridge cluster, in which many firms would have their roots in the regional universities and where many academics would maintain active contacts to the business sector and to former colleagues now being working in private firms. This specific entrepreneurial-academic spirit would decisively ease the emergence and the success of these forms of cooperation (Interview EoE 4: 3ff.).

7.2.2.2 Association Based Mechanisms of Knowledge Transfer

The second dimension of knowledge transfer is represented by the association based mechanisms, namely employers-, business-, and trade associations, chambers of industry and commerce, and binding and common technical standards. Building on the literature on VoC and NIS, the theoretical assumptions have been derived that the first two of them should not play any role for knowledge transfer processes in the UK regions, while the latter should have a medium range importance. These assumptions have basically been confirmed by the interviewed experts when looking at the East of England as a whole.

Employers-, Business-, and Trade Associations/ Chambers of Industry and Commerce

With regard to both, *employers-, business-, and trade associations*, as well as *chambers of industry and commerce*, the interview partners consistently confirmed that these institutions would not play any role for knowledge transfer processes in the East of England (Interviews EoE 1: 3ff.; EoE 2: 7f.; EoE 3: 5ff.; EoE 4: 14f.). Associations and chambers would traditionally not engage in the topics of innovation and knowledge transfer, but, rather, in interest representation and other aspects of business activities. In addition, associations and chambers would represent more the less knowledge intensive sectors, which might be more important in other UK regions or the rural parts of the East of England, while many of the knowledge intensive service providing and manufacturing firms characterizing the East of England's economy would not be their members, as the CEO of *Cambridge Network* explains:

> "It depends where you are. E.g. within the UK the biggest employers association would be the *CBI*, which is the business industry group, or the *chambers of commerce*, which tend to be the small businesses. Neither of them is very strong in their membership in Cambridge itself, because the employer interests that they represent are very old economy. It has to do with logistics, very basic level things about levels of the IT, levels of taxation in general, it is the very lowest common denominator. They do not address very closely the particular concerns of high-tech firms. So within Cambridge many of the companies would be members of the *Cambridge Network* but they probably would not be members of the Chamber of Commerce. So I do not see much activity in this respect. The activity that I see from chambers of commerce or from the CBI are dinners or breakfast meetings which are very basic, so general things like how to be better organized, and lobbying for a better transport in the region." (Interview EoE 3: 5f.)

Binding and Common Technical Standards

Furthermore, associations and chambers would also not play an important role in the definition, regulation and implementation of *binding and common technical standards*, and especially not in the regions medium-high and high-technology sectors (Interview EoE 3: 7). In these sectors, standard setting would be a highly competitive race among already existing or upcoming market leaders to first bring their products to the markets and, with that, to implement them as technical standard. Nevertheless, even if not implemented by associations or chambers but by market dynamics, these common standards would play a significant role in knowledge transfer processes when established, as drivers of innovation activities, as a joint reference point for firms operating in related fields, and as indicators for the direction of further technological developments. In addition, these standard setting processes would be explicitly non regional, but global processes, although also East of England firms would have been successful in the past in globally implementing technical standards, as the CEO of *Cambridge Networks* remarks:

> "[W]e have both, people that are developing and elevating in standards, who are creating standards like *Bluetooth*, etc. [...] The creation of standards or creating de facto standards of people that are in use, is very much an international activity within high-tech. There are local companies who are very influential in defining standards, especially the larger players have people who are contributing to standard bodies at very high levels, but there is a very international forum. Compared to e.g. physical equipment to do manufacturing and creating standards for a physical piece of equipment, these are very global markets. I think that this creates the difference. It will be hard for a local grouping to set a standard that would be accepted." (Interview EoE 3: 4, 7)

Association Based Mechanisms in the East of England's Automotive Sector

Concerning the role of associations and chambers in *automotive*, the interviewed experts basically confirmed the results obtained for the region as a whole. While chambers would not play any role at all for knowledge transfer and innovation in the sector, the associations would be slightly more active, although by far not decisive actors in these fields (Interview EoE 5: 3f.). Most of the automotive associations like the *Society of Motor Manufacturers and Traders* (SMMT) or the *Motorsport Industry Association* (MIA) would be nationally organized and would not have a specific focus on the UK regions in general and the East of England in particular. Moreover, their contribution to innovation and knowledge transfer would mostly be indirectly, through offering training and education activities, networking events and conferences. In addition, also their interest representation activities towards central government and the EU have been mentioned in this regard, where they would contribute to the development of strategic goals and common technical standards, like e.g. emission standards concerning automotive, especially in relation to the low carbon topic. However, one expert explicitly highlights that these associations would only be one voice in the sector among others, and that in the decisive arenas like the *UK Automotive Council* the larger firms would also defend their interests individually. Overall, this expert also remarks that the traditionally weak industrial relations in UK automotive and a lack of concerted interest intermediation would be seen as decisive competitive disadvantages towards the country's major competitors by many automotive actors, especially in the sector's transition to the low carbon era (Interview EoE 5: 4f.).

Association Based Mechanisms in the East of England's Biotechnology

For *biotechnology*, on the other hand, both chambers and associations would not play a role in knowledge transfer activities at all, especially because of the global orientation characterizing the sector, as one interviewed expert for the East of England's biotech sector points out:

> "In the biotechnology I do not really see them. As I told you before, a lot of mergers and acquisitions are done with companies outside the region, and so the local chambers are often not working at the level that a lot of these companies are working at. Not only in the biotech industry you have the situation that your customers will not only be local. The usual clients of the chambers in the UK will be local supermarkets and car dealers, e.g. The biotech industry is global and nobody of the people running these companies will see the local chambers or networkers being hugely important to him. […] And, moreover, the biotech firms are much more mobile than others, so before engaging with government for a better environment they might just go somewhere else." (Interview EoE 4: 14f.)

Finally, none of the interview partners could report on the role of common and binding technical standards in the East of England's biotech sector.

7.2.2.3 Organization Based Mechanisms of Knowledge Transfer

As organization based mechanisms of knowledge transfer in regional innovation systems the transfer of technical and scientific staff, purposeful headhunting and business start-ups have been identified. Building on the literature on VoC and NIS, it has been assumed that all of these mechanisms should be centrally important for knowledge transfer processes in the regional innovation systems of the UK's LME. These assumptions have basically been confirmed by the interview partners when looking at the East of England region as a whole.

Transfer of Technical and Scientific Staff

With regard to the *transfer of technical and scientific staff*, the interview partners consistently pointed to highly fluid labor markets especially in the Greater Cambridge area, but less intensive transfer activities in other parts of the East of England region (Interviews EoE 1: 7f.; EoE 2: 11; EoE 3: 10f.; EoE 4: 5). While around Cambridge there would be intensive pull-effects from the business sector using the transfer of academic staff as major channel of interaction with the region's scientific institutes, and while the entrepreneurial university model characterizing Cambridge University would generate push-effects to the academic personnel to transfer to the business sector, outside Cambridge different mentalities and logics of action between the two worlds would make these transfers less likely, as the director of innovation at *EEDA* describes:

> "There is quite a lot of movement of staff around, and particularly in Cambridge, because it gets a lot of pull from businesses, and in UEA [University of East Anglia; MO] there is probably none, at least not into local businesses. People will stay at the university as long as they can and if they get funded they stay staff there. But generally there is a force for university staff, because not many people have permanent tenure. What happens in Cambridge is that people will move out to spin-outs or you will get senior university staff who is involved in companies, but still retain their academic position for a long time. There are many companies where the professor who ran the group from which some of the work comes out is associated with a director of the business, maybe even spends one or two days a week in the business, but is still a member of the university. Cambridge is probably much more flexible in this then the most other universities." (Interview EoE 2: 11)

To enhance personnel movement activities also in other areas of the East of England, central government would offer a support program mediated by EEDA, called *Knowledge Transfer Partnerships (KTP)*, which has been consistently evaluated as a quite successful tool by all interview partners (Interviews EoE 1: 7f.; EoE 2: 11; EoE 3: 10f.), and even as the "most successful government scheme in technology transfer" (Interview EoE 3: 10) by the CEO of *Cambridge Network*. A general rationale underlying this scheme would be the fact that normally most personnel transfers would be those of university graduates and doctoral students, and that experienced senior researchers would only scarcely (and even then in most cases only part time) change to industry. Therefore, the interview partners acknowledged it as a specific strength of the KTP program to also indirectly involve the respective supervisors of the transferred graduates or doctoral students, as an exemplary statement of an interview partner from the *Business Link East* illustrates:

"Through the *Knowledge Transfer Partnership* process what you have is that a graduate is [temporarily; MO] recruited, but has an academic supervisor, who has a very close link to the graduate working within the business. Therefore, the knowledge is transferred directly into the business. What they say is that the business itself benefits hugely, because everybody working around that process actually adopts skills as part of that. So it is hugely valuable. That is one way in which the supervisor from the university is directly working within that business. [T]hey have regular meetings and the partnership is based around one particular project that is of value for the business. So the knowledge is generated and supported within the business itself." (Interview EoE 1: 7f.)

Purposeful Headhunting

Also personnel transfers within the business sector and *purposeful headhunting* activities would be very common and regularly realized, especially in the Greater Cambridge cluster (Interviews EoE 2: 11f.; EoE 3: 11f.; EoE 4: 6). In this, major recruiting pools would be e.g. the previously discussed network initiatives like *Cambridge Network*. However, as the director of innovation at EEDA accentuates, fluctuation and headhunting would still be less volatile than in comparable US-clusters, pointing with that to important differences within the group of LMEs:

"Yes, this happens. Particularly in the cluster I am sure it happens all the time. Although I do not think that it happens as much as in *Silicon Valley*, where people move around every year. So there is a different attitude here. […] And in the USA it is still much more driven by the rewards, so people look at their options and they might decide quite quickly if they could do more money in another company or not. In the UK it would also be driven by the options, but I do not think that they will be comparing if in one year they could go somewhere else. For them it might be more important if they are in an interesting company and they like to work there and if they have an interesting product. So I would not say that there is a huge movement, but there is enough." (Interview EoE 2: 11f.)

Business Start-Ups

With regard to *business start-ups* and academic *spin-offs*, the interviewees consistently emphasized that these would be vital mechanisms of knowledge transfer in the East of England, and especially pointed to the Greater Cambridge cluster and other universities in the region as the UK's and Europe's most entrepreneurial places (Interviews EoE 1: 9f.; EoE 2: 12; EoE 3: 12; EoE 4: 6f.). In particular, the region's scientific parks and business incubators, also with support of EEDA, and business R&D-centers would be essential drivers of start-up activities in the East of England, as one interview partner remarks:

"In Cambridge it is very common, it is probably the most entrepreneurial cluster in Europe. And in Cambridge it is not only spin-outs from the university, but also people who come up with business ideas and set up their companies. In the rest of the region this is less. You get hot spots, and we are encouraging that through these scientific science parks, where there will also be spin-outs from there. There are big R&D-centers from *Unilever* e.g. etc. and research institutes, from where we also have a lot of spin-outs. So there are pockets for spin-outs around the Cambridge Cluster." (Interview EoE 2: 12)

One explanation of these pointedly high start-up activities would lie in the region's specific history of industrialization and its particular structural conditions, as the CEO of *Cambridge Network* illustrates:

"Historically, the East of England was an agricultural region, where there was not much historical industry. That means that because the rents are very high and a lot of the land is controlled by the university or by colleges, you have to create very high margins perhaps. And so, because of the peculiar conditions here, we have many *science parks* close to Cambridge and it creates an environment where there are many start-ups." (Interview EoE 3: 12)

Concerning the availability of seed- and venture capital for these start-up activities, as could be assumed, the experts pointed to a mature ecosystem of investors and business angels in the East of England region (Interviews EoE 1: 9f.; EoE 2: 12ff.; EoE 3: 12f.; EoE 4: 7f.). While the provision of venture-capital for later stages of the business life cycle and angel support would be sufficiently available, some shortcomings in terms of seed-capital for the early phases of business creation would exist, which would, however, be among other things addressed by EEDA providing small amounts of initial and proof-of-concept funding.

Organization Based Mechanisms in the East of England's Automotive

Turning to *automotive*, staff transfers and purposeful headhunting would occur, but only to a limited extent, and therefore, on a lower level than in the region as a whole (Interview EoE 5: 5ff.). On the one hand, automotive would be regarded as a less attractive employer by many highly qualified persons and graduates, and overall a too low share of students would opt for engineering studies in general. On the other hand, the firms in the sector could mostly not offer competitive wages and career options to highly qualified staff, not least because of wage pressures imposed by international competition and business headquarters located overseas, as well as the lack of OEM headquarters in the East of England. In addition, also a certain disregard of the automotive firms themselves to invest more in training and education schemes would contribute to this comparatively low mobility in the sector. However, in the strong R&D-sector of the East of England's automotive industry, intensive transfers of academic staff from universities and institutes to the business R&D-facilities would take place, pointing again to the specific and distinguishing academic ecosystem in the region. Finally, also start-up activities would only occur to a very limited extent, above all because of the mature and consolidated character of the sector (Interview EoE 5: 7).

Organization Based Mechanisms in the East of England's Biotechnology

For *biotechnology*, both the transfer of academic staff to industry and headhunting activities would be actively used and usual mechanisms of knowledge transfer. In contrast to other sectors, also the converse direction from industry to academia would not be exceptional. However, even in this sector and in the Cambridge cluster these transfers would not totally be free fly, since distinct logics of action and working routines would markedly differ between the two worlds and would constitute a considerable barrier for qualified staff to switch sides (Interview EoE 4: 5f.). In addition, and at first sight surprising, business start-ups and spin-out activities would only play a limited role in the East of England's biotech sector. Apart from being an already mature sector with tendentially lower start-up and spin-out activities, especially a lack of veritable large scale venture-capital would lower these activities. As a consequence, the businesses anyhow created would be considerably smaller than in the past, and many firms would increasingly outsource large parts of their activities. It would then be predominantly foreign, and especially US investors creating life science businesses in the region, often employing East of England staff, but not purely basing on East of England IP, as one interview partner from *Cambridge Enterprise* critically highlights. These firms could, therefore, not be considered as spin-offs from the region's universities (Interview EoE 4: 6ff.).

7.2.2.4 State Induced Mechanisms of Knowledge Transfer

The state has been defined to intervene in knowledge transfer processes in regional innovation systems above all through mechanisms such as public programs, differentiated by national or regional policy measures, through (larger) state-induced innovation projects, the creation of scientific parks and technological centers, as well as the maintenance of technology transfer offices or companies at the universities and public research institutes. Basing on the discussion of the literature on national production and innovation systems, it has been assumed that in the UK's regional innovation systems all these mechanisms should only play a subordinated role for knowledge transfer activities, according to the liberal, *laissez-faire* and market creating and preserving role exercised by the state in the UK's LME, taking an arms' length approach to business and labor and only playing a limited role in setting rules and moderating conflicts and interaction. Only scientific parks, technological centers and technology transfer agencies/companies should be slightly more important. None but the last of these assumptions has been confirmed by the interviewed experts, ascribing the first five of them a significantly higher importance for knowledge transfer processes as assumed.

National and Regional Government Measures

With regard to *national and regional government measures*, the interview partners consistently highlighted first of all the lack of regional governments in the English regions. Except Scotland, Wales, Northern Ireland and the Greater London region (see below), the rest of the UK regions would not have far reaching regional policy competences in the sense of own regional parliaments with larger legislative and fiscal room to maneuver, but would depend strongly on central government action. Therefore, national government measures would be clearly the decisive level of action directed towards the fields of innovation and knowledge transfer (Interviews EoE 1: 11f.; EoE 2: 3ff.; EoE 3: 14f.; EoE 4: 15f.). However, since the late 1990s innovation and technology policy in the UK has experienced considerable regionalization processes, especially through the establishment of Regional *Development Agencies* (RDAs) in the UK regions, *Government Offices* for the regions and *Regional Assemblies*. Through these measures, especially implementation of national science, technology and innovation policy has partly been devolved to the regional level. In this connection, particularly the work of the RDAs, and in this case of EEDA, has been consistently evaluated as highly valuable in initiating and supporting innovation and knowledge transfer, as well as in creating the respective framework conditions in the East of England.

Nevertheless, and again consensual, the interviewees assessed that the room for EEDA to develop a unique regional innovation and knowledge transfer policy, specifically tailored to the sectoral conditions and structural needs of the East of England region, would be very limited by the strict political requirements set up by central government (Interviews EoE 1: 11f.; EoE 2: 3ff.; EoE 3: 14f.; EoE 4: 15f.). Two factors would be major reasons for this limited scope of regional policy action in the East of England, of which the first would be the detailed central government predefinition of policy measures to apply. In this, a major impact on innovation and knowledge transfer would have the *Business Links*, a business support program directed to small businesses, and, above all, the *Solutions for Business* portfolio, a package of publicly funded business support products and services to help firms in the different phases of their business life cycle. These programs and packages would include measures like e.g. the KTPs, innovation vouchers, start-up support, collaborative R&D, networking for innovation and innovation advice, which would then be implemented and managed by EEDA and the other RDAs in the UK, or would be delegated to the *Business Links* in the regions like *Business Link East*. Although all interview partners acknowledged these measures as successful and efficient supporting tools in innovation and knowledge transfer, they would basically be the same across all UK regions and for all industrial sectors and only leave marginal space for regional priority setting by EEDA.

The second factor mentioned is the redistributive nature of innovation and knowledge transfer support through the RDAs, leaving fewer resources to the economically better performing regions like the East of England (section 7.2.1.1) to realize stronger strategic support of the region's key sectors, as the CEO of *Cambridge Network* describes contrasting this to parts of the German innovation policy:

> "The answer would be very different in the UK compared to Germany. In Germany they have the *"Spitzencluster" program*, which concentrates resources in regions which are meant to be highly innovative and, though, the regional funding which is then available is significant. By contrast, our *Regional Development Agencies* in the most innovative regions as the East of England have less money than the least innovative regions. So it is a taxation system for moving money from the regions which are rich and are competing successfully in high tech markets to old regions with dying industries like ship-building or coal. Thus, we have a different approach and our regional funding is not significant [...]. It is set up to be a redistributive taxation system. The Eastern region has taken the decision to focus more or less exclusively on clean tech, where they think that they can have some impact by supporting the setup projects in clean tech. Within that sector, there will be very significant funding, but in general, for most sectors central government would be much more significant in terms of the amount of funding." (Interview EoE 3: 14f.)

The protagonism in innovation and knowledge transfer support for businesses would, therefore, be clearly left to central government (Interview EoE 3: 14f.). However, the director of innovation at EEDA evaluates these points less critical, since he recognizes specific opportunities in the intensified concentration on clean tech and low carbon sectors, as the East of England region would have particular strengths in many of these sectors like renewable energies and automotive (Interview EoE 2: 2ff.).

Scientific Parks and Technological Centers

In addition, especially EEDA's strong focus on supporting *scientific parks*, *technological centers* and business incubators would be a tribute to the region's specific characteristics as a location of high-tech industries and scientific excellence, and would differentiate it from other UK regions. In this regard, the high importance and central role of these scientific parks and technological centers in knowledge transfer processes, especially in terms of start-ups, spin-offs and collaborative R&D-activities in close physical proximity, have also been affirmed by the other interviewed experts (Interviews EoE 1: 4; EoE 2: 2ff.; EoE 3: 2).

(Larger) State Induced Innovation Projects

Concerning the role of *larger state induced innovation projects* as mechanisms of knowledge transfer, the interview partners ascribed them a high importance, although strongly dependent on the sector. While the state would be a major initiator and driver of innovation projects in the low carbon fields, also setting up incentives and creating favorable regulatory framework conditions to this, it would also be a major customer and initiator of large scale innovation projects of the defense industry, which would in the East of England among others be represented by the aerospace sector (Interview EoE 3: 14).

Technology Transfer Agencies

Finally, the role of the *technology transfer agencies and companies* at the region's universities and research institutes has been evaluated as important and valuable for knowledge transfer processes (Interviews EoE 1: 10f.; EoE 2: 14; EoE 3: 13f.; EoE 4: 1ff.). Although the interview partners also pointed to differences in the management capacities of these institutions, and to their difficulties in fulfilling an extensive switchboard function, especially the role of those at the larger universities like *Cambridge Enterprise* at Cambridge University would be highly professional, proactive and efficient. Nevertheless, also in the East of England these transfer agencies would not necessarily be the decisive nodal points in the knowledge transfer streams between academia and business, although because of different reasons than in the other investigated European regions. As the CEO of *Cambridge Network* illustrates giving the example of patenting activities, many transfer activities would occur passing by these transfer agencies, not least because of the anyway strong ties between academic institutions and business sector, constituting a major difference also to other UK regions:

"If you contrast *Imperial College* or *Oxford* with Cambridge, the largest level of patenting by far was Cambridge, beating Oxford by factor five. Within Cambridge, *Cambridge Enterprise* is only a tenth of the total volume in terms of patenting activity. I also compared the means of everybody who is patenting in Cambridge and I compared it with a staff telephone listing to try to find out whether Cambridge inventors are going into the markets through local companies, given that there are many local companies patenting. [O]nly about 30% of all the patenting is done by university academics. That means that there is a huge connection between the university and the local high-tech cluster. If you contrast that with Oxford, *ISIS* is probably 60% of all the patenting in Oxford." (Interview EoE 3: 13)

State Induced Mechanisms in the East of England's Automotive Sector

From a sectoral perspective, both automotive and biotechnology in the East of England do not decisively deviate from these results for the region as a whole. On the one hand, also for the East of England's *automotive* industry central government would be the decisive policy level, and EEDA would be the most important regional policy body (Interview EoE 5: 9f.). Both would address a wide range of measures, strategies and instruments towards automotive, starting from general subsidies by central government for business support mainly coming from EEDA, but also R&D- and start-up funding and skills development etc. As one expert for the East of England's automotive sector stresses, this public action would be helpful in enhancing innovation and knowledge transfer activities, although more through creating adequate framework conditions and less through direct intervention. In particular, this public action in automotive would increasingly be part of a broader policy strategy directed to the low carbon challenge affecting many related sectors. Apart from the opportunities connected to this technology and policy shift, the interview partner also mentions concerns of automotive actors to be pushed out of policy focus stronger than before, and also points to different roles of public action when comparing the UK with its European peers:

"[E]specially business leaders claim that automotive is more and more out of policy focus. But as I said, EEDA and central government are both recently strongly pushing forward low carbon innovation. This effects many sectors, and automotive is definitely one of them, so we will see how auto companies will profit from that. [A]t least that creates an environment in which innovation and knowledge transfer can occur. [I]n facing the low carbon challenge it is indispensible that central government provides active support, and surely that will enhance innovation […]. But if you compare this to Germany or France, the focus is more on the framework conditions in which innovation will take place, not so much on basic inputs. I think in the UK the links between the larger firms and central government are just not so dense as overseas. In terms of innovation and technology transfer what is most important here is the supply chains and inputs coming from customers and suppliers, and, of course, our research institutes and universities, especially in the East of England." (Interview EoE 5: 9f.)

Larger state induced innovation projects would, therefore, only play a role where the state would have a certain protagonism in fostering technological change and pushing innovation, like in the low carbon field, and where it would act as major customer of respective products, like in the defense branch of the aerospace sector, what indeed would have effects on the East of England's aerospace cluster (Interview EoE 5: 9). Moreover, as a comparatively R&D-intensive automotive region also scientific parks and technological centers would be actively used and important mechanisms of knowledge transfer. The most renowned would be, apart from those around Cambridge, *Cranfield University Technology Park*, hosting the *Nissan R&D-center*, and the *Hethel Engineering Center*, where *Lotus* would be engaged, and where a technology park is being created as well. Finally, this high R&D-orientation in the sector would also lead to intensive interaction with academia through technology transfer agencies, especially in the larger and R&D-intensive firms, but less in the smaller and more production or service oriented companies (Interview EoE 5: 7f.).

State Induced Mechanisms in the East of England's Automotive Sector

On the other hand, also the East of England's *biotech sector* would mirror most features identified for the region as a whole: Central government as the major policy level; *EEDA* as central policy actor at the regional level supporting the sector's knowledge transfer activities through network promotion (ERBI) and the mechanisms of the *Solutions for Business* portfolio; scientific parks, technological centers and business incubators as important physical places of interaction between academia and business and breeding grounds for start-ups and spin-offs; as well as technology transfer offices being major channels of patenting, licensing, IP-management, academic consultancy and spin-off support. Only state-induced innovation projects would not play a role at all in East of England's biotech sector (Interview EoE 4: 1ff.).

7.2.2.5 Market Based Mechanisms of Knowledge Transfer

Market based mechanisms of knowledge transfer have been defined as mechanisms through which knowledge is "bought" and "sold" via the market. To investigate this market dimension of knowledge transfer, the interviewed experts have been asked for the role of external consultancies, mergers, acquisitions and hostile takeovers, as well as the buying, selling and licensing of patents in knowledge transfer processes. The assumption derived from the literature on VoC and NIS is that all of these mechanisms should play a central role in the regional innovation systems of the predominantly market action driven LMEs like the UK.

External Consultancies

Overall, these assumptions have fully been confirmed by the interviewed experts when looking at the East of England region. *External consultancy* services would be used intensively in the region throughout all stages of the business life cycle and as important connecting agents between the scientific world and the business sector (Interviews EoE 2: 15; EoE 3: 16f.; EoE 4: 10f.). In this, it would be the large multinational and national consultancies like *PPT* or *PA*, and also larger and smaller regional and local consultancy firms like *Cambridge Consultants* which would be decisive tools in managing innovation and knowledge transfer activities in both larger and smaller firms, as the CEO of *Cambridge Network* explains:

"The strongest local consultancies as *Cambridge Consultants*, *PPT*, *PA* etc. have a very strong track record of creating spin-offs and of applying university technology. So yes, they are actively used. […] For the larger companies they would be contracted for product development work and for small companies they would be contracted for a proof of concept. [S]till it is a very similar need which is to contract with somebody to develop a particular devise in a time frame. That is the typical thing. […] A very similar story would be a start-up with very small scale. It does not have the broad range of capabilities and *Cambridge Consultants* contracted with a local company that wanted to build a washing machine that washes without water by using little plastic chips. And that was something that *Cambridge Consultants* could do for them very quickly, because they have the labs and they have the consumer product expertise about how to create a devise that looked nice and was easy to use. So I think the motivation is pretty similar both for small and for large companies." (Interview EoE 3: 16f.)

In addition, also individual academics from the region's universities and scientific institutions would provide consultancy services to industry, representing one more aspect in the close links between both worlds in the East of England and especially in the Greater Cambridge area, as an interview partner from *Cambridge Enterprise* adds:

"We also have a service by which our members of staff can be *consultant* to industry. As a member of the university they can either consult as a private individual, and there is no limit on how much time they spend doing consultancy in Cambridge. If they do it as private individuals, they are not allowed to use university facilities [...]. Or they can use our services, whereby *Cambridge Enterprise* would put in place all of the contractual arrangements and they can also access university facilities to conduct their consultancy service. They are also covered by the university's insurance policy. What we are trying to do with that is to promote the engagement with industry on the basis of *consultancy* through the university." (Interview EoE 4: 2)

Mergers, Acquisitions and Hostile Takeovers

Furthermore, also *mergers, acquisitions and hostile takeovers* would be, as expected, actively used as mechanisms of knowledge transfer in the region (Interviews EoE 1: 13f.; EoE 2: 15; EoE 3: 17f.; EoE 4: 11). Yet, this would depend strongly on the sector, with the high-tech sectors showing the highest activities in this regard. In addition, a large share of these activities would be with larger multinationals from abroad and especially from the US acquiring East of England firms.

Buying, Selling and Licensing of Patents

Finally, this would also be true for the *buying, selling* and especially *licensing* of *patents and IP*, which would be important channels of knowledge transfer in the East of England (Interviews EoE 1: 14; EoE 2: 15; EoE 3: 18; EoE 4: 12f.). Again, the impact of these mechanisms would be limited by a markedly international rather than regional dimension of IP-trading, making licensing and patenting activities strongly transcending the regional environment. And, second, as the CEO of *Cambridge Network* points out, also the trade of IP would highly depend on the respective sector and would, from his perspective, overall be less significant than other mechanisms, like e.g. consultancy:

"Licensing is a very important mechanism for some companies, like *ARM* e.g., which is a licensing company. But relative to the volume of technology that is being transferred to consultancies, it is much smaller. That will be true for all regions in the whole world. You will not find that licensing is one of the primary methods. There is a lot of licensing around Cambridge, and there are really successful technology companies like *Illumina*, which is a very successful local spin-out licensing designs. It is a very typical model for the really high-techs, because it enables them to create an international take up without having to do the full delivery service. But compared to the level of impact in society which is generated through consultancy, I think it is a much larger volume of technology transfer realized through consultancy, at least in the Eastern region." (Interview EoE 3: 18)

Market Based Mechanisms in the East of England's Automotive Sector

From a sectoral perspective, the use of these market based mechanisms of knowledge transfer in *automotive* slightly differs from the results of the region as a whole and appears to be generally lower. External consultancies, and above all specialized engineering consultancies would indeed be used, but would not be vital for knowledge transfer processes in the sector. Mergers, acquisitions and hostile takeovers would frequently occur, but more as a part of a consolidation and shrinking process characterizing the sector, and also as part of a selling-out of manufacturers and suppliers to foreign companies, in which indeed knowledge transfer would be one motive among other strategic considerations. Finally, the trade of IP would also play a role, and especially in the R&D-intensive automotive firms in the East of England and for the firms maintaining their R&D-centers in the region. However, substantial IP-transactions in the sector would be directed to the larger firms overseas and less to the smaller automotive firms in the Eastern region (Interviews EoE 5: 11f.; EoE 3: 17f.).

Market Based Mechanisms in the East of England's Biotechnology Sector

As was to expect, the results for the region as a whole also appeared when looking at the *biotechnology* sector. External consultancies would be used intensively, and the broad supply of these services in the East of England has been evaluated as highly valuable for the regional biotech cluster to lift its innovative capacities and enhance knowledge transfer activities. More than in any other investigated region and more than for other sectors the interviewed experts reported that mergers, acquisitions and hostile takeovers would be channels of knowledge transfer, since being acquired by another (mostly larger pharmaceutical) firm would be the explicit business model of many biotech firms in the region, and the overall mature character of the sector would already have produced a high number of acquisitions in the past. Finally, also in biotech the trade of IP would be an important, but not dominant, and international rather than regional mechanism of knowledge transfer (Interview EoE 4: 10ff.; EoE 3: 17f.).

7.2.2.6 Community Based Mechanisms of Knowledge Transfer

To address the community dimension of knowledge transfer processes, the interviewed experts have been asked for their estimation on the importance of three knowledge transfer mechanisms: Informal contacts, regional platforms and additional (regional) institutions, which would allow for establishing trust-based relationships and base on a common regional identity, culture and tradition. Building on the theoretical discussion on national systems of production and innovation, it has been assumed that these community based mechanisms of knowledge transfer should be of a medium range importance within the institutional framework of the UK's LME, which basically relies on market coordination of economic activities. This market coordination has been supposed to be only limitedly complemented by informal or community action.

Informal Contacts

These assumptions found broad confirmation by the experts when looking at the East of England as a whole. *Informal contacts* would positively complement most of the already discussed dimensions of knowledge transfer, and especially within the high-tech business-university cluster of the Greater Cambridge area this informal streams of knowledge transfer would be particularly intensive and enabling (Interviews EoE 1: 13; EoE 2: 16; EoE 3: 18f.; EoE 4: 9f.). However, the most substantial parts of knowledge transfer activities in the East of England would be formalized, and there would also be pointed differences in the relevance of informal knowledge flows when looking at comparable clusters in the US e.g., as an interview partner from *Cambridge Enterprise* illustrates:

> "I […was; MO] working in *MIT's technology licensing office*, and that is a proper cluster. Here [in Cambridge; MO] everything is distributed out into *science parks*. But in Boston my friends used to say that it takes them an hour and a half to get a sandwich at the corner shop and come back, because they will meet somebody they know. Cambridge does not work like that, we are not quite on top of each other. But it is a very small and tight community in that regard. If you look at the size of Cambridge, it is actually a small town and I am surprised if I go into the center of town on weekends and I do not meet somebody that I know. So that does work." (Interview EoE 4: 9f.)

Regional Platforms and other (regional) Institutions

Regional platforms would especially be provided by the region's universities and research institutes, by EEDA, e.g. through regional round tables, sectoral business associations, the already mentioned network initiatives and, more on the local than the regional level, by the local assemblies and chambers of commerce (Interviews EoE 1; EoE 2; EoE 3; EoE 4; EoE 5). Furthermore, some interview partners also named the *East of England Science and Industry Council* as an *additional regional institution*, a regularly meeting reunion of high rank science and industry leaders elaborating strategic concepts for the economic and scientific development of the East of England region, which would have an important driving stake in the regional innovation and knowledge transfer policy.

The question if a strong and commonly shared regional identity would be a driving factor behind knowledge transfer through informal interaction and regional platforms, has been neglected clearly by all interviewed experts (Interviews EoE 1: 13; EoE 2: 16; EoE 3: 18f.; EoE 4: 9f.; EoE 5: 11). Two explanations have been provided for this: First, most of the nine *English Government Office Regions* would be perceived as artificial, since they would not represent historically grown regional structures and would rather be statistical units established at the end of the 1990s to create UK regions covering the statistical NUTS-1 category. In the East of England case, regional identities would, therefore, rather be found on the sub-regional level, in East Anglia, in Essex, Bedfordshire and Hertfordshire. And especially the Greater Cambridge area with its universities and high-tech industry would be an important point of reference and self-identification, creating a dense structure of interaction and knowledge flows, even across the other parts of the East of England. However, and second, even in the Greater Cambridge cluster, the main orientation of most actors would be on international rather than regional networks, markets and actors. As most interview partners accentuate, the main driving force of interaction on the regional level would then be geographical proximity, and not regional culture and identity.

Community Based Mechanisms in the East of England's Automotive and Biotech Sectors

From a sectoral perspective, all these points would also apply to both investigated sectors, biotechnology and automotive (Interviews EoE 4: 9f.;

EoE 5: 10f.). What additionally resulted when focusing on the two sectors is the cross-regional dimension of knowledge transfer contexts and cluster-building, connecting the East of England with other UK regions, like the *One Nucleus* life sciences network with Greater London, and the *Motorsport Valley* with the southern and eastern UK regions. This tendency of increasingly creating national super-clusters also demonstrates a strong national dimension of innovation and knowledge transfer processes complementing regional and local contexts in the East of England. In addition, and pointedly stronger in biotechnology than in automotive, the interviewed experts highlighted the strong international orientation in terms of knowledge transfer and innovation, rather than towards the region (Interviews EoE 4: 9f.; EoE 5: 10f.).

7.2.3 Summary

In sum, these results provide broad correspondence with the theoretical assumptions derived from the literature on comparative political economy and innovation system studies (Table 12). In this, the case of the East of England region resulted to be a well matching example of a regional innovation and knowledge transfer system within a LME context, showing full confirmation of the assumptions concerning the association, organization, market and community based mechanisms of knowledge transfer, and only minor deviations from these assumptions on the network based, but stronger on the state induced dimensions of knowledge transfer. Thereby, many features of the East of England's knowledge transfer system have been traced back to the region's specific historical-industrial development path, coming from a strong agricultural and feudalistic tradition, experiencing a comparatively late industrialization process and developing a pointedly high-technology dominated industrial and services model in the present. This economic development has been complemented and enforced by a traditionally strong academic sector, emerging out of a region being pointedly shaped by colleges and universities, and giving decisive impulses for this economic development path, leading into one of Europe's most prominent high-tech and research business-university clusters.

In the analysis, in all investigated dimensions of knowledge transfer, these specific characteristics of the East of England resulted to be shaping the structure of the region's knowledge transfer system and the concrete mechanisms applied in it, and especially its strong links between business and academia: Vital start-up and spin-off activities from the business sector and from academia as innovative impulse givers to the region's high-tech industries, highly volatile job markets for qualified technical and scientific staff, collaborative R&D-activities of firms and scientific actors as major incentives for R&D-intensive firms to locate in the region and for researchers to mobilize funds within the entrepreneurial university model, numerous and intensively used scientific parks and technological centers as places of close interaction of business and academia in physical proximity, external consultancies as connecting actors between firms and scientific institutes managing innovation and knowledge transfer, and intensive trading activities of IPRs as commercialization strategy and business model of R&D-intensive scientific institutes and universities as well as high-tech companies constitute the central channels of knowledge transfer in the East of England's regional innovation system.

The highly networked structure of this business-university cluster also explains the slightly higher significance of business networks in the region than assumed, even if, overall, the links between firms and academia appeared to be generally stronger than those to competitors or customers and suppliers, independently of being formalized or informal in nature. What resulted to not play a role for the East of England's regional knowledge transfer system have been business associations and chambers, which, as expected, lack the tradition, as well as the organizational and structural resources to do so. However, especially for the majority of the region's high tech firms they appeared to be obsolete, since innovation and knowledge transfer do not depend on associational or corporatist intermediation and circumvent these actors in the system. In addition, it came out that this dense interaction of academic and business actors also functions without being driven and held together by a resilient historically grown and commonly shared regional cultural identity, but rather by the common fixing point of the region's excellent scientific institutes and universities. This aspect is even reinforced by the comparatively low self-identification of regional actors with the East of England region as a whole, being perceived as an artificial and primarily administrative and statistical, but not emotionally binding geographical unit.

Finally, the largest deviations from the theoretical assumptions appeared with regard to the state induced mechanisms of knowledge transfer. These resulted to be throughout more important than assumed, with public support to innovation and knowledge transfer in many fields, like e.g.

network-building, skills development, personnel transfer, larger state induced innovation projects e.g. in the defense industry, or general subsidies. However, the difference to the German and, above all, the Spanish cases is, that the state has by far not been evaluated as the central actor in facilitating, organizing and funding knowledge transfer activities, but only as one actor among others, creating the framework conditions to innovation and knowledge transfer and further enhancing already existing and functioning channels of knowledge transfer. This reinforcing rather than intervening role of public action has to be seen as particularly valuable within the East of England context with its diverse and well functioning links between business and academia.

Thereby, central government has been evaluated as the decisive policy level in the fields of innovation and knowledge transfer, due to the strongly limited autonomy and competences of regional policy in the English regions. However, approaches to regional devolution and to bringing national policy closer to the English regions through the RDAs like EEDA, Government Offices for the Regions and Regional Assemblies have been consistently evaluated as positively and effective. What has been criticized in this regard is, above all, the insufficient autonomy to develop regional policy instruments specifically tailored to the particular needs of the region. This points to a whish of regional actors in the East of England to transform parts of central government support into more region specific tools taking into account the region's individual structural and economic characteristics.

What must, therefore, be added in this context is that actual political developments in the UK are moving clearly towards a different direction. Shortly after having conducted the interviews in the UK regions, with the shift of government from *Labour* to the coalition of *Tories* and *Liberal Democrats* in May 2010, the central pillars of regional devolution policy have been successively taken back, because of both political and financial rationales, giving room to a reinforced re-localization of policy. As a consequence, the *Government Offices* for the regions have been closed in March 2011, *Regional Assemblies* already in April 2010 and the RDAs including *EEDA* will stop their operations in March 2012 and will be presumably be closed in summer of that year, developments which could not be foreseen by the experts when they were interviewed in early March 2010. It will be to observe and be left to future research which consequences this recent policy shift will have for the innovation and knowledge transfer policy in the East of England.

As in the other investigated regions, also in the case of the East of England the two sectors of automotive and biotechnology showed distinct structures and mechanisms of knowledge transfer and, therefore, distinct strategies in dealing with the available structural conditions in the region. Thereby, the East of England's automotive industry resulted to be a traditional industrial sector being shaped by the decline of this industry across the whole country in the last decades. The lack of own OEMs in the region, foreign owned manufacturers and suppliers, tendencies of supply chain hollowing out, a strong specialization on niche markets like luxury vehicles, engine construction, motorsport and aerospace, are some of the consequences of this specific situation, also influencing innovation and knowledge transfer in the East of England's automotive sector. These features are at least partly compensated by a strong orientation on R&D-activities and high-end automotive technology, which is in line with the region's general character as a high-tech and science region.

However, the East of England's automotive industry also showed many of the structures and mechanisms of knowledge transfer being typical for this sector, but differentiating it from the generally dominating structures shaping the region's knowledge transfer system: A pointedly higher significance of supply chain relationships, collaborative R&D-activities and business associations than in the region as a whole, comparatively strong strategic guidance and financial support by the state, markedly less volatile labor markets and start-up activities, and an only medium range importance of the market based mechanisms represent characteristic features of knowledge transfer processes in the sector. On the other hand, the analysis also provided some evidence for regional specificities pointing to inadequate structural support mechanisms for an incrementally innovating industrial sector like automotive: Less stabilized horizontal network structures as channels of knowledge transfer, a lack of highly and adequately qualified staff, an insufficient associational intermediation and traditionally low industrial relations have been evaluated as major shortcomings for automotive in the East of England when compared to its continental competitors, and could also be interpreted as explanatory factors for the generally less successful performance of the sector in the last decades.

On the contrary, biotechnology in the East of England almost parallels the results obtained for the region as a whole, and, as an example of a high-technology sector, it appears to find a well fitting environment for innovation and knowledge transfer activities in this region. Where it deviates from these results, it mainly does so because of exhausted flows of large venture capital rounds, and, above all, its mature and consolidated character, differentiating it not only from many other high-tech sectors in the East of England, but also from the other investigated biotech-regions in this study. Therefore, horizontal network relations are comparatively strong and show well-connected sectoral actors, start-up and spin-off activities are markedly lower and public support is less significant, also when compared to automotive.

Both sectors share a relatively low orientation towards the region they are operating in. Apart of a strong international focus, characterizing above all the biotech industry, both sectors are part of cross-regional networks and clusters, like *One Nucleus* in life sciences and the *Motorsport Valley* in automotive, which connect the regional actors to cross-regional contexts of innovation and knowledge transfer. One could interpret this as a manifestation of the generally lower historical, cultural, social, political and economic tradition in the English Government Office regions, making regionally defined and contextualized industries less likely than in the investigated German and Spanish regions, where regional policy competences are much stronger and where much more specific and differentiating contexts of innovation and knowledge transfer can be created, limiting the incentives to broaden cluster-building activities across regional boundaries. Overall, this points to the East of England as a regional knowledge transfer system dominated by high-tech industries, streams of analytical knowledge, market mechanisms and dense interaction of business and academia, providing fruitful contexts for innovation and knowledge transfer to radically innovating sectors like biotechnology. However, it provides a less adequate environment for innovation and knowledge transfer to incrementally innovating traditional industrial sectors like automotive, which rely much stronger on synthetical stocks of knowledge, what, as a consequence, enhances niche-sector specialization and a strong focus on R&D-activities. These results are also summarized in Table 10. If these findings on the regional knowledge transfer system of the East of England are exemplary for other UK regions, or if they represent a unique structural and institutional configuration, will be investigated by contrasting them with the results of the qualitative case study on the Greater London metropolitan region, which will be discussed in detail in the following sections.

Table 10: Qualitative Relevance of Knowledge Transfer Mechanisms in the East of England

MECHANISMS	UK*	EoE	Auto	Bio
NETWORK BASED MECHANISMS				
business networks	–	O	O	+
informal networks	O	O	O	O
institutionally stabilized networks	O	O	O	+
customers and suppliers	O	O	+	O
cooperative inter-firm R&D-projects	–	–	O	–
cooperative R&D-projects with scientific institutions and universities	O	+	+	+
ASSOCIATION BASED MECHANISMS				
employers-, business-, and trade associations	–	–	O	–
chambers of industry and commerce	–	–	–	–
binding and common technical standards	O	O	O	???
ORGANIZATION BASED MECHANISMS				
transfer of technical and scientific staff	+	+	O	+
purposeful headhunting	+	+	O	+
business start-ups	+	+	O	O
STATE INDUCED MECHANISMS				
national government measures	–	+	+	+
regional government measures	–	O	O	O
(larger) state-induced innovation projects	–	+	O	–
scientific parks	O	+	+	+
technological centers	O	+	+	+
technology transfer agencies/companies	O	O	O	+
MARKET BASED MECHANISMS				
external consultancies	+	+	O	+
mergers and acquisitions, hostile takeovers	+	+	O	+
buying, selling and licensing of patents	+	+	O	+
COMMUNITY BASED MECHANISMS				
informal contacts	O	O	O	O
regional platforms	O	O	O	O
additional (regional) institutions	O	O	O	O

+ *Reported (*Hypothesized) as crucial mechanism of knowledge transfer*

O *Reported (*Hypothesized) as important but not decisive mechanism*

– *Reported (*Hypothesized) as not important at all*

??? Not reported by the interview partners

[dark grey] = *Important deviation from hypotheses*

[light grey] = *Modest deviation from hypotheses*

[white] = *No deviation from hypotheses*

7.3 The Greater London Case

7.3.1 Locating the Greater London Region in the UK Context

7.3.1.1 Selected Facts on Greater London's Economic Structure

The Greater London metropolitan region is the major urban agglomeration area in the UK and contains the country's capital city of London. It is located in the south east of the British Islands, bordering to the East of England and the South East Region, and is one of the nine *English Government Office Regions* (ONS 2003). Through the Thames River, which crosses the region, Greater London is connected to the North Sea. Historically, this specific location along the Thames was decisive for the region's urban and economic development. Starting with the Roman bridging of the river, London was positioned at the center of national and international trade routes, and could develop into an important center of trade and commerce in the Middle Ages, bringing together merchants from Northern and Southern Europe, especially for the trade of wool and woolen products. From the 16th century on this tradition of open trade expanded to the British colonies, converting London into the political and economic center of the British Empire, and to one of the world's most important trading hubs. As a consequence, large trading companies emerged, accompanied by an early services sector supporting international trade and colonial expansion, like banking, financial and (marine-) insurances services. At the end of the 19th century, London had not only become the world's largest city, but also its undisputed financial center and leading market place and service provider of world trade (GLA 2010: 25, 69ff.).

This specific economic history also led the region into a particular path of industrialization. In contrast to other British regions like the North or the Midlands, Greater London never really experienced the development of a large scale industrial factory system and a specialization on a few heavy industry sectors relying on natural resources as could be observed in other UK settings like Manchester. Much more, Greater London's role in the country's industrialization process was that of a large consumer market for industrial products and luxury goods, as a reloading site for maritime transport, as a location of docking industries and ship-building, and as a hub for capital, information and expertise in banking, finance, shipping and insurance. However, the region has also been hosting a highly diversified and flexible manufacturing sector (contrasted to large scale industry) at the edge of technological development, addressing the specific and diverse needs of a dynamically growing metropolitan bourgeois population (GLA 2010: 25, 73ff.). This created an economic geography encouraging specialization through interdependent networks of skills, resources and knowledge, as a recent study on Greater London's economic history pointedly formulates:

"In this sense the idea of a close-knit series of industrial districts is a better way of conceptualizing London's manufacturing base, than seeking evidence of the factory system proper. Flexibility gave London a leading role in the emergence of an enterprise economy, where circuits of informal information flow and casual methods of employment meant entrepreneurs could respond quickly to changes in taste and consumer demand by varying batch production of both high and low value goods. Drawing upon the wide variety of skill and tradition of innovation in the metropolis, together with a large, flexible workforce, the dynamism of London's Victorian economy stood in sharp contrast to the emphasis on mass production of standardized capital and consumer goods in the classic industrial cities of the period." (GLA 2010: 74)

With an explosion of consumer demand and the rise of Fordist production models in the beginning 20th century also larger industrial production plants appeared in the region, producing consumer goods like vacuum cleaners, light bulbs, televisions, radios or *Vauxhall* cars. Many of these plants based on US-investments, as in the cases of *Gillette* and *Hoover*, and were located in first business parks outside the city. With the increasing standardization of industrial production, however, most of these consumer goods producing firms located their production plants away from Greater London, but left in many cases their business headquarters in the city, giving rise to a further tertiarization of the region's economy, especially in the fields of highly qualified services like management consultancies, advertising agencies and publishing (GLA 2010: 25, 73ff.).

Today, (2009) with a population of about 7.7 million Greater London is, after the South East, the UK's second largest region in terms of inhabitants, leaving behind the North West and the East of England, and represents 12.5% of the UK's total population, but only 0.7% of its territory, pointing to the expectable high density of population in the country's major metropolitan agglomeration area. Greater London is also the UK's most powerful economic region, with a total GVA of about 265 billion Pounds, representing 21.5% of the UK's total GVA (ONS 2010b: 12). The region is, together with New York, one of the world's two leading financial places, is integrated into the network of the world's global cities, and is the

UK's undisputed powerhouse in the fields of national government and administration, business services, commerce, culture, media and tourism (GLA 2010: 20ff.).

The economic structure of Greater London markedly differs from the national average and also from the one of the East of England, showing all typical characteristics of a major metropolitan area. Its primary sector is marginal with a share of only 0.03% of the regional total GVA (UK: 0.8%; EoE: 1.3%), and also its industrial sector (without construction) is minor, ranking with 5.7% considerably lower than the national average of 14.0% or the East of England with 14.7% and the South East with 11.9%[91] (Figure 44; ONS 2010b: 20ff.). This reflects Greater London's previously described economic history and also a reinforced tertiarization process in the region in the last decades, which is further illustrated by a downturn of employment in manufacturing from 23% in 1971 to less than five percent in 2007 (GLA 2010: 18).

Fig. 54: Greater London's Industrial Structure

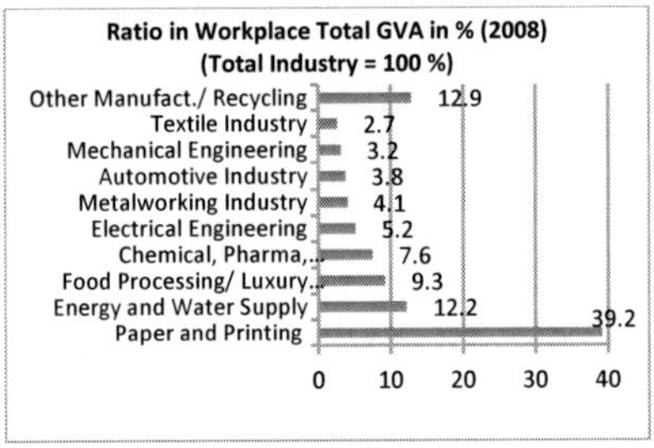

Source: Own Calculations and Illustration based on Data from ONS (2011g)

Fig. 55: Sectoral R&D-Performance in Greater London's Manufacturing

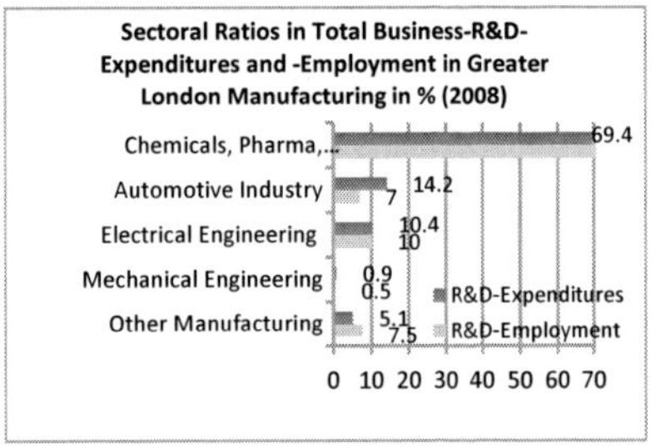

Source: Own Calculations and Illustration based on Data from ONS (2011e)

What is clearly dominating Greater London's economic structure is, therefore, the service sector, accounting for 90.2% of the regional total GVA, a value which is not only pointedly leaving behind the UK average of 78.8% and the value for the East of England of 76.1%[92], but also the values of any other region investigated in this study, including the Madrid Community (76.5%)[93] (Figure 44; ONS 2011b: 20ff.). In this, Greater London's

91 Own calculations based on ONS (2010b): 20ff., data for 2008

92 Own calculations based on ONS 2011b: 20ff., data for 2008

93 Own calculations based on data from IESTADIS 2011, data for 2008

service sector shows the typical structure of western metropolitan agglomeration areas and national capital city regions, with a comparatively elevated share of highly qualified services. Measured as a share in the regional total GVA, the most important sectors are real estate, renting and business services (29.8%), financial services (19.5%), trade services (8.7%), transport and communication services (8.0%), health and social services (5.3%), and public administration (4.5%)[94] (ONS 2011b: 23).

Looking at the structure of Greater London's small industry sector, a strong concentration of manufacturing activities in paper and printing appears, which accounts for 39.2% of the region's industry's total GVA, pointing to the strong role in media production and publishing activities the region plays nationally and internationally. Energy and water supply represent 12.2% of the regional industry's total GVA, followed by food processing and luxury foods (9.3%), chemicals (7.6%), electrical engineering (5.2%), metalworking (4.1%), automotive (3.8%), mechanical engineering (3.2%), and textiles (2.7%)[95] (Figure 54).

Greater London's business structure, on the contrary, almost parallels the national values, above all in being markedly dominated by micro firms with less than nine employees, which represent 96.4% of the total number of firms in the region's private sector (UK: 95.7%). Within the category of micro-firms the group of self-employed persons (autonomous) without employees stands out: This group represents 81.6% of the micro firms in the region and 78.7% of the total number of firms in Greater London's private sector, values that are even above those of the East of England (77.6% and 74.5%). Slightly lower is the region's share of small businesses (10–49 employees) and medium sized firms (50–249 employees) of 2.9% and 0.5%, also when compared to the national average of 3.6% and 0.8%, respectively. Finally, the share of larger firms with more than 250 employees of 0.2% slightly surpasses the average values for the UK and the East of England[96]. In this, the micro firms employed 28.0% of all employees in the private sector, while the small, medium sized and large firms employed 10.4%, 9.5% and 52.0%, respectively. Overall, this points to a markedly higher average firm size of the large firms in Greater London when compared to the national average (40.8%) or the East of England

94 Own calculations based on ONS 2011b: 23, data for 2008

95 Own calculations based on data from ONS 2011g, data for 2008

96 Own calculations based on data from ONS 2009, data for 2008

(45.4%), and highlights again its role as location of large national and multinational firm's headquarters[97] (Figures 56 and 57; ONS 2009).

Fig. 56: Business Structure Greater London I

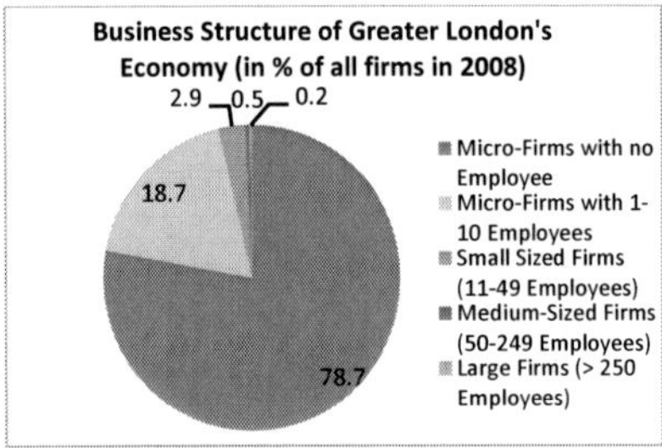

Source: Own Calculations and Illustration based on Data from ONS (2009)

Fig. 57: Business Structure Greater London II

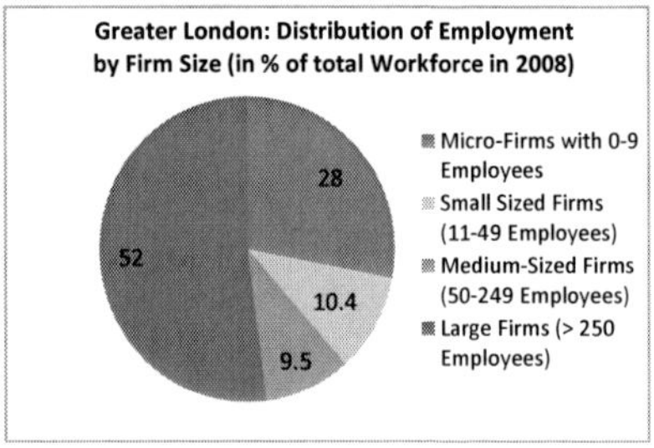

Source: Own Calculations and Illustration based on Data from ONS (2009)

Another important structural indicator describing the region's knowledge generating capacity is its R&D-activities. When looking at Greater London's R&D-intensity (gross domestic expenditure on R&D as percentage of total workplace based GVA) of about 1.2%, it becomes obvious that the region is one of the less R&D-intensive UK regions, ranking considerably lower than the national average (2.0%), and, especially than the East of England (4.9%)[98] (Figure 52). Thus, despite the region's high number of universities and research institutes, the share of R&D-activities in Greater London's economic performance in comparatively low, even if compared to other national capital city areas like Madrid (2.0%). This is further underlined by the region's business sector R&D-efforts: In 2008, Greater London had total business R&D-expenditures of about 1.1 billion Pounds, representing a share of 6.7% of the total business R&D-expenditures in the UK. This is only the fifth largest contribution to these expenditures among the UK regions, markedly falling behind the East of England (26%), the South East (21.6%), the North West (14.4%), and the South West (8.6%)[99] (ONS 2011c: 18). In terms of the region's business sector's R&D-intensity (business R&D-expenditure as percentage of total workplace based GVA), Greater London even is the lowest ranking among all UK regions: With a

97 Own calculations based on data from ONS 2009, data for 2008

98 Own calculations based on data from ONS 2011f, data for 2007

99 Own calculations based on data from ONS 2011c: 18, data for 2008

value of 0.4% it only reaches one third of the national average value of 1.2%, and not even one ninth of the value of the East of England as the UK's leading region in this category (3.7%)[100] (Figure 52). This points to a comparatively less R&D-intensive metropolitan and services economy in the Greater London region, which markedly differs in this regard from the high and medium-high technology oriented business sector in the East of England.

The strong service orientation of Greater London's economy is also revealed when taking a sectoral perspective on the business R&D-activities in the region: 49.1% of the total business R&D-expenditures and about 60% of the total business R&D-employment in Greater London are covered by service firms, almost doubling the average values for the UK with 24.7% and 28.3%, and also considerably surpassing the values of the East of England (36.7% and 31%). Anyway, 50% of business R&D-expenditures and 40% of business R&D-employment is concentrated in Greater London's small manufacturing sector, with some of the region's most important manufacturing sectors also being the most R&D-intensive ones: The chemical industry (including pharmaceuticals and biosciences), accounts for 69.4% of manufacturing R&D-expenditures and about 75% of its R&D-personnel, and is, with that, by far the region's most R&D-intensive industrial sector, while automotive (14.2% and ca. 7%), electrical engineering (10.4% and ca. 10%), and mechanical engineering (0.9% and ca. 0.5%)[101] follow with large distance on the subsequent ranks (Figure 52).

7.3.1.2 On Greater London's Marginalized Automotive Industry

A closer focus on automotive shows that it has a comparatively low share in Greater London's manufacturing sector, accounting for only 3.8% of the total GVA in the region's manufacturing[102]. Given the region's overall weight in the UK's economy, Greater London also has a relatively low share in UK automotive, with the automotive firms located in the region covering a share of 9.5%[103] of the sector's total sales volume (IMI 2007: 18). In 2008, there were around 5,900 people employed in Greater London's automotive sector, representing about 4% of the UK automotive

100 Own calculations based on data from ONS 2011f, data for 2008

101 Own calculations and estimations based on data from ONS 2011e, data for 2008

102 Own calculations based on data from ONS 2011g, data for 2008

103 Own calculation based on data from IMI (2007): 18; data for 2007

workforce, and not even half of the number of employees of its East of England counterpart (SEMTA 2011: 29f.). In this, the sector consisted of around 255 companies, representing 4.9% of all UK automotive firms. The sector's business structure is characterized by a larger share of SMEs and large firms when compared to the region as a whole, as well as by a markedly lower proportion of micro-firms. While the latter (0–9 employees) represent 80.4% of all firms in the sector (of which 85.4% have less than four employees), the small (10–49), medium sized (50–249) and large (>249) firms account for 13.7%, 3.9% and 2.0% of all firms in Greater London's automotive sector, respectively[104].

Nevertheless, automotive presents itself as research intensive sector, accounting for 14.2% of Greater London's manufacturing R&D-expenditures and about 7% of its R&D-personnel, being with that the second highest contributor to the region's manufacturing R&D-activities behind chemicals[105]. Like the rest of the country, also Greater London lacks of own automotive OEMs. The only larger assembly plant in the region is the one of *Ford* in Dagenham, which is specialized in engine construction, and which with its about 4,000 employees is the largest automotive employer in Greater London. Apart from that, it is especially smaller and medium sized local suppliers shaping the structure of Greater London's automotive sector.

7.3.1.3 Greater London's Biotech Industry

Looking at Greater London's biotech industry, it can be stated that it has much less significance for the regional economy than its East of England counterpart. In 2008, the sector consisted of about 135 biotechnological and biopharmaceutical companies, including a broad spectrum of sectoral business activities like drug discovery and development, contract research and contract manufacturing firms, suppliers, drug delivery, platform technology, and diagnostics firms, as well as bio- and chemo-informatics companies. These firms include facilities of the largest and most renowned international companies in the field, like *AstraZeneca*, *Amgen*, *Beck Pharma*, *Glaxo Smithkline* and *Zanofi Aventis UK* (One Nucleus 2011).

Within the UK, Greater London ranks in the middle fields of business activities in biotechnology, concentrating 8.4% of the total turnover, 6.5%

104 Own calculations based on data from ONS 2011g, data for 2009

105 Own calculations and estimations based on data from ONS 2011e, data for 2008

of the employees and 9.5% of the total number of companies in the sector. With that, it ranks clearly behind the leading regions of the East of England with 27% of the total turnover, 24% of the employees and 24% of the total number of companies in UK biotechnology, and the South East region (26%, 21%, 14%)[106] (BIS 2010c: 35ff.). Greater London's biotech sector is, however, extraordinarily R&D-intensive, accounting together with the related sectors of chemicals and pharmaceuticals for 69.4% of manufacturing R&D-expenditures and about 75% of its R&D-personnel[107].

However, apart from basic research, in particular in the field medical biotechnology, the development of new products, and especially of new pharmaceuticals, requires long clinical test phases (phase I–III) of up to ten years or more, before a new pharmaceutical is approved for commercial distribution. In total, in 2008 Greater London biotech firms had 93 drug candidates in phases of clinical trials, and another 93 candidates in pre-clinical phases. Looking at the distribution of Greater London drug candidates in the three clinical phases, 34.4% of the drug candidates were in phase I, 47.8% in phase II, 17.8% in phase III and 3.3% in the status of approval[108]. With that, and in comparison with the national average and the mean values of all other European countries active in biotechnology, the Greater London biotech firms have comparative strengths in drug candidates in phases I and especially III, but are slightly behind in drug candidates in phase II, pointing to the overall matureness of the sector with many businesses being in later phases of their business life cycles, but also to a dynamically evolving sector, with many firms developing new candidates in earlier stages (One Nucleus 2011; Ernst & Young 2010: 42).

7.3.2 Structures and Mechanisms of Knowledge Transfer in the Greater London Region: Six Governance Dimensions

7.3.2.1 Networks as Mechanisms of Knowledge Transfer

Basing on the theoretical assumptions developed in section 3.2 and on the discussion of quantitative indicators characterizing Greater London's regional economy in the preceding section, in the following sections the re-

106 Own calculations based on BIS 2010c: 35ff., 47; data for 2010

107 Own calculations and estimations based on data from ONS 2011e, data for 2008

108 Own calculations based on Ernst & Young 2010: 45, data for 2009

sults of the empirical qualitative case study of Greater London's knowledge transfer system will be presented. The first dimension of knowledge transfer mechanisms that has been investigated in this regard is the network dimension, for which the interviewed experts broadly confirmed the theoretical assumptions.

Horizontal Business Networks

In this, *horizontal business networks* as mechanisms of knowledge transfer would be used only to a limited extend (Interviews LON 1: 2f.; LON 2: 4ff.; LON 3: 3f.; LON 4: 2f.; LON 6: 2f.). On the one hand, Greater London's economic and business structure with its myriads of small and micro firms as well as other institutional actors would create a confusing environment for businesses operating in the region, making business networks in the first place important points of reference and orientation. On the other hand, this environment would make it difficult to oversee potential cooperation partners and to develop awareness of operating in a larger economic context or cluster, a fact which would be a barrier to more intensive network-building. Therefore, central government as well as the regional political bodies like the *London Development Agency* (LDA) or the *Mayor of London* would actively support the creation of business networks in Greater London, like e.g. *London Technology Network*, *London Business Network*, or the *London Biotechnology Network*. The head of business research at LDA formulates this discussing the example of life sciences:

"We funded e.g. the *London Biotechnology Network* four or five years ago in various tranches of funding to do various things. We did not create the network, but we provided the majority of its funding in its early years, to develop it into a self-sustaining network that it is today. In terms of money that the LDA has invested in the area of knowledge transfer, I think creating something in London's case having a network like that is absolutely critical. It really helps people to understand that they are in a cluster, not only in biotech, but also in the whole life sciences. London is such a big city that people do not perceive that they are part of something bigger, because you have so many universities. They do all collaborate and work together, but they are very independent as well. And because there is competition as well as collaboration between them, having a forum that brings people together plays a very important role. For business [...] one of the big problems that people have with London is that they do not know where to go. [...] But having a network like the London Biotechnology Network just makes it much easier for that flow of information and sharing of information to happen." (Interview LON 3: 3)

In this regard, the interview partner also points to an important difference when comparing Greater London to the East of England: While the latter would provide a clear anchor point to the regional business sector with Cambridge University and the high-tech cluster around it, Greater London's economic and innovation landscape would be harder to oversee:

> "In London the common place of the universities is very important as well, because we have got so many of them, whereas in the East of England everybody wants to go to Cambridge University and wants to talk to them. You come to London and people get stuck. So there is an additional drive for providing an access point into that cluster in London as well." (Interview LON 3: 4)

The majority of these networks would be *formalized* private membership networks, but interaction and knowledge transfer within these networks would mostly occur on an *informal* basis and around topics like the framework conditions of business activities, funding, public subsidies, access to information, etc., and would only in a limited number of cases lead to deeper collaboration in formalized consortia. Apart from these pan-London networks, a London-specific element in terms of networks would be borough- or district-specific business networks focusing on promoting business activities and knowledge transfer in the region's different sub-spaces, as it would be the financial sector in the City of London. These networks would in some cases be historically grown and have a long tradition, and would even stronger rely on informal interaction and flows of knowledge than the pan-London ones (Interviews LON 1: 3; LON 2: 5f.; LON 3: 4; LON 4: 3; LON 6: 3).

Vertical Supply Chain Networks

Vertical supply chain networks would, on the contrary, not be dominant mechanisms of knowledge transfer in Greater London, although strong sectoral differences have been identified in this connection by the interviewed experts (Interviews LON 1: 14; LON 3: 5; LON 4: 17; LON 6: 4). The networks between customers and suppliers would, if anything, only play a role in the region's small manufacturing sector, which would have a marginal share in Greater London's economy as a whole, and would, therefore, only lowly coin the region's knowledge transfer system. On the other hand, and because of this minor dimension of manufacturing in the region, supply chains would often be incomplete, and firms would rather be part of cross-regional supply chain contexts than London specific networks.

Cooperative Inter-Firm R&D-Projects

In addition, like in the East of England case, *cooperative inter-firm R&D-projects* were ascribed only a low importance as mechanisms of knowledge transfer for Greater London firms (Interviews LON 1: 3f.; LON 3: 6f.; LON 4: 3f.; LON 6: 6). These cooperations would only in a very few cases occur purely out of market dynamics, and, if anything, take the forms of publicly, and especially EU-funded R&D-consortia, e.g. through the European *Framework Programme.* Most firms in the region, however, would be too small and lack the resources to cooperate in R&D-activities with other firms, or would just not operate in R&D-intensive fields.

Cooperations between Business and Academia

More important would be, though, *cooperations between firms and academia*, which would especially profit from the broad supply of scientific institutions in Greater London, with more than 40 higher education institutes, as well as a high number of research oriented institutes and hospitals (Interviews LON 1: 2ff.; LON 2: 3ff.; LON 3: 7ff.; LON 4: 4; LON 6: 6f.). This huge academic knowledge base would be a major differentiating characteristic in Greater London's knowledge transfer system when comparing it to other UK regions, and it would create variegated links between business and academia, which would also be supported by central government and the LDA. However, these links would not be that vital as in the East of England e.g., above all because of a generally less R&D-intensive business sector, as the head of business research at the LDA accentuates:

"We here in London probably provide more support getting universities and businesses to work together. The advantage is that London has this big research knowledge base in the universities and in the hospitals. Actually it is trying to gain economic impact from that and therefore there is more activity on that level. […] It happens more in London than it happens anywhere else, and that is partly because we have more universities than anywhere else. But in terms of business innovation and skills they are doing an annual report on the level of commercial activity that these universities do. They measure how much contract research they do and how many spin-out companies they have formed, how many licenses they have got, how many patents they have applied for. London is first place in this. But if you look at the business R&D-spending in London, we are very small. [W]e do not have these big R&D-intensive companies. *Glaxo* will not put itself in the centre of London, because it is too expensive. However, the opposite is true in the public side of it, so in the university R&D London is hugely dominant. We spend more in this than the South and the East of England put together, which includes Cambridge and Southampton etc. So London has this huge advantage in the knowledge base compared to anywhere else." (Interview LON 3: 7f.)

This would refer to a pointed disequilibrium between public and business R&D-efforts in the region, for which the interview partner provides two explanations: First, a lack of large R&D-intensive firms, and, second, "hidden" business R&D-activities conducted by the larger service sector firms in the region, what would in the eyes of the interview partner lead to an underestimation of Greater London's business R&D-activities and also of the links between these firms and scientific institutes:

"Partly it is because of what is measured by business R&D. I think in reality it is not so low. E.g., the South and the East of England have massive amounts, but in simple terms they have got companies like *Glaxo*, *Pfizer* and […] *BA Systems* etc., so they have got some big MNCs who are R&D-intensive and are spending thousands millions of Pounds. So that is very easy. But if you come along to London, our economy is dominated by the financial sector. If you go to *Goldman Sachs* and ask how much they spent on R&D, they would say that they are a bank and they do not spend on R&D. But actually if you look a little closer they do spend money on R&D in terms of programming or on what the stock markets are actually doing and they sometimes come up with an algorithm that saves them half a second in terms of saying what is going to happen to a particular stock in comparison to their competitor. That can turn into a big advantage. They actually spend money doing this sort of things. But they do not see that as research and so it does not get catched. In a sense I think that London has got an artificially low business R&D measure, because of the nature of its economy." (Interview LON 3: 7f.)

To support interaction between business and academia, especially the LDA would apply a broad variety of mechanisms, like creating consortia of London universities to ease the access to them through a *one-front-door-approach*, but also knowledge transfer partnerships, innovation vouchers and, as a London specialism, the *Knowledge Connect* project, which would help SMEs in interacting with universities and would assume a kind of switchboard service for Greater London firms in this regard (Interviews LON 3: 8f.; LON 2: 2ff.).

Network Based Mechanisms in Greater London's Automotive Sector

From a sectoral perspective, *automotive* would not completely correspond to these results. In general, both horizontal and vertical business networks would be important channels of knowledge transfer in the sector, but its small size and the low number of firms in the region would not represent a complete value added chain and would, therefore, not make the Greater London region a central point of interaction for Greater London automotive firms. Rather, network relationships would occur cross regional, especially with the East of England, the South East and the Midlands. In this, Greater London's automotive industry would have a strong focus on engine construction, and cross-regional network-building and supply chain relationships would, therefore, mainly occur within this particular sub-sector, and would include both formal and informal interaction. In addition, one interviewed expert emphasized that the lack of larger OEMs in the region, which could function as focal points around which stable networks could emerge and develop, would be a major barrier to stronger London specific networking activities in automotive. Not relevant would be, furthermore, cooperative inter-firm R&D-projects, especially because there would be no larger automotive R&D-department in the region, and most of the Greater London automotive firms would either be pure assembly plants or just too small to conduct own R&D or to cooperate in R&D with others. Cooperative R&D-projects with research institutes and universities would occur more often, although still only to a limited extend and not exclusively with regional universities and institutes, but also with more specialized scientific institutions in other regions (Interview LON 5: 2ff.).

Network Based Mechanisms in Greater London's Biotechnology Sector

The *biotech sector*, on the other hand, would not considerably deviate from the results obtained for the region as a whole. Horizontal business networks would be important and used intensively, although not vital mechanisms of knowledge transfer. Especially the *London Biotechnology Network* would be an important London-wide arena of interaction for the Greater London biotech sector, providing opportunities for formal and especially informal interaction. Vertical supply chain networks in a classical sense would not play any role in the sector. Only when considering the biotech firms as suppliers to the pharmaceutical and chemical industry this would be of relevance in terms of knowledge transfer. Cooperative inter-firm R&D-projects would regularly occur and would be important transfer channels, in many cases building upon public funding of R&D-consortia by the EU, central government and LDA. However, this type of cooperation would always be strictly formalized and would require the highest awareness of the accompanying project management with regard to IP-protection and knowledge contamination. Vital, again, would be R&D-cooperations with scientific institutes and universities, which would be motors of the R&D-driven biotech industry (Interviews LON 3: 3ff.; LON 4: 2ff.; LON 6: 2ff.).

7.3.2.2 Association Based Mechanisms of Knowledge Transfer

Concerning the second dimension of knowledge transfer through association based mechanisms like employers-, business-, and trade associations, chambers of industry and commerce, as well as binding and common technical standards, the interviewed experts fully confirmed the theoretical assumptions when looking at Greater London as a whole.

Business Associations and Chambers of Industry and Commerce

Therefore, both *business associations* and *chambers* would not play a significant role for knowledge transfer in the region (Interviews LON 1: 14f.; LON 2: 6ff.; LON 3: 5f.; LON 4: 17f.; LON 6: 5). Basically, knowledge transfer and innovation would not be among the traditional topics covered by these players, and they would not be explicitly engaged in them. However, in contrast to the East of England, the experts at least ascribed them an indirect function in bringing actors together through events, conferences and meetings, as well as in transmitting (but not implementing) certain policy measures to their members. In addition, and as a London specific feature, they would fulfill an important function of concerting the region's ethnic business communities and enhance interaction among them, as well as between them and other communities (Interview LON 2: 6ff.).

Business Associations and Chambers in Automotive and Biotechnology

Divided by sectors, these results would also apply for *automotive*. Regional automotive business associations would not exist in Greater London, and, if anything, national sectoral associations would be used by London automotive firms as platforms for interaction with other automotive firms. The *London Chamber* would provide events and meetings with the same goal, but neither specifically on the field of automotive nor on innovation and knowledge transfer (Interview LON 5: 3f.). In *biotechnology*, the role of associations has been, on the contrary, estimated as higher (Interviews LON 4: 17f.; LON 6: 5). Although their main contribution would also be events, conferences and meetings, one interviewed expert for the biotech sector evaluated this as a very important driver of knowledge transfer (Interview LON 4: 17f.). The *London Chamber*, on the other hand, would not have recognized biotechnology as one of their topics yet.

Binding and Common Technical Standards

Especially in *biotechnology*, the impact of business associations in formulating and implementing *binding and common technical standards* would be high. In this, they would be actively lobbying the respective regulatory bodies and influence regulation and industry guidelines. Since these would frequently even become law, their role in fostering knowledge transfer activities among firms has been evaluated as positive and essential (Interview LON 4: 18f.). In *automotive*, technical standards would also be important facilitating mechanisms of knowledge transfer, although the role of associations and chambers in influencing this regulation would be low, and the protagonism would mainly be left to central government and the respective EU-bodies (Interview LON 5: 4).

7.3.2.3 Organization Based Mechanisms of Knowledge Transfer

Like for the East of England case, also for Greater London the interviewed experts fully affirmed the theoretical assumptions on the organization based mechanisms of knowledge transfer like the transfer of technical and scientific staff, purposeful headhunting and business start-ups when looking at the region as a whole.

Transfer of Technical and Scientific Staff

As assumed, the *transfer of technical and scientific staff* has been evaluated as a central mechanism of knowledge transfer in Greater London, being reflected by a highly fluid labor market for qualified personnel (Interviews LON 1: 5f.; LON 2: 11f.; LON 3: 9ff.; LON 4: 4ff.). These strong transfer activities in the region would especially be nourished by its strong equipment with universities and research institutes, but also with knowledge intensive financial and business services companies, as well as business headquarters. This talent pool would be a major motive for many firms to locate themselves in the region, and the region's character as a global city and international trading and communication hub would attract highly qualified staff worldwide, what would be even further supported by Greater London being an English language region. To also provide smaller and less knowledge intensive firms access to the region's academic talent pool, also Greater London would offer the already discussed mechanism of *Knowledge Transfer Partnerships* through the LDA, supporting these firms to temporarily employ university graduates or doctoral students and, with that, to also access the expertise of the respective supervisors. However, as the head of business research of the LDA points out, in Greater London this specific scheme would be used much less than in the other UK regions, since transfer activities would be intensive anyway (Interview LON 3: 9ff.).

Purposeful Headhunting

This specific metropolitan labor market with its broad talent base and intensive demand for expertise generated by knowledge (but not necessarily R&D-) intensive firms would also lead to a high competition in recruiting the highest qualified experts. *Purposeful headhunting* activities would, therefore, be essential channels of knowledge transfer in the region, and a differentiating characteristic when comparing Greater London to other UK regions (Interviews LON 1: 6; LON 3: 10f.; LON 4: 6). As the head of business research at the LDA emphasizes, this would especially hold true for the City of London, but less for Outer London:

> "One of the reasons that London is on top of this is that we have got the universities and we create so many graduates, we have got a pool of talents and access to people of the university. So we attract companies to London who want to be based here because of this pool of expertise. And then you have got the financial people as well, and the spin-out companies from the universities. [...] Yes it [purposeful headhunting; MO] is happening a lot, and I think London has a quite dynamic economy in that sense. [...] From that base there is a turnover of staff and a demand for it. And in terms of the talent in London, in the centre of London the qualifications that people have are very high compared to outside. But because of the high number of high skilled people you also have a high level of competition. [...] You have got this special situation I described in the centre of London, but outside of this, London is not so much different from the rest of the country. Outside the center of London the economy and the businesses are much more similar to what you would find in any other part of the UK." (Interview LON 3: 9f.)

Business Start-Ups

Also *business start-ups* and *academic spin-offs* have been evaluated as central and frequently used mechanisms of knowledge transfer in the region (Interviews LON 1: 6f.; LON 2: 12ff.; LON 3: 11f.; LON 4: 6ff.; LON 6: 8f.). Two of the interview partners reported that its start-up activities would be, in absolute terms, the highest in the UK, although in relation to Greater London's knowledge base they would still be expandable. Therefore, a lot of public support measures would encourage entrepreneurial activity in the region, and both central government through LDA and the Mayor of London would provide expertise and funding to potential entrepreneurs, like e.g. through the *Business Link London.* However, all experts consistently emphasized their critical view on start-ups and spin-offs as mechanisms of knowledge transfer, since these would be risky, costly and inefficient ways of bringing knowledge and innovation to the market, especially because of their high failure rates. Public funding should, therefore, rather be used to promote alternative channels of knowledge transfer, like e.g. licensing. Anyway, enough seed- and venture capital would be available for start-up activities, above all in terms of smaller amounts of funding, and for large scale venture-capital (Interviews LON 1: 7; LON 2: 14f.; LON 3: 11f.; LON 4: 7ff.; LON 6: 8f.). For the medium sized quantities of funding, however, an "equity gap" would exist in the UK and even in London as one of the world's major financial places, as the head of business research at the LDA illustrates:

> "I would point to the equity gap in the UK when we are talking about technology businesses in particular. […] You can get some money to get your business started up, people can either raise funds themselves or via business angels etc. So you can find amounts of hundreds of thousands of Pounds, but if you need millions then it gets much harder. To find five or ten million Pounds is difficult, but if you are a firm that needs even more, then the venture capital firms are getting in there and certainly we have seen a number of the venture capitalists that some years ago have been looking at starting businesses are now looking for investing in more established businesses. So this gap in funding in early stages has become bigger." (Interview LON 4: 12)

Organization Based Mechanisms in Greater London's Automotive Sector

Divided by sectors, Greater London's *automotive* industry would have much less fluid labor markets than the region as a whole (Interview LON 5: 5).

The small scale of the sector in the region and a lack of larger employers which could offer top-management, -engineering or research positions to highly qualified persons would make transfers of staff and purposeful headhunting less likely instruments of knowledge transfer, as one interviewed expert for Greater London's automotive sector points out:

> "People change their employment. That is nothing exceptional. But for the whole country we currently observe a lack of qualified staff in this field. This will particularly be true for London, since we are not a classical automotive region with a grown pool of qualification in this field. And if you talk about the best qualified experts, especially in London many would prefer to go to one of the large banks and insurances or consultancies anyway, which might offer better opportunities. And I would also not see aggressive headhunting activities going on, just because we do not have those large firms which would invest in that and which would have top-management positions to offer." (Interview LON 5: 5)

Also business start-ups would be comparatively low in Greater London's automotive sector, especially because of its mature and consolidated character, but also because of a general economic downturn in UK automotive. In addition, the expert did not recognize insufficient supply of seed- and venture capital, but a lack of innovative business ideas in the sector, which could, however, be improved through the highly funded low carbon strategy of central government (Interview LON 5: 5f.).

Organization Based Mechanisms in Greater London's Biotech Sector

In *biotechnology*, on the contrary, all organization based mechanisms would be equally important than in the region as a whole, with strong transfer and headhunting activities (Interviews LON 4: 4ff.; LON 6: 7f.). Also academic spin-offs, but especially start-ups coming from the pharmaceutical industry would be central and frequently used mechanisms of knowledge transfer (Interviews LON 4: 6f.; LON 6: 8). What would limit even more elevated entrepreneurial activities would be a lack of management skills of most scientists creating a business, as well as a lack of understanding for the specific business culture in biotechnology, as a technology consultant for the life sciences at *London Technology Network* explains:

> "The trouble with knowledge transfer is that you must also bring the scientists out of academia into the spin-out to take it forward and they are probably not the best to do a spin-out anyway. Just because they are creative in a university environment, it does not mean that they are going to be successful in a commercial one. This can

become very disappointing, because they can very quickly get off an advantageous position, and the company is pushed into the bankrupt, or professional managers are bought in to bring the company forward." (Interview LON 4: 6f.)

Finally, also for the biotechnology sector the interviewed experts did not see seed- and venture capital provision as a major barrier to these entrepreneurial activities, not least because especially in London there would be a broad supply of venture-capital firms specialized in the life sciences sector (Interviews LON 4: 7ff.; LON 6: 8f.).

7.3.2.4 State Induced Mechanisms of Knowledge Transfer

Stronger deviations from both the theoretical assumptions made for the UK as a LME and the results obtained for the East of England, appeared when looking at the dimension of state induced mechanisms of knowledge transfer.

National and Regional Government Measures

Consistently, all experts reported both *national and regional government measures* to be important drivers of innovation and knowledge transfer in Greater London (Interviews LON 1: 10ff.; LON 2: 18ff.; LON 3: 17ff.; LON 4: 14ff.; LON 5: 8). In this, and contrasting the other English Government Office Regions including the East of England, Greater London would be the only English region with an own regional government represented by the *Mayor of London* in support of the *London Assembly*. In addition, in its work also the LDA would be much more dependent on coordination with this regional government, bringing it into a unique regional policy constellation between central government and the Mayor of London, and differentiating it strongly from the RDAs in the other English regions, as the head of business research at LDA emphasizes:

"London is slightly different, because we have a Mayor [...], our funding comes from Central Government, but the person who writes the *economic development strategy* or whose economic development strategy we deliver is the Mayor. [We have; MO] to deal with two agendas: We have the Mayor's economic development strategy agenda and we have the Central Government economic development strategy, which may not always perfectly align. Other Regional Development Agencies also develop their own economic development strategies, but theirs would probably be seen more as a direct flow from the Central Government, whereas ours has a different value. [...] Our regional government is the Mayor and he has a local government assembly. In EEDA there is no local government that would oversee the operations of the East of England. They have local approaches underneath them, but there is no body above them apart from reporting straight into Central Government. We have this intermediary, and the Mayor is the person we would really report into. Slightly we are in between these two." (Interview LON 3: 17f.)

Decisive in this connection would be the regional *economic development strategy* developed by the Mayor in collaboration with the LDA, and the national *Solutions for Business* portfolio transmitted through the LDA, which have both been evaluated as positive and efficient policy agendas in the fields of innovation and knowledge transfer by all interview partners (Interviews LON 1: 10ff.; LON 2: 18ff.; LON 3: 17ff.; LON 4: 14ff.; LON 5: 8). With that, all experts also acknowledge the LDA as central policy player in the fields of innovation and knowledge transfer. Although the flexibility to develop particular regional policy measures specifically tailored to the structural conditions of Greater London would be limited by the framework of the *Solutions for Business* portfolio and by the fact that most funding would be provided by central government, the regional policy level with the Mayor would pointedly be involved in the strategic agenda and priority setting, not least in innovation and knowledge transfer. In this, one major objective would be a stronger connection between Greater London's research base and the business sector, through mechanisms like the *KTPs*, *Knowledge Connect*, *Innovation Vouchers*, the *Business Link London* and start-up support. In addition, also the support of stronger collaboration of businesses, along sectoral value added chains as well as cross-sectorally, would be addressed by the regional innovation and knowledge transfer policy, like in the cases of the *London Biotechnology Network*, *London First* or the *London Technology Network*.

Scientific Parks and Technological Centers

Furthermore, the LDA would recently have invested larger funds in the creation of business incubators at three of London's universities, addressing with that a lack of *scientific parks* and *technological centers* in the region's metropolitan university landscape. As mechanisms of knowledge transfer, these institutions would, therefore, not play a decisive role yet (Interviews LON 3: 8f.; LON 4: 15; LON 5: 7; LON 6: 10). This would constitute a major difference to other UK regions like the East of England, but not necessarily a disadvantage, as the director of biotechnology and healthcare at *London First* remarks:

"I guess where London is quite different to some regions is the nature of business hubs at its ease. [W]hen we do events, a number of companies and advisors and investors more come in from areas of the right site of London, just because it is a fairly major business home to come and do business here. I suspect that we do not tend to have such sort of localized science parks etc., which are always built for the closer net community in other regions like Cambridge etc. That is the nature of London versus a fairly rural science community. I think in the current world where the industry is growing up, then most of the business will be done through major business hubs, it will not be done through local networks." (Interview LON 6: 10)

(Larger) State Induced Innovation Projects

Larger state induced innovation projects would only play a minor role in Greater London's regional knowledge transfer system (Interview LON 1: 10; LON 3: 15; LON 5: 7; LON 6: 10). On the one hand, the region would lack firms in sectors addressed by these projects, like defense or aerospace, and, on the other hand, the business R&D-level of Greater London firms would mostly be too low to participate in these projects. Therefore, it would more likely be the region's universities and research institutes which would participate in them.

Technology Transfer Offices

Finally, the role of *technology transfer offices and companies* at Greater London's universities has been evaluated as ambivalent by the experts: For the innovative and knowledge intensive firms these agencies would be important hubs in the interaction with universities, since most IP-related transactions would have to pass through them. For the majority of the low technology and less knowledge intensive firms in the region, however, these offices would be less important (Interviews LON 1: 8f.; LON 2: 16ff.; LON 3: 13f.; LON 4: 9ff.; LON 6: 9). In addition, the interviewees remarked large differences in the professional quality of these offices, as an exemplary statement of the head of business research at the LDA demonstrates:

"You have got something like *Imperial Innovations*, which is a technology transfer office that spinned-out a lot of companies from a world leading university, which has very professional staff, the company is quoted on the stock market, they are very professional and they need to be so. Down to, you have some relatively small inexperienced universities which may have a handful of people working on technology transfer. There you sometimes have a lack of management capacities and so you have a huge range in London. Some relatively small universities may have some important pieces of technology or skills etc., but might not have the access to do the way across." (Interview LON 3: 13)

This would frequently lead to a disequilibrium of negotiating skills, giving advantage to larger firms in accessing academic stocks of knowledge, as a technology consultant for the life science sector at *London Technology Network* illustrates, who also points to the fact that the technology transfer offices would apply a broad variety of knowledge transfer tools apart from patenting and licensing:

"It is not working [...]. Now the selling out process is a lot better than it was. Historically, there has been an underinvestment in the offices, not in number of people, [...] but in the quality of people. The quality of people is not high enough to do this kind of job. [R]emember that you can be negotiating a license with somebody like *Glaxo Smith Kline* which has a bunch of business development managers on the other side and each of them has 20 or 25 years of licensing experience. And you have some university licensing managers a couple of years after their PhD. Now that is changing. University business development offices are now employing better and higher caliber people. [T]hese people will make a difference. But remember that they are not only licensing, they are also transferring knowledge in other ways, through collaborative research, consultancy, contract services, like contract testing, training, executive education and many other ways that the universities are transferring their knowledge and add value." (Interview LON 4: 9f.)

State Induced Mechanisms in Greater London's Automotive Sector

Divided by sectors, *automotive* would almost mirror these results. Regional and national government measures would be important drivers of innovation and knowledge transfer in the sector, partly compensating for incomplete industry and network structures in Greater London (Interview LON 5: 8). Larger state induced innovation projects would not play a role, since most firms in the region would not be operating in the respective subsectors like defense or aerospace. However, one interviewed expert for the sector expects that the low carbon strategies developed by both central government and the Mayor of London will also have positive effects on automotive in the region and that projects funded through these strategic plans will enhance innovation and knowledge transfer in the sector (Interview LON 5: 7). Scientific parks and technological centers in automotive related fields would not exist in the region, a fact that has been evaluated as major shortcoming in comparison to other UK regions like the East of England. In addition, also technology transfer offices at the universities would not be frequently used by Greater London automotive firms, demonstrating again their low R&D-intensity and the respective barriers between automotive firms and universities (Interview LON 5: 6f.).

State Induced Mechanisms in Greater London's Biotechnology Sector

Biotechnology, on the other hand, would also profit strongly from both national and regional government measures. Especially the support of network initiatives like the *London Biotechnology Network* or the *London Technology Network* has been evaluated as highly valuable in enhancing innovation and knowledge transfer in the sector (Interviews LON 4: 14ff.; LON 6: 1ff.). Larger state induced innovation projects, however, would not play a role in this particular industry (Interview LON 6: 10). Also for biotechnology related fields there would be no scientific park or technological center located in the Greater London region. Nevertheless, these installations would be indispensible tools for biotech firms to interact with academia in close physical proximity, so that many Greater London biotech firms would also use science parks outside the region (Interview LON 4: 15f.). Finally, and like in the East of England, also Greater London biotech firms would use intensively the services provided by the technology transfer offices at the universities and access through this channel university knowledge bases, especially in form of IP-transactions (Interviews LON 4: 9ff.; LON 6: 9).

7.3.2.5 Market Based Mechanisms of Knowledge Transfer

With regard to the market based mechanisms of knowledge transfer like external consultancies, mergers, acquisitions and hostile takeovers, as well as the buying, selling and licensing of patents, the experts did not fully confirm the theoretical assumptions, but evaluated them as less important than assumed.

External Consultancies

External consultancies would be used only to a limited extend as brokers of knowledge transfer activities, since for the majority of the smaller and medium sized firms in Greater London it would not be economic to use their services (Interviews LON 1: 13; LON 2: 22f.; LON 3: 15f.; LON 6: 10f.). In addition, knowledge transfer would not be one of the main activities being outsourced to external consultants by Greater London firms. The consultancy services used would then be those provided by the universities, by smaller and specialized consultancies, as well as those offered by the regional government through the *British Library*, which offers expertise and consultancy services to firms in IP-related transactions. Furthermore, many of the large firms in the region would also have internal consultancy departments exercising the function of managing innovation and knowledge transfer.

Mergers, Acquisitions and Hostile Takeovers

Also *mergers, acquisitions and hostile takeovers* would not be usual and frequently used mechanisms of acquiring knowledge (Interviews LON 1: 13f.; LON 3: 16; LON 4: 13; LON 6: 11f.). In those cases they would be used in this sense, it would be strongly sector dependent, and it would mostly be the larger firms using them to incorporate specific stocks of knowledge and technologies, while the smaller firms would lack the financial and organizational resources to access knowledge via this channel and would also fear the risks involved.

Buying, Selling and Licensing of Patents

Not in line with the theoretical assumptions as well as with the results obtained for the East of England is also the only medium high significance the experts ascribed the *buying, selling and licensing of patents* as mechanisms of knowledge transfer (Interviews LON 1: 14; LON 2: 23f.; LON 3: 17; LON 4: 14; LON 6: 12). For the technology based firms in the region this trade of IPRs would indeed be essential, whereby for the majority of the firms operating in other (less knowledge intensive) sectors their use might be an exception. Furthermore, the trade of IPRs as a knowledge transfer mechanism would be especially relevant for the universities, but less for the business sector, which would use patenting more likely as an instrument to protect their own know-how, as the head of business research at the LDA remarks:

"I think it seems important for universities, and the government spent a lot of time and money ten years ago trying to increase technology transfer from universities in the UK. We were lacking behind the USA in this in the 80s. I have now some good examples where universities have developed technologies that were making large amounts of revenues, either from spin-outs or from licensing. And of course the RDAs would be supporting that as well. I think licenses etc. are seen as more important as a tradable commodity from the universities standpoint than I have seen it from the businesses. Businesses have more used it as a way of protecting their value, their developments and their technology, so that the whole company could not be sold. We can observe some licensing going on, but as a wider protecting of the technology we develop here, whereas the universities very much by the nature of the creator of a lot of patents have a lot of businesses that allow them to develop these technologies and taking them forward." (Interview LON 3: 17)

Market Based Mechanisms in Greater London's Automotive Sector

Also for the low scale and lowly innovative Greater London *automotive* sector the use of external consultancies as agents of knowledge transfer would only be minor (Interview LON 5: 10). Mergers, acquisitions and hostile takeovers would have been important accompanying mechanisms in the decline and consolidation process the sector experienced in the past, although in the present firms in Greater London would not have been frequently involved in these processes (Interview LON 5: 10). Finally, as in the region as a whole, the automotive firms in the region would mostly use patents to protect own stocks of knowledge, but less as strategic instruments of acquiring knowledge, what has been traced back again by one interviewed expert to the low R&D-intensity in the sector. More likely, London automotive firms would, however, resort to licensing as mechanisms of knowledge transfer, even if only in a limited number of cases (Interview LON 5: 11).

Market Based Mechanisms in Greater London's Biotechnology Sector

In the Greater London *biotechnology* sector, external consultancies would only scarcely be used (Interview LON 6: 10f.). First, the firms would not outsource knowledge transfer activities, and second, they would still be small enough and their senior managers would still be close enough to academia to exercise these functions themselves. Like in the East of England, also in Greater London mergers, acquisitions and hostile takeovers would especially occur when a biotech firm is acquired by a larger pharmaceutical firm, while within the sector itself these processes would be an exception (Interviews LON 4: 13; LON 6: 11f.). Finally, and much stronger than in the region as a whole and as in automotive, the trade of patents and licenses would by an absolutely vital channel of knowledge transfer in Greater London biotechnology (Interviews LON 4: 14; LON 6: 12), as the director of biotechnology and healthcare at *London First* accentuates:

> "The whole sector is driven by patents, by selling and licensing them. Without them this sector would not exist. [...] This is a very important mechanism of knowledge transfer. A lot of that knowledge transfer happens as a result of someone licensing another company's or academic institution's intellectual property. In this sector that is more often packed up in patents. For individual reasons, in these collaborations we talked about there will still be intellectual property secured within it. So I would say it is pretty much the key mechanism." (Interview LON 6: 12)

7.3.2.6 Community Based Mechanisms of Knowledge Transfer

Lastly, with regard to the community dimension of knowledge transfer processes and the respective mechanisms of informal contacts, regional platforms and additional (regional) institutions allowing to establish trust-based relationships basing on a common regional identity, the interviewed experts did not fully confirm the theoretical assumptions made in section 3.2.

Informal Contacts, Regional Platforms and other (Regional) Institutions

In this, *informal contacts* and formalized interaction have been reported to go hand in hand, without any of them dominating the other (Interviews LON 1: 3ff.; LON 2: 5ff.; LON 3: 4ff.; LON 4: 3ff.; LON 6: 3ff.). In any case, the informal dimension of knowledge transfer in Greater London has been evaluated as important precondition to establish further reaching, then more formalized forms of interaction. In addition, Greater London would offer a broad variety of opportunities to meet and interact in informal ways, much more than other UK regions. Not least the unique density of institutional infrastructure and the numerous *regional platforms* in a metropolitan environment would provide broad opportunities to interact and exchange knowledge (Interviews LON 1: 3ff.; LON 2: 5ff.; LON 3: 4ff.;

LON 4: 11ff.; LON 6: 3ff.). Especially the borough- and district-specific business platforms, the different ethnic communities with their respective organizations as well as the sector based networks like the financial industry in the City of London would exercise these functions.

Nevertheless, all experts consistently neglected the existence of a strong and commonly shared *regional cultural identity* in the Greater London region (Interviews LON 1: 12f.; LON 2: 21f.; LON 3: 19ff.; LON 4: 19f.; LON 6: 10). In this, the region would not differ from the other English regions including the East of England, which would all not be characterized by pointed regional identities. These could only be found in the nation-based regions of the UK, like Wales, Northern Ireland and Scotland. Greater London, however, would have never had developed a regional identity, but rather a national and international perspective, as one of the world's global cities and trading hubs. Therefore, also a strong preference to interact with other Greater London actors because of a shared regional identity has not been identified by the experts. Rather, they pointed out that physical proximity would be perceived as an advantage in knowledge transfer activities by Greater London actors, but only as one factor among others. In case specific expertise and stocks of knowledge would be needed, the regional actors would look with the same intensity for cooperation partners outside the region and internationally.

Community Based Mechanisms in Greater London's Automotive and Biotechnology Sectors

From a sectoral perspective, the two investigated sectors of automotive and biotechnology would basically fit into this picture (Interviews LON 4; LON 5: 8f.; LON 6). Like in the case of the East of England, also the focus on these two sectors in the Greater London region additionally revealed a strong cross-regional dimension of knowledge transfer contexts and cluster-building, connecting the regional actors in Greater London to other UK regions, like the *One Nucleus* life sciences network with the East of England. This tendency of increasingly creating national super-clusters also demonstrates a strong national dimension of innovation and knowledge transfer processes complementing regional and local contexts in the Greater London region. Finally, and pointedly stronger in biotechnology than in automotive, the experts highlighted the strong international orientation in terms of knowledge transfer and innovation, rather than towards the region.

7.3.3 Summary

Overall, these results provide both some correspondence with many of the theoretical assumptions derived from the literature on comparative political economy and innovation system studies, but also essential deviations from others. In this, the Greater London region resulted to be a regional innovation and knowledge transfer system which is only partly fitting into a LME context as defined by the theory, showing broad confirmation concerning the association, organization and community based mechanisms of knowledge transfer, and larger deviations from these assumptions on the network and market based, as well as on the state induced dimensions of knowledge transfer. Thereby, a majority of features shaping Greater London's regional knowledge transfer system have been traced back to the region's character as a major metropolitan agglomeration area as well as to its specific historical-economic development path, conserving a strong tradition in trade and its related services, experiencing an incomplete and low scale industrialization process coined to the particular needs of a bourgeois consumer market, and developing in the present an economic model being markedly dominated by knowledge intensive business and financial services as well as creative industries, and being complemented by a marginalized manufacturing sector operating in less R&D-intensive fields. As the UK's capital city area, Greater London is, furthermore, equipped with a disproportionately high number of universities, research institutes and R&D-oriented hospitals, creating a huge academic knowledge and expertise base, which, however, does not find its correspondence in a highly R&D-intensive business sector.

When analyzing the Greater London case, in all investigated dimensions of knowledge transfer these specific characteristics of Greater London as a metropolitan region resulted to be shaping the structure of the region's knowledge transfer system and the concrete mechanisms applied in it: A large talent pool coming from the region's numerous universities, knowledge intensive firms and business headquarters creating a highly fluid metropolitan labor market for scientific and technical personnel and strong competition for qualified staff through headhunting activities, a high number of universities and research institutes and an attractive business environment motivating intensive business start-up and academic spin-off activities in the region, as well as the pointed density of institutional infrastructure, regional platforms and business actors allowing to establish a broad variety of informal and formalized network relationships, especially between academia and business, constitute the main channels of knowledge transfer in Greater London's regional innovation system.

Other dimensions of knowledge transfer, however, have been strongly effected by the region's specific economic and business structure, with a vast majority of small and micro firms, a strongly services-dominated regional economy, a very low share of manufacturing sectors in the region's total economic activity, and an overall low technology and business R&D-level characterizing Greater London's business sector. As a consequence, complete and densely linked supply chain networks are rare, firms only scarcely cooperate in R&D-activities, formal and informal network relationships of Greater London actors mostly have to transcend regional boundaries to find adequate cooperation partners, and the technology transfer offices at the region's universities and research institutes are only used by a minority of firms. In addition, and as expected, also in Greater London the association based mechanisms of knowledge transfer like business associations and the London Chamber resulted to not play a significant role in innovation and knowledge transfer activities, since they lack the tradition to explicitly engage in these fields and have not yet recognized them as some of their main missions.

Larger deviations from the theoretical assumptions appeared with regard to the market dimension of knowledge transfer mechanisms. The respective mechanisms of external consultancies, mergers, acquisitions and hostile takeovers, and the trade of IPRs resulted to be surprisingly low for a metropolitan region within a LME, what has been traced back again to the region's peculiarities in terms of economic and business structure, as well as to its lowly R&D-intensive business sector. Further deviations are rooted in Greater London's metropolitan university structure: In contrast to more rural and industrial university landscapes like in the East of England, Greater London's universities mostly lack related scientific parks and technological centers as physical loci of interaction between business and academia in close spatial proximity, strongly limiting the impact of this rich university setting on the business sector and innovation activities.

Another striking feature in Greater London's regional knowledge transfer system is the unexpectedly high impact the interviewed experts ascribed to the distinct policy levels. Regional and national policy measures support innovation and knowledge transfer activities in a broad variety of fields, like network creation and maintenance, transfer of staff, interaction of business and academia, public consultancy services, start-up support, and general subsidies. Even if in this more active role of the state Greater London not substantially differs from the East of England case, it must be stated that, overall, the role of public policy has been evaluated as slightly more interventionist in Greater London, explicitly addressing many of the shortcomings discussed before, but especially the intensification of the links between business and academia. However, also in Greater London the policy focus of the different government levels is clearly on further enhancing and reinforcing already existing and functioning channels of knowledge transfer and creating and improving the respective framework conditions, and is, with that, far away from being the central actor in facilitating, organizing and funding knowledge transfer activities as it is in the Spanish, and, although to a less degree, German regions.

One important difference to the East of England region in this regard is the existence of a regional government, represented by the Mayor of London, which, in dense collaboration with the LDA, creates an explicit regional policy level with far reaching agenda and strategy setting autonomy, allowing to adapt more policy instruments to the specific needs of Greater London and its metropolitan structural conditions. This higher degree of regional policy autonomy resulted to be evaluated throughout positively by the experts, as well as the central role the LDA assumes in co-developing and, above all, implementing Greater London's policy agenda. What must, however, be added for the Greater London case, is that also this region is affected by the recent policy shifts already discussed above, pretending a turning back of regional devolution tendencies in the UK, including the closure of the LDA in March 2012. Therefore, it will be to observe and be left to future research which consequences these developments will have for the innovation and knowledge transfer policy in Greater London. Nevertheless, most of the functions concerning innovation and knowledge transfer will then be folded into the organizational structures of the Mayor of London and the Greater London Authority, so that one can expect that the impact of these changes on Greater London might be lower than on other UK regions, since at least one essential part of regional policy making will be preserved.

Similar to the other investigated cases, also in Greater London the two sectors of automotive and biotechnology showed distinct structures and mechanisms of knowledge transfer and, therefore, distinct strategies in dealing with the available structural conditions in the region. In this, Greater London's automotive industry resulted to be a traditional industrial sector being shaped by the decline of this industry across the whole country in the last decades. What remained, is a very small sector without any own OEMs, some foreign owned manufacturers and suppliers, only one larger assembly plant, an incomplete automotive value added chain, a low sectoral R&D-intensity with no larger R&D-department located in the region, absent regional sector-specific business associations, and some specialization in engine construction, which, overall, has only a marginal share in the region's manufacturing sector. As a consequence, horizontal and vertical customer-supplier networks are comparatively weak and frequently have to transcend regional borders, and cooperative inter-firm R&D-projects are relatively rare. In addition, the interaction between automotive firms and academia is considerably lower than in the region as a whole and as in the East of England, which is reflected in a few R&D-cooperations with universities, infrequent start-up and spin-off activities, as well as low use of scientific parks, technological centers and technology transfer offices at the universities.

Nevertheless, Greater London's automotive industry to a certain degree also showed some of the structures and mechanisms of knowledge transfer being typical for this sector, slightly differentiating it from the structures shaping the region's metropolitan and service sector dominated knowledge transfer system: More significant vertical supply chain relationships than in the region as a whole, comparatively strong strategic guidance and financial support by the state, markedly less volatile labor markets, and an only medium range importance of the market based mechanisms represent characteristic features of knowledge transfer processes in the sector. On the other hand, the analysis also demonstrated that, overall, the structures and mechanisms of knowledge transfer applied in Greater London's automotive sector do not differ much from those of the region as a whole. This could be interpreted in the sense that the sector is too marginalized within Greater London's economic landscape to create and maintain own modes of knowledge transfer, a fact that resulted to be considerably stronger with regard to its East of England counterpart.

Table 11: Qualitative Relevance of Knowledge Transfer Mechanisms in the Greater London Region

MECHANISMS	UK*	GL	Auto	Bio
NETWORK BASED MECHANISMS				
business networks	–	O	O	O
informal networks	O	O	O	O
institutionally stabilized networks	O	O	O	O
customers and suppliers	O	–	O	O
cooperative inter-firm R&D-projects	–	–	–	O
cooperative R&D-projects with scientific institutions and universities	O	+	O	+
ASSOCIATION BASED MECHANISMS				
employers-, business-, and trade associations	–	–	–	O
chambers of industry and commerce	–	–	–	–
binding and common technical standards	O	O	O	+
ORGANIZATION BASED MECHANISMS				
transfer of technical and scientific staff	+	+	O	+
purposeful headhunting	+	+	O	+
business start-ups	+	+	–	+
STATE INDUCED MECHANISMS				
national government measures	–	+	+	+
regional government measures	–	+	+	+
(larger) state-induced innovation projects	–	–	O	–
scientific parks	O	–	–	–
technological centers	O	–	–	–
technology transfer agencies/companies	O	O	–	+
MARKET BASED MECHANISMS				
external consultancies	+	O	O	O
mergers and acquisitions, hostile takeovers	+	O	O	+
buying, selling and licensing of patents	+	O	O	+
COMMUNITY BASED MECHANISMS				
informal contacts	O	O	O	O
regional platforms	O	O	O	O
additional (regional) institutions	O	O	O	O

+ *Reported (*Hypothesized) as crucial mechanism of knowledge transfer*

O *Reported (*Hypothesized) as important but not decisive mechanism*

– *Reported (*Hypothesized) as not important at all*

??? Not reported by the interview partners

[dark grey box] = *Important deviation from hypotheses*

[light grey box] = *Modest deviation from hypotheses*

[white box] = *No deviation from hypotheses*

However, it also appeared that the Greater London metropolitan knowledge transfer system provides inadequate structural support mechanisms for an incrementally innovating industrial sector like automotive: Lowly stabilized horizontal network structures, incomplete and only loosely linked value added chains, a lack of highly and adequately qualified staff, insufficient associational intermediation and traditionally low industrial relations resulted to be important weaknesses in Greater London's automotive sector in comparison to the investigated continental automotive regions. This allows for concluding that a major metropolitan agglomeration area like London might not be a fruitful environment for automotive to exchange knowledge intensively and to be innovative, what could also be seen as an explanatory factor for the generally less successful performance and the decline of the sector in the last decades.

Considerably stronger knowledge transfer and innovation activities characterize Greater London's biotech sector, which seems to find better matching conditions within the region's structural setting. In this, and as a high-technology sector, the biotech industry can make use of the broad scientific knowledge base provided in the region, and cooperate intensively with the region's academic institutions in manifold ways, but especially in collaborative R&D-projects, personnel transfers, business start-ups and academic spin-offs, trade of IPRs and the use of the technology transfer offices. However, Greater London seems not to allow for establishing network relationships in the same density as in a high-technology region like the East of England. One major reason for this is the absence of any sector related scientific park and technological center in which Greater London biotech firms could interact with other firms and academic actors in close physical proximity. Another reason might be the institutional landscape in a major urban agglomeration area with its myriads of actors generally being hard to oversee and which might also effect automotive and other sectors than those being closer investigated in this study.

Instead, both investigated sectors maintain intensive cross-regional network and cluster relationships, as it would be the *One Nucleus* life sciences network connecting Greater London with the East of England. Overall, this points to the fact that both sectors have a comparatively low orientation on Greater London as a region and are integrated into region-transcending contexts of innovation and knowledge transfer. In addition, also the international focus of the Greater London actors, and especially of those in biotechnology, must be highlighted in this regard. The region's long history as national capital city area, as center of the British Empire and its openness to national and international trade activities might be major explanatory factors for this, above all in the region's few larger manufacturing sectors as well as in the dominating financial and business services sector. However, and demonstrating the strong contrasts within a major agglomeration area, many small manufacturing firms and service providers are strongly rooted in the region and even in its single districts and boroughs, building on a historically grown business district culture. All these results are summarized in Table 11.

7.4 Lessons from the UK Case

When confronting these results with the two research questions motivating this study, the qualitative case studies on the East of England and the Greater London region showed a mixed picture, providing broad correspondence of the findings to the predicted structures and mechanisms of knowledge transfer for the UK case of a LME in some investigated dimensions, as well as considerable variation from these theoretical assumptions in others. In this, the two investigated UK regions are characterized by markedly different structural conditions shaping and limiting the concrete forms their respective regional knowledge transfer systems take, but apply similar approaches in dealing with these conditions, creating despite all structural differences more coherent and homogeneous innovation and knowledge transfer systems than one could have expected.

Nevertheless, the analysis revealed two strongly diverging types of regional innovation and knowledge transfer systems, with the East of England as a prime example of a high-technology industry region with highly R&D-intensive academic institutions and firms, and Greater London as a metropolitan region with a strong knowledge intensive service sector but predominantly low-technology firms. In this, the East of England's knowledge transfer system resulted to be mainly driven by the dense links between the academic research and high-tech-business-cluster around Cambridge, generating strong flows of knowledge especially in the direction from academia to business, and being the nodal point in most streams of knowledge in the region. As such, it very much resembles the ideal ERIS-type of a regional knowledge transfer system within a LME context as derived from the literature, showing most of the respective attributes: Strong networking activities leading to formal and informal modes of interaction, dynamic start-up and spin-off activities, fluctuation of qualified personnel as carriers of expert knowledge, intensive collaboration of firms and the scientific world in close physical proximity within scientific parks and technological centers, and strong trading activities of IPRs reflect a strong auto-dynamically learning and knowledge transferring regional economy, which needs only low associational intermediation and few intervention by public policy (Figure 58).

Diverging from this ideal model resulted to be Greater London's regional knowledge transfer system, which, overall, can be described as being predominantly talent and less institution driven, since especially the transfer of highly qualified staff, purposeful headhunting activities, business start-ups and academic spin-offs are the main channels of transferring knowledge within the region. In this, regional actors especially make use of the large pool of expertise available in Greater London coming from universities, business headquarters and knowledge (but not R&D-) intensive firms. This is especially true for the knowledge intensive services sector dominating the region, which has, though, not fully been in the focus of this study. However, the metropolitan character of Greater London as a major urban agglomeration area with its myriads of academic, business and institutional actors makes the available opportunities for knowledge transfer hard to oversee and hinders network creation and consolidation in all sectors, but especially in the few small scale manufacturing sectors in the region. This brings public policy into a slightly more intervening role, trying to address these structural barriers to innovation and knowledge transfer where possible (Figure 59).

This role of the state and the different levels of government in the UK needs some more attention in this regard: On the one hand, in both investigated UK-regions public support measures resulted to be important drivers for innovation and knowledge transfer, contrasting with that the theoretical assumptions on the LME context, which would suggest a marginalized role of public policy in knowledge transfer processes. On the other hand, these public measures are quite similar across both investigated regions due to the UK's centralized policy structures, provoking that in the East of England case they develop a reinforcing and in the case of Greater London a more intervening character. However, this centralized policy structure overall contributes to a certain homogenization among the UK's regional knowledge transfer systems, even if the degree of regional policy autonomy differs between both regions.

In this regard, the structures of regional innovation and knowledge transfer policy in the UK regions resulted to currently be in a flux. Regional devolution tendencies since the late 1990s with the establishment of the Government Offices for the Regions, Regional Assemblies and, above all, the RDAs in the UK regions, have introduced a regional governance perspective across the country giving regional actors a certain room to maneuver in implementing and developing more regionally adapted policy instruments in the fields of innovation and knowledge transfer. Despite the fact that in both investigated regions all experts consistently regarded this as a highly valuable process and concordantly argued that regional policy autonomy in these fields should even be increased, the recent shift of government from Labor to Tories and Liberal Democrats led to a fundamental change of policy strategy in this regard, with the consequence that the previously mentioned regional devolution measures are basically taken back and that innovation and knowledge transfer policy in the UK is recentralized. However, even if both analyzed regions are affected by this policy shift through the closure of their respective RDAs, it is to be expected that its impact will be considerably stronger on the East of England, since it lacks, in contrast to Greater London with the Greater London Authority and the Mayor of London, a regional government which could take over the respective policy competences. Further analyses in the upcoming years will be necessary to observe the exact consequences for both investigated regions as well as for the UK as a whole.

Furthermore, another striking finding for the case of the UK refers to the market based dimension of knowledge transfer. This has been assumed to be particularly significant within a LME context, an assumption which has indeed been confirmed for the East of England. However, it has not been confirmed for the case of Greater London, what must be seen as surprising when taking into account that the region is a major financial place and trading hub, being as such a center of liberal capitalism. What is demonstrated by this deviation from the theoretical assumptions is, above all, that the concepts of national production and innovation systems are too closely built around single empirical regional cases, like *Silicon Valley* or *Route 128* in the USA, when describing the LME model as a whole. Taking this into account, it is not surprising that the knowledge transfer system of the East of England as a high-technology region fits well into this category, and makes intensive use of market based mechanisms like external consultancies, mergers and acquisitions, hostile takeovers and the trading and licensing of patents, since it has a similar industrial focus like those regions from which these theories are mainly derived and where these mechanisms are important preconditions for the respective high-technology sectors to acquire knowledge. But for other regions which are not characterized by a focus on high-technology industries, these mechanisms appear to be much less significant, which is demonstrated clearly by the case of Greater London. And, even if speculating here, it seems plausible to assume that also other UK regions which have a stronger focus on traditional industries on a lower technology level might not fully fit into this category.

In addition, the case of the UK also demonstrates that the knowledge transfer systems of Greater London as a metropolitan and the East of England as a high-technology industrial region differ due to distinctly oriented university systems. While both regions are characterized by a comparatively high number of universities and research institutes, the East of England has a pointedly stronger focus on technology related disciplines and technological development, which is not least manifested in a high number of scientific parks and technological centers as physical loci of collaborative R&D-activities and knowledge transfer. Greater London, on the other hand, has a much less technology oriented university landscape, is stronger focused towards higher education and training, and within its metropolitan university setting lacks such scientific parks and technological centers. Interaction and collaboration in technology oriented fields between business and academia are, therefore, less frequent, a fact which is further reinforced by a markedly lower business R&D-intensity in the region. As a consequence, Greater London represents much less than the East of England the entrepreneurial university model which is generally regarded as typical for a LME context. Furthermore, it must be added that this specific model is not least rooted within an institutional specialism at Cambridge University: Here, a large share of researchers is allowed to own the IPRs from their inventions and discoveries, whereby in all other UK regions including London it is the universities to own them. This shows that it might also be differences lying within single institutions that might decisively coin the knowledge transfer systems of particular regions, at least if they possess structural relevance.

What is further shown by the analysis is that in the case of the UK commonly shared and historically grown regional identities and cultures do not play a significant role in facilitating and enhancing knowledge transfer activities. In both regions these identities resulted to be only weakly developed and regional actors transfer knowledge independently of strong regional communitarian ties, but by using the advantages of close spatial proximity. In this, especially the English regions in the UK resulted to lack explicitly regional traditions and identities, and the boundaries of the English Government Office Regions are in many cases perceived as artificial demarcation lines by the actors involved in innovation and knowledge transfer processes. Therefore, the UK is even stronger than the other investigated countries characterized by cross-regional contexts of knowledge transfer, with horizontal and vertical networks often transcending regional boundaries and a tendency to create rather national than regional superclusters, as has been shown e.g. for the two sectors of biotechnology and automotive.

As was to expect, these two closer investigated sectors showed diverging structures and mechanisms shaping their knowledge transfer processes from the two regions they are respectively operating in, pointing also for the case of the UK to specific sectoral needs and modalities of knowledge transfer within regional systems (Figures 58 and 59). However, these sectoral differences appear to be relatively low when compared to the other investigated countries, what must be interpreted in the sense that the institutional framework in the UK does much less allow single industrial sectors to develop their own regional and sectoral knowledge transfer systems than in the cases of Germany and Spain. In this regard, it is not surprising that it is biotechnology which best fits to both the theoretical assumptions on the LME context and the results obtained for the two investigated regions, but especially those for the East of England, since as a high-technology, radically innovating and analytic knowledge based sector it should find the better framework conditions within them. Automotive, on the other hand, indeed deviates slightly stronger from the theoretical assumptions and the regional results, although much less than one could have expected for a traditional industrial sector relying on incremental innovation processes and synthetic stocks of knowledge within the context of a LME, pointing to the strong coining impact this macro-institutional context exercises. As a consequence, one can conclude that the structural weaknesses of the sector and its comparatively low performance might basically been traced back to these inadequate institutional framework conditions to innovation and knowledge transfer.

Fig. 58: Regional and Sectoral Variation from the UK's Liberal Model: The East of England

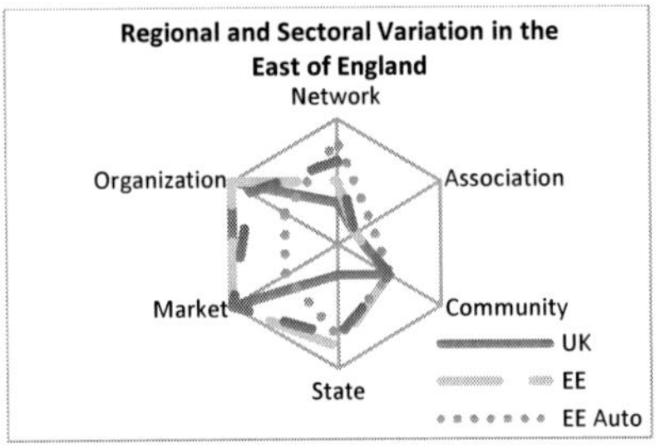

Source: Own Illustration

Fig. 59: Regional and Sectoral Variation from the UK's Liberal Model: Greater London

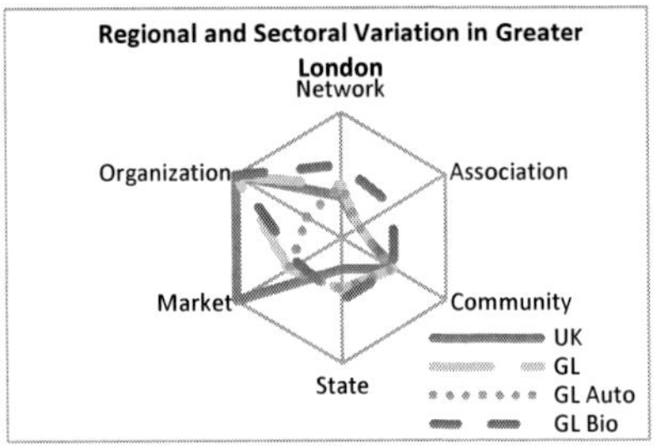

Source: Own Illustration

The analysis revealed across both investigated regions a UK automotive industry in decline, being shaped by a lack of own OEMs, foreign own manufacturers and suppliers, incomplete value added chains and comparatively low R&D-activities. The most important sources for innovative impulses and knowledge transfer in the sector, like dense and stable customer-supplier relationships, horizontal networks, collaborative R&D-activities, highly qualified staff with durable work contracts, and associational intermediation supporting interaction and knowledge transfer are not fully developed, incomplete or insufficiently available, so that concerning all these channels of knowledge transfer major shortcomings have been stated in both regions by the practitioners in the sector pointing to important competitive disadvantages in terms of innovation towards other European automotive regions. However, the analysis also demonstrated that apart from these commonly shared features the respective regional sectoral specialisms must be taken into account, which further influence the structures and mechanisms of knowledge transfer applied in the sectors.

In this, the East of England's automotive industry is strongly specialized in high-value and more R&D-intensive subsectors of automotive, like aerospace, luxury vehicles, engine construction and motorsport. With that, it efficiently uses the available structures of knowledge transfer in the region and its high-tech cluster, focusing on R&D-activities and the sector's higher technology segments in close interaction with the region's universities and research institutes. This corresponds strongly to the strategy of *subsector-specialization* described by Soskice (1999) for biotechnology in Germany's CME. Greater London's automotive sector, on the other hand, lacks this opportunity to focus on the high value segments. It is to be described as a marginalized manufacturing sector within a region in which manufacturing per se is marginalized, and the general picture resulting from the analysis is that it must be characterized as lowly innovative and only weakly knowledge transferring sector. In conclusion, this can be interpreted in the sense that automotive needs a critical mass of firms and sector related institutional actors to create resilient structures and mechanisms of knowledge transfer, which is not available in Greater London anymore. Thus, it appears that the sector does not find adequate support structures in Greater London's metropolitan knowledge transfer system which could compensate for these structural deficits.

The analysis of biotechnology in the East of England and Greater London, on the contrary, showed a mature and consolidated high-technology sector, which is highly innovative and transfers knowledge intensively through a broad variety of channels, and especially through those linking it with the academic world. As such, it also finds further supporting structures in both regions, as are, above all, formalized business networks, specialized external consultancies, venture-capital providers and business angels. In this, especially the East of England resulted to be the national and one of the international poles of excellence in biotechnology, offering favorable conditions to the sectoral actors to be highly innovative and to transfer knowledge intensively, something which, to a certain degree, is also true for Greater London. One of the East of England's specific strengths is to be seen in the dense ties between high-tech business firms and academia, which are characterized by a pointedly open and collaborative culture of interaction. In this form, this does not exist in Greater London, where it resulted to be much more difficult for the biotech firms to oversee the available cooperation partners and to establish stable network relationships to other business and academic actors, among other things due to the lack of scientific parks and technological centers in the region. And again, there has been only very few evidence, as e.g. publicly initiated and funded networks, that the actors in biotechnology could compensate this within the framework of a purely metropolitan knowledge transfer system like Greater London, but, rather, that they circumvent these structural barriers by creating and engaging in cross-regional network structures and using extra-regional scientific parks and technological centers.

Table 12: Qualitative Relevance of Knowledge Transfer Mechanisms in two UK Regions and Sectors: The East of England and Greater London

MECHANISMS	UK*	EoE	Aut	Bio	GL	Aut	Bio
NETWORK BASED MECHANISMS							
business networks	–	O	O	+	O	O	O
informal networks	O	O	O	O	O	O	O
institutionally stabilized networks	O	O	O	+	O	O	O
customers and suppliers	O	O	+	O	–	O	O
cooperative inter-firm R&D-projects	–	–	O	–	–	–	O
cooperative R&D-projects with scientific institutions and universities	O	+	+	+	+	O	+
ASSOCIATION BASED MECHANISMS							
employers-, business-, and trade associations	–	–	O	–	–	–	O
chambers of industry and commerce	–	–	–	–	–	–	–
binding and common technical standards	O	O	O	???	O	O	+
ORGANIZATION BASED MECHANISMS							
transfer of technical and scientific staff	+	+	O	+	+	O	+
purposeful headhunting	+	+	O	+	+	O	+
business start-ups	+	+	O	O	+	–	+
STATE INDUCED MECHANISMS							
national government measures	–	+	+	+	+	+	+
regional government measures	–	O	O	O	+	+	+
(larger) state-induced innovation projects	–	+	O	–	–	O	–
scientific parks	O	+	+	+	–	–	–
technological centers	O	+	+	+	–	–	–
technology transfer agencies/companies	O	O	O	+	O	–	+
MARKET BASED MECHANISMS							
external consultancies	+	+	O	+	O	O	O
mergers and acquisitions, hostile takeovers	+	+	O	+	O	O	+
buying, selling and licensing of patents	+	+	O	+	O	O	+
COMMUNITY BASED MECHANISMS							
informal contacts	O	O	O	O	O	O	O
regional platforms	O	O	O	O	O	O	O
additional (regional) institutions	O	O	O	O	O	O	O

+ *Reported (*Hypothesized) as crucial mechanism of knowledge transfer*

O *Reported (*Hypothesized) as important but not decisive mechanism*

– *Reported (*Hypothesized) as not important at all*

??? *Not reported by the interview partners*

[dark grey box] = *Important deviation from hypotheses*

[light grey box] = *Modest deviation from hypotheses*

[white box] = *No deviation from hypotheses*

Overall, these results illustrate that also the UK's LME model allows for considerable regional variation and cannot fully be seen as coherent national institutional configuration. Apart from the well-fitting high-tech region of the East of England, also other knowledge transfer models exist within the framework of the UK economy, as has been shown for the case of Greater London's metropolitan knowledge transfer system. In contrast to the coordinated or the mixed models in Germany and Spain, however, the UK allows for less sectoral variation, especially with regard to incrementally innovating sectors. Obviously, the respectively required structures and functional needs characterizing these sectors are hard to create in a LME environment, especially those concerning non-market coordination through stable networks and associational intermediation. This, as well as the high degree of policy-centralization, should be taken into account when developing a more varieties-centered research agenda on innovation and knowledge transfer, as will be discussed in detail in the following, concluding chapter. An overview on the results for the UK is provided in Table 12.

8. Conclusions

Taking these empirical results together, numerous empirical and theoretical conclusions can be drawn from the discussion with regard to the research questions. The first of these questions asked, in which way and to which degree the liberal, coordinated or mixed institutional arrangements at the level of national systems of innovation and production are also reflected in the regional level. In other words: Is it possible to identify in the regional innovation systems of national production models the corresponding liberal, coordinated and mixed structural mechanisms in the field of knowledge transfer, which the VoC-approach describes for the national level? Furthermore, the second question asked how exactly and why these structures and mechanisms are adapted at the regional level. Which are the specific institutional structures and mechanisms facilitating knowledge transfer at the regional level, and through which driving factors can regional variation be explained?

The respective empirical and theoretical implications for the distinct levels of analysis (national, regional, sectoral, conceptual) will be discussed in this concluding chapter. Due to the complex design of the study, with its multiple embedded cases and its distinct interwoven analytical levels, the different levels of analysis will be discussed in distinct sub-sections: Section 8.1 summarizes the empirical results from a comparative cross-case perspective and focuses on cross-case similarities to discuss deeper insights on the nature of knowledge transfer processes in the context of innovation as elements of a more sophisticated theoretical approach on knowledge transfer. Section 8.2 draws conclusions concerning the degree and the nature of regional variation in the three investigated national production and innovation models, as well as its major explanatory factors. Sections 8.3 and 8.4 focus on two specific drivers for this regional variation, namely the antagonism of traditional industrial and metropolitan regions as well as sectoral specialisms. Sections 8.5 and 8.6 discuss the theoretical implications for the concepts of VoC, NIS and RIS and claim both the need to develop a more "varieties-oriented" research agenda on innovation and knowledge transfer, and a more innovation and knowledge transfer oriented focus in the analysis of capitalist diversity. Section 8.7 finally discusses policy recommendations and possible future developments derived from the analysis.

8.1 Towards a Theory of Knowledge Transfer in Innovation Systems

In the first place, the region- and nation-specific results of the empirical case studies presented above also allow for drawing some general conclusions on the nature of knowledge transfer processes in the context of innovation. Taking the conceptual shortcomings of the approaches of RIS, NIS and VoC regarding knowledge transfer into account, these conclusions can be seen as further steps towards a more explicit, encompassing and systematic understanding of these processes and an advance in theory-building on the issue. Starting with the general conceptual level, knowledge transfer has been described as a *systemic process* which is *institutionally embedded* in specific *models of social order* and is coordinated by particular *modes of governance*. With the *heuristic model of knowledge transfer systems* developed in this study a useful *toolbox* has been introduced to catch the basic elements of these systems and to reconstruct their underlying logic of governance. This has been proved in all six regional and three national settings, in which all central mechanisms of knowledge transfer could be addressed and for which the respectively predominating modes of coordination could be identified. In this connection, it resulted to be highly valuable to differentiate the coordination of knowledge transfer by the six dimensions of governance, to systemize the numerous possible channels and mechanisms of knowledge transfer clearly and to evaluate them analytically.

From confronting this heuristic model with empirical evidence from the distinct regional and national settings, on the one hand, *case specific insights* into knowledge transfer processes in distinct institutional contexts have been derived, which have already been extensively discussed above. They demonstrate to which degree, how and why structures and mechanisms of knowledge transfer vary across nations and regions. On the other hand, going beyond these case specific results, taking a *comparative cross-case perspective*, and focusing on *cross-case similarities*, allows to reveal deeper insights on the nature of knowledge transfer processes, which can be major elements for the construction of a more sophisticated theoretical approach on knowledge transfer in the context of innovation. In addition, this enables to identify *good practice models*, which can serve as benchmarks for policy actors concerned with constructing and reforming knowledge transfer systems at the national or regional level. These cross-case similarities, as well as the resulting policy recommendations will be discussed in the following paragraphs for the single analytical dimensions of the heuristic model.

Networks

Concerning the *network* dimension, the results indicate that network based mechanisms are most important in innovation systems in which incrementally innovating sectors dominate, which require intensive, stable, and formalized interaction of the involved actors, in which large and small firms are complemented by a diversified medium sized business sector, in which dense collaboration of customers and suppliers is necessary, in which complete value added chains exist, in which collaborative R&D-activities are required, in which a strong sense of a commercial, technological or professional community exists, and in which a cooperative rather than individualistic business culture dominates. As expected, networks as non-market coordination instruments are used most intensively in the German region of Baden-Württemberg, but they also play an important role in the East of England, where a strongly high-tech-oriented business sector maintains intensive network relations around the fixing pole of Cambridge University, and where technological rather than purely commercial ties predominate. In other cases, in which the sectoral focus would require network based forms of knowledge transfer, but in which the structural conditions are not conductive to this, as it would be the case in the MR HBGW, Catalonia or the Madrid Community, public policy resulted to be at least partly compensating these shortcomings by initiating networking activities top-down, especially through providing funding and organizational structures, as has been shown among other things in the examples of the *Innovationsnetzwerk Niedersachsen*, the *Landesinitiativen* or the *Regional Growth Concepts*, the TECNIO and NUCLIS programs, or *Madrid Network* and the *madri+d* system. However, the analysis also demonstrated that the use of these "constructed" network structures as channels of knowledge transfer is considerably lower than the one of their "grown" counterparts in Baden-Württemberg and the East of England, since structural barriers like incomplete value added chains and individualistic business cultures are hard to bypass.

Network based cooperations between business and academia resulted to be generally more important than assumed, except for the case of Catalonia. This refers to an increased importance of scientific knowledge in all production processes as well as a certain need to use scientific infrastructure in business R&D-processes. These R&D-oriented networks between science and industry are strongest in high-technology dominated regional settings like the East of England, in which firms rely strongly on analytic knowledge bases generated by the scientific world, and where an entrepreneurial university model motivates academia to interact intensively with industry. They are lower, but still important, in settings like Baden-Württemberg, in which the sectoral structure requires intensive inter-firm R&D-activities rather than intensive business-academia interaction, and they are almost marginal in settings like Catalonia, where strong mental and cultural barriers impede cooperation of firms and scientific institutes, and where the dominating incentive structures hinder academia to engage with industry.

In this regard, especially the investigated metropolitan regions face particular structural barriers to networking activities, as are, above all, the hard to oversee cooperation and networking options available due to the high number of actors (Madrid, London), or multipolar regional structures (MR HBGW) in urban agglomeration environments: In these regions many actors are just not aware of their potential options and find it hard to establish adequate and stable network relationships. However, this study also displayed policy answers to these barriers which have been evaluated as positive and highly efficient by the interviewed practitioners, as the *Innovationsnetzwerk Niedersachsen*, *Madrid Network* or the *madri+d* system, which share a certain centralization of intermediation activities and public support to network-building. Especially the centralization and stronger coordination of the different intermediary actors can be seen as a good practice model, which could help to overcome networking related barriers not only in other metropolitan regions like Greater London, but also in other regional settings. Making the regional actors visible, as well as creating a central access point to the regional knowledge transfer systems appears to be one central challenge for improving knowledge transfer activities in many regions, but especially in those characterized by metropolitan agglomerations.

Association

Second, the results on the *association based mechanisms* of knowledge transfer indicate that also associational intermediation is especially important in regions dominated by incrementally innovating sectors to moderate and stabilize knowledge transfer activities, and where corporatism plays a role in economic activities, like in Germany's CME and Spain's MME. They do not play any important role in the investigated UK regions. However, the results also demonstrate that to efficiently exercise this role chambers and business associations need to be well organized in terms of financial and human capital and to be regionally-wide coordinated, as has been the case with the *Madrid Chamber*, *CEIM* or the *Baden-Württemberg Chambers* with their lead management principle. It is only then when corporatist actors are seen as efficient service providers and drivers of knowledge transfer. Improving the chambers' and associations' internal organization and coordination, as well as a stronger acknowledgement of innovation and knowledge transfer as important corporatist missions are, therefore, recommendations which can be derived from these results. However, the results also showed that where a strong tradition of chambers and associations to take this responsibility does not exist, and where their internal structure is historically fragmented, their abilities to intervene in these fields is limited and also their perception as competent partners in innovation and knowledge transfer is rather low. One efficient solution to compensate these structural shortcomings is realized in the MR HBGW with the *Innovationsnetzwerk Niedersachsen* which also includes chambers and associations. This network enables them to better coordinate their activities region-wide, not only among each other, but also with other intermediary actors, giving their services a certain "catalyzing effect." Especially for the region of Catalonia this could be a good practice model to improve associational intermediation.

Organization

Third, the *organizational dimension* of knowledge transfer, as expected, resulted to be most important in the UK regions, where strong fluctuation of staff and intensive start-up as well as spin-off activities are vital channels through which knowledge is transferred. Especially in the "talent driven" knowledge transfer system of Greater London this appeared to be particularly strong. But also in the regions of Germany's coordinated model these mechanisms play a surprisingly important role, pointing to an increasingly deregulated and fluid labor market, as well as an increasing need of including scientific knowledge into innovation processes even in incrementally innovating sectors. This is manifested in remarkably vital transfers of scientific and technical staff. Where these transfer activities are lower, the reasons are similar across the investigated regions: Small firms are less involved in transfers of academic staff than larger companies, since they mostly lack the organizational capabilities to conduct intensive recruiting activities; firms with lower technology levels transfer less qualified staff than those with higher technology levels because of stronger mental and socialization related barriers to do so. Many policy measures address these issues, especially through providing support schemes for SMEs to recruit academic staff and graduates, as it would be e.g. voucher schemes or KTPs, which resulted to be well accepted by the practitioners and can be expected to be an important initial start to lower barriers to get into contact with scientifically educated personnel.

Pointed start-up and spin-off activities as mechanisms of knowledge transfer are strongly sector and business demography dependent, with regions dominated by younger and radically innovating sectors showing higher start-up rates, like in the East of England and Greater London, and regions characterized by incrementally innovating and consolidated sectors with lower start-up rates like Baden-Württemberg and the MR HBGW. In addition, also particular entrepreneurial cultures resulted to influence this dimension of knowledge transfer strongly, with markedly developed entrepreneurial activities in the investigated UK regions and pointed risk aversion in the Spanish and, above all, German regions, impeding higher business creation rates. Moreover, Spain and Germany also share a lack in professional and specialized seed- and venture-capital provision. Even if enough financial resources are available in the investigated regions in the eyes of the interviewed experts, the decisive bottlenecks are insufficient management capacities and expertise of banks, venture-capital firms and business angels in the field of technology related business creations. Therefore, policy measures providing seed- and venture capital and fostering the creation and consolidation of business angel networks exist in all investigated regions, although their success and sustainability is seen throughout critically. Apparently, this particular element of knowledge transfer activity is especially hard to support by isolated policies, but requires more systemic reform processes as well as economy wide learning and mentality changes.

State

Fourth, the results on the *state dimension* of knowledge transfer processes indicate that in all types of market economy models and in all regional varieties within them the state plays an important role in knowledge transfer processes. This role varies, however, both between and within national models. In Spain, the state is the most central actor in *intervening* in all stages and fields of knowledge transfer processes, while in Germany it exercises an important role as *bystander and enabler* to these processes. Even in the UK's LME, the state resulted to play a surprisingly important role as *catalyzer* of knowledge transfer processes. Across the investigated regions, the role of the state varies between predominantly state driven knowledge transfer systems like Catalonia, systems with pointed state intervention but also strong corporatist elements like in the Madrid Community or the MR HBGW, systems corresponding to the enabling and bystander model, like Baden-Württemberg, and models with a more catalyzing state like the East of England and Greater London. In addition, regional variation in the role of the state resulted to be strongest where regional autonomy is most developed and where central government has less impact into regional knowledge transfer processes, so that strongest regional variation has been observed in federalist Germany, followed by the Spanish regions with high autonomy rights but strong central government impact, and the lowest variation in the central government driven political system of the UK.

In any case, these results suggest that across national and regional models knowledge transfer in all its dimensions requires an active role of policy, however this might be defined. Improving the learning and knowledge transfer capacity of regions and whole national economies is, therefore, also to be seen as a *collective challenge*, and competition on innovation is also a competition on the *most successful and best fitted policy models* in these fields.

In this regard, the results of this study indicate that it is especially the interface of publicly organized academia and the business sector which is one of the most critical points in all knowledge transfer activities. A central role in all knowledge transfer systems would thus correspond to the *technology transfer agencies and offices* at the scientific institutes and universities, which should be the main interlocutors between both worlds. However, in all investigated regions these agencies resulted to be inadequately missioned, organized and equipped, so that in none of these cases they could fulfill this function. The challenge for all investigated regions is, therefore, to ascribe these agencies a more proactive and less administrative mission as central *marketplaces* for exchange processes between business and academia, to professionalize their knowledge and innovation management capacities, to better equip them with highly qualified and experienced staff, and to increase their funding to create measures to actively approach science-averse actors like especially SMEs and to intensify their marketing efforts to become more visible. Also the stronger inclusion of these agencies into wider networks of intermediary actors to leverage their impact appears to be promising in this regard, as, once more, the policy practice models of the *Innovationsnetzwerk Niedersachsen* and the *madri+d* system showed.

Market

Fifth, the results for the *market based dimension* of knowledge transfer indicate that across the investigated regions and national economies, the use of external consultancies, mergers and acquisitions, hostile takeovers and the trade of IPRs is strongly dependent on the predominating sectoral structures and technology levels, with regions dominated by radically innovating sectors and higher technology levels showing more intensive use of these mechanisms than those with incrementally innovating sectors and lower technology levels. Consequently, in the East of England this dimension is the strongest, followed by Greater London and the MR HBGW, and is comparatively low in Baden-Württemberg and the two Spanish regions. One major shortcoming characterizing the investigated German and Spanish regions in this regard is a *lack* of supply of *specialized private consultancy services* in the fields of innovation and knowledge transfer. Even if this is partly compensated by consultancy offered by chambers and associations, as well as by public consultancy, the analysis displayed a noticeable demand for these external services, pointing to a growing market potential in this field. With that, the results basically confirm the assumptions derived from the VoC-approach that market based mechanisms are the strongest in the LMEs like the UK, although even here considerable regional variation could be observed pointing to the uniqueness rather than representativeness of the East of England's high-tech economy. What these results do not confirm is, then, the assumption that market based mechanisms would be particularly strong in the investigated metropolitan regions which are supposed to be nodal points in global networks, and would, therefore, be places where possible tendencies of institutional change towards the liberal model would be to observe.

In addition, mergers, acquisitions and hostile takeovers can be clearly attributed to be only central mechanisms of knowledge transfer in purely high-tech environments, where even knowledge bound to particular firms has substantial analytic elements and can, therefore, more easily be integrated into another organization structure, where highly transparent and dynamic market structures predominate the economic landscape and where a respective business culture of acquiring competitors exists. Interesting in this connection is the finding that even in the German and Spanish regions neither institutional nor structural barriers have been reported to exist by the interviewed experts, but, rather, functional and cultural ones. This indicates that in the continental regional and national economies these particular mechanisms of knowledge transfer can be supposed to remain marginal.

This dense relation of market based mechanisms and high-tech regional economies also appeared when pointing to the trade of IPRs, with two particular points to detail on: First, even though also in incrementally innovating and non-high-tech sectors and regions the trade of IPRs played and plays a certain role, this role resulted to decrease, especially in the German regions. Shorter product cycles and difficulties to avoid product imitation in transparent global competitive environments make precise patent descriptions and long term licensing periods less attractive or even counterproductive, and foster alternative strategies like technology leadership. Second, the intensity of IPRs-trading activities is also related to its institutional embeddedness: In all investigated regional cases the IPRs on scientific results are held by the respective institutions and universities, making the technology transfer agencies and offices central players in their commercialization. With the already discussed shortcomings of these institutions especially in the German and Spanish cases, however, also the intensity of trading IPRs is limited. An alternative model is realized at Cambridge University, where large parts of faculty can hold the IPRs from their work by their own, and where the commercialization of IPRs is quite intensive. Yet, it can be doubted that this model would also work as benchmark for other environments, since it is part of an overall more entrepreneurial university structure which is not comparable to other regional settings, especially not to those in Germany and Spain. What policy can do, however, would be to create more incentives for researchers to invest in the protection and commercialization of their IPRs, giving transfer activities in addition to publishing a higher relevance for their scientific careers. Especially for the Spanish case, where patenting and IPR-trading activities are almost marginalized, this could help to overcome a transfer- and innovation impeding culture of seeing the results of public research as public rather than marketable goods.

Community

Finally, concerning the *community dimension* of knowledge transfer, the results demonstrate that the corresponding mechanisms like informal contacts and regional platforms are most important in the German regions, and especially in Baden-Württemberg. This confirms that these non-market channels of knowledge transfer are the strongest in a CME environment, in which incrementally innovating sectors require stable, long term and trust based interaction of economic actors. The nature of this dimension of knowledge transfer is informal, and across all investigated cases the most important value of these transfer channels resulted to be creating and stabilizing a *sense of community*, might it be technological or commercial, and, above all, to initiate then *more formalized and substantial* forms of interaction and transfer. Strong and historically grown regional cultures, however, resulted to not be the main drivers of this community based transfer activities, but, rather, spatial proximity, *questioning with that large parts of the literature on RIS* which see regional cultural identities as important location factors and motors of learning processes in regions. Regional knowledge transfer communities can, thus, more likely be described as *spatially rather than culturally integrated communities*. An exception among the investigated regions might only be Baden-Württemberg, where not primarily a strong regional identity, but a *highly associative business culture* could be identified, as well as a pointedly strong commercial community as a reference point of self-identification with considerable preferences to cooperate among each other.

Altogether, this analysis demonstrated to provide a useful basis to further develop the theoretical approaches of NIS, RIS and VoC in making processes of knowledge transfer explicit and systematically understanding them as basal processes underlying innovation and determining competitiveness in knowledge based economies. This advanced understanding of the nature of knowledge transfer in the context of innovation can be seen as highly valuable, not only in theoretical conceptual terms, but also for the involved actors and policy makers concerned with respective reform processes. In this connection, the applied heuristic model can be seen as a helpful analytical basis for reconstructing further knowledge transfer systems in other national and regional settings and with that broadening up the empirical basis for our understanding of knowledge transfer processes per se. However, as the analysis also showed, this theory-building process must necessarily take into account driving causes for national and regional variation of innovation and production systems, which are central explanatory factors behind distinct institutional models of knowledge transfer. Therefore, in the following paragraphs, conclusions concerning the question on regional variation in national systems will be drawn.

8.2 The Nature of Regional Variation in VoC

With regard to the question of regional variation in national systems of innovation and production, the results of the qualitative case studies demonstrate clearly that national market economy and innovation models are by no means homogeneous institutional configurations, at least not in the field of knowledge transfer, but are, rather, to be seen as *regionally diversified institutional settings*. In all three investigated national market economies, the two analyzed regional knowledge transfer systems showed at least some *variation to the assumed national models*, as well as *among each other*. Conceptualizing national production and innovation systems as homogeneous and coherent entities, like broad parts of the literature on VoC and NIS do, should, therefore, be reconsidered. Rather, knowledge transfer processes should be thought as being institutionally embedded into specific regional governance systems, which create particular structures and support specific mechanisms through which knowledge is transferred in given regional settings.

However, with the same explicitness in which the analysis demonstrated the existence of regional variation in national knowledge transfer systems, it also showed that the room to develop region-specific institutional models of knowledge transfer is *limited by the surrounding national institutional frameworks*. Even if all investigated regional cases deviated in many aspects from their respective national models, this variation has been punctual rather than systemic, and the degree of variation has only been in a few instances fundamental. None of the investigated cases provided evidence for completely deviant logics of coordination on the regional level which would allow for speaking of fundamentally contradicting coordination logics within one national model. This indicates that the VoC-framework indeed captures the dominant institutional structures and emphasizes the principal mechanisms of interaction of national market economy systems as well as of their respective national knowledge transfer subsystems (Figure 60).

What must be added to an encompassing view on national systems of production and innovation is, therefore, the *degree* to which they allow for regional variation, as well as the specific *explanatory factors* determining this regional variation, for what this study provided empirical evidence in both dimensions. This empirical evidence demonstrates clearly that it is above all *regional economic structures, business structures, sectoral specializations, actors' constellations, policy approaches and business cultures* which are major explanatory factors for regional variation and deviating governance structures within national models. Understanding the complexity of national innovation and knowledge transfer models in depth, using these models as benchmarks for political reform processes and regional economic policy making requires, thus, taking into account these regionally variegated appearances of national governance models.

In this, the extensive analysis of the empirical case studies revealed specific evidence on regional variation in the three investigated national market economies. In the case of Germany's CME, regional variation in the two analyzed regional cases resulted to be the strongest (Figure 61). The two regions of Baden-Württemberg and MR HBGW showed clearly diverging structures and mechanisms of knowledge transfer, demonstrating that the Baden-Württemberg model of a regional knowledge transfer system should neither necessarily be taken as paradigmatic for the totality of the German regional economies, nor for the coordinated model of capitalism as a whole. What resulted to be the major differentiating factors in the German case are, first, distinctly developed *business cultures*, giving the MR HBGW's knowledge transfer system a more individualistic and distanced character and hindering knowledge transfer, while Baden-Württemberg is characterized by more cooperative business attitudes, giving network based and collaborative channels of knowledge transfer substantial weight. Second, distinct *actors' constellations*, especially with regard to the chambers of commerce, rooting in diverging organizational structures, distinct degrees of competition and cooperation among the different local chambers, and different traditions and self-understandings, make the MR HBGW a much less corporatist and associationally intermediated knowledge transfer system than Baden-Württemberg.

Third, the *role of the different policy levels* differs to a considerable degree among both regions, with the state exercising a more decentralized and bottom-up approach in Baden-Württemberg, leaving more initiative to the economic and corporatist actors and relying strongly on their respective coordination capabilities, while the Lower Saxon innovation policy takes a stronger top-down orientation, assuming more responsibility in initiating and coordinating the activities in the fields of innovation and knowledge transfer. In this connection, it is especially Germany's federalist political system which gives its *Länder* sufficient autonomy to create specifically tailored regional innovation policies, which is manifested in the markedly more centralized coordination of the Lower Saxon innovation policy, above all through the participation of most of the chambers, associations and other intermediary actors in the *Innovationsnetzwerk Niedersachsen*, partly compensating the organizational shortcomings characterizing many of these actors per se. Finally, also *distinctly strong developed regional identities* due to specific historical experiences, but also *diverging degrees of regional coherence and competition* among the different regional sub-spaces, account for regionally variegated knowledge transfer systems: While in the MR HBGW knowledge transfer is basically organized according to a "sub-spatial logic" in *regional growth concepts*, Baden-Württemberg relies more on a sectoral organization of these processes around specific value added chains in *cluster initiatives*.

Fig. 60: Varieties of National Knowledge Transfer Systems in Germany

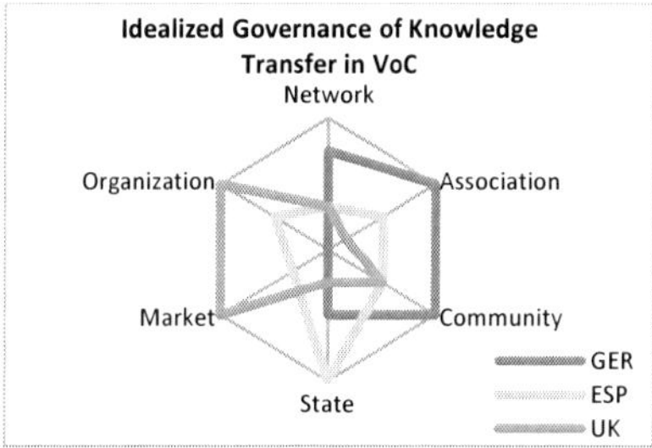

Source: Own Illustration

Fig. 61: Regional Varieties of Knowledge Transfer Systems: Germany, Spain and UK

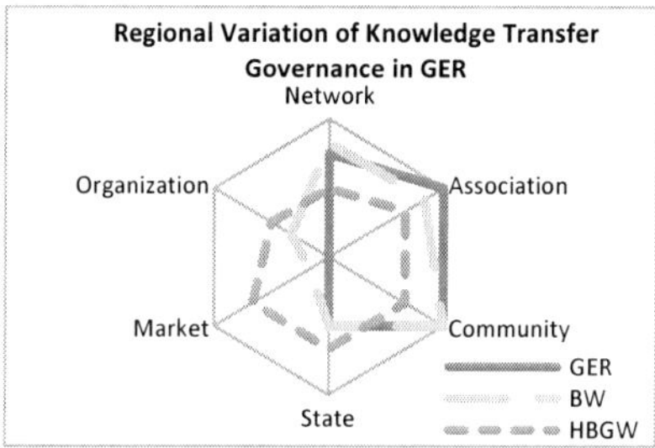

Source: Own Illustration

Fig. 62: Regional Varieties of Know-ledgeTransfer Systems in the UK

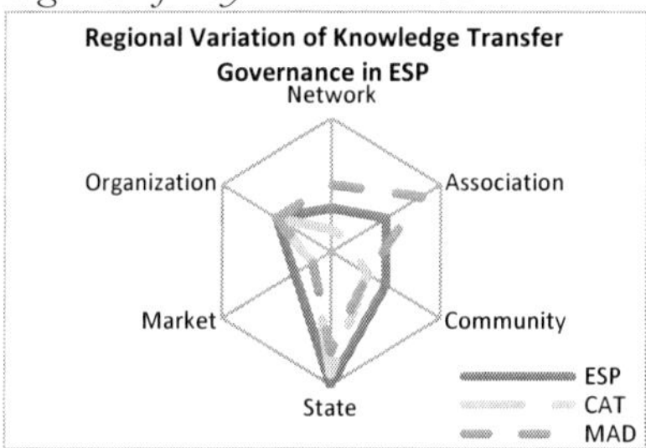

Source: Own Illustration

Fig. 63: Regional Varieties of Know-ledge Transfer Systems in Spain

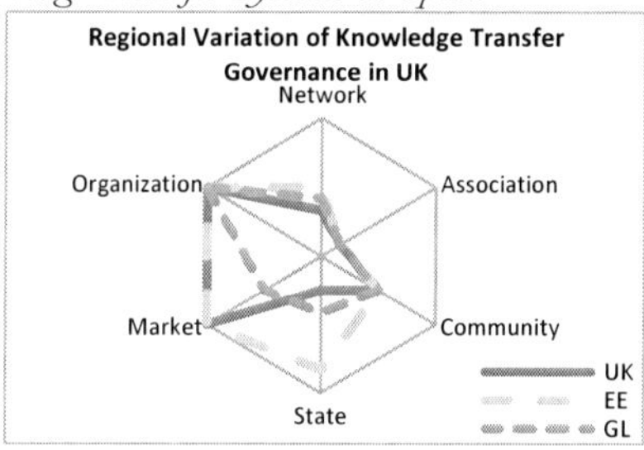

Source: Own Illustration

What we learn from the analysis of the German case for regional variation in national systems of innovation and production is that, apart from all homogenizing effects exercised by national institutional frameworks, considerable degrees of freedom to create regional governance systems deviating from these models exist in Germany's CME, being able to address particular structural, sectoral, cultural and actor-related conditions. Many of these deviations from the presumed national structures can be traced back to specific regional deficits with regard to the very much idealized coordinated model as defined by the VoC-approach and the respective compensation efforts at the regional level. This flexibility should be seen as strength of national systems rather than weakness as suggested in the orthodox version of the VoC-approach, raising the knowledge transferring ability of regional economies. However, the question to which extent these "creative incoherencies", as positive they might have been evaluated from a qualitative point of view, can lead to institutional equilibriums which are similarly successful than the ideal coordinated model, should be object to further, more quantitative and econometric research designs. In any case, the direction this variation from the ideal model takes in the German context is, indeed, in parts towards the liberal model, with elevated relevance of the market dimension and lower relevance of network mechanisms, but it also tends towards the state driven model of the mixed countries, so that it is hard to speak of a clear confirmation of the liberalization thesis (see 2.4.2.5.3; 4.1.2) in the German case.

Regional variation resulted to be lowest developed in the case of Spain's MME (Figure 62). Spain illustrates a case of knowledge transfer systems in a predominantly state driven market economy environment, in which regional variation of these systems depends strongly on different regional *policy approaches*. On the one hand, the strong dependence of both regions on policy intervention and the considerable autonomy rights to create own policy concepts are important explanatory factors for regional variation in the investigated Spanish regions. In this regard, the much stronger focus on *network creation* in the Madrid Community, as well as the much more *centralized* policy approach with the ACCIÓ-Agency in Catalonia are major differences among both systems. On the other hand, the strong role exercised by central government in the fields of innovation and knowledge transfer, especially through providing the most important shares of funding for specific programs, resulted to have a highly homogenizing effect across the Spanish regions in the relevance of particular mechanisms.

While the different levels of policy in the investigated Spanish regions have to address almost similar structural conditions (and deficits), as fragmented business structures, low technology levels or incomplete value added chains, two major specialisms stand out, which also explain the main differences between both regions: First, the Madrid Community is characterized by a *much stronger role of the corporatist actors* like the Madrid Chamber or CEIM, which are active policy partners of regional government and allow for a more *pluralist, self-regulating and network centered policy approach* than in Catalonia. Efficient organization structures and a strong tradition to focus not only on the region as a whole but also on the topics of innovation and knowledge transfer, basically explain this *actors' constellation* and substantially differentiate Madrid from the situation in Catalonia. Second,

Catalonia appears to be characterized by an (even for the Spanish context) *extremely individualistic business culture* impeding almost all associational and collaborative forms of knowledge transfer, and requires, thus, a more *centralized and interventionist policy approach*, which especially addresses the basic elements of knowledge transfer through, e.g., raising the visibility of potential cooperation partners (TECNIO) and the funding of initial forms of cooperation (NUCLIS).

Overall, the Spanish case discloses that also the mixed model of knowledge transfer systems is characterized by regional variation, even if the degree of this variation is pointedly limited by national institutional frameworks and especially by central government intervention. As the results demonstrate clearly, the direction this regional variation takes in the Spanish case is clearly towards the *coordinated model*, with elevated relevancies of the network and association based dimension of knowledge transfer. No evidence appeared, however, for variation towards the liberal model, which is manifested in the market and organization based dimensions being only as or even less important than expected. In this regard, it can be assumed that the particular structural conditions and especially the numerous structural deficits in the Spanish model allow much more for network based interaction and associational intermediation than they would permit the development of a strong market and organization based dimension of knowledge transfer. One possible interpretation for this could be that these forms of non-market coordination require less auto-incentives especially by the firms, which can be expected to be better compatible with the previously described structural deficits in the Spanish regions.

For the investigated cases of the UK's LME, regional variation came out to be slightly higher pronounced than in Spain, but lower than in Germany (Figure 63). The two analyzed regions of the East of England and Greater London show similar institutional structures and mechanisms of knowledge transfer, even if they are characterized by pointedly distinct structural conditions. The UK's *highly centralized political system* with only *marginal regional political autonomy* and the absence of a tradition to create institutional structures and policies at the level of the region can be seen as major *homogenizing factors* in this regard. However, also the two UK regions showed some variation, which are, on the one hand, rooted in the differences of Greater London's *metropolitan and higher education dominated university setting* standing in the center of a *talent driven regional knowledge transfer system*, and the *rural and research oriented university landscape* of the East of England being the major driver of a *technology oriented regional economy* maintaining close links to industry, not least through technological networks, technology centers and scientific parks.

On the other hand, important differences among both regions exist in the relevance of the market dimension of knowledge transfer, which is considerably more important in the East of England. This indicates that the strongly capital oriented market based mechanisms are basically bound to explicitly high-technology oriented economic environments, and might, even for a LME context, be rather exceptional than paradigmatic. One can expect, therefore, that to the degree to which other regions in the UK or other LMEs do not have such a pointed focus on high-technology, they might also not fully correspond to the ideal types of such economies. In this regard, the VoC-approach can indeed be criticized for developing its liberal ideal type focusing too strong on high-technology regions and *underestimating* the broader spectrum of possible production models within this ideal type.

What the results on the UK regions further demonstrate is that within the group of LMEs as defined by VoC not only regions, but also whole national cases can deviate from this ideal type. The case of the UK shows clearly a *general underestimation of the state induced dimension* of knowledge transfer, what might basically be explained with the fact that VoC is basically oriented towards the USA when defining the liberal model, and in this especially towards a few extremely specialized high-technology regions, like *Silicon Valley* or *Route 128*. In the investigated regional knowledge transfer systems in the UK, however, the state resulted to be even of *high* relevance, above all in its role as a *catalyzer* of knowledge transfer processes, making the UK regions much more policy driven knowledge transfer systems than assumed. Yet, this role of the state in innovation processes came out to be very homogenous across the UK regions, above all, again, due to the strong dominance of central government in the UK's political system.

8.3 Traditional Industrial vs. Metropolitan Knowledge Transfer Settings

Other explanatory factors for regional variation are rooted in the distinct structural conditions shaping traditional industrial regions and metropolitan regions. The investigated traditional industrial regions all shared to a certain degree historical path-dependent and industrial trajectories leading to established practices and business structures dominated by SMEs, complemented by MNEs standing at the head of comparatively dense networks of customers and suppliers. They also shared institutional frameworks specifically tailored for the needs and the support of the particular industrial focus, as well as entrepreneurial mentalities of a commercial or technological community. The investigated metropolitan regions, on the other hand, are all characterized by a high density of institutional actors involved in knowledge transfer processes, more service oriented economic structures, more diversified sectoral structures, pronounced gateway-functions with strong inter-regional and international channels of knowledge transfer, control-functions with a high concentration of headquarters of larger multinational companies and central financial service institutions, as well as learning-functions with dense infrastructures of universities and research institutions being characterized by a strong focus on higher education and training.

Fig. 64: Cross-National Comparison of Traditional Industrial Regions

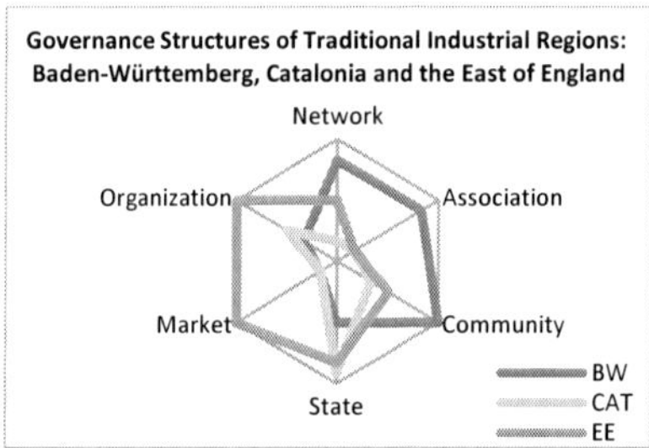

Source: Own Illustration

Fig. 65: Cross-National Comparison of Metropolitan Regions

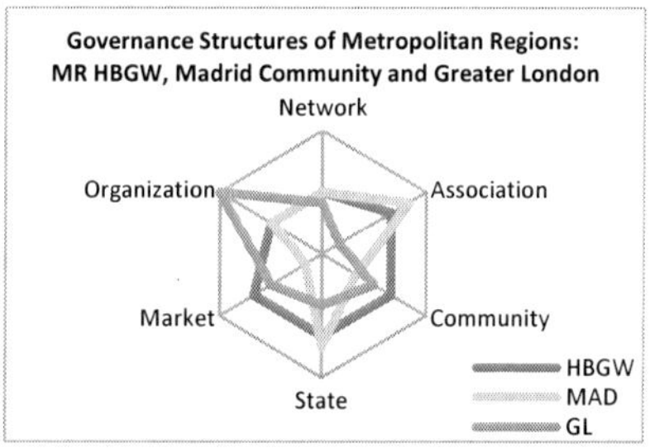

Source: Own Illustration

However, the analysis did not reveal cross-nationally consistent metropolitan or traditional industrial regional knowledge transfer models (see also Figures 64 and 65). Instead, the results demonstrate clearly that *across the national cases the traditional industrial regions correspond much more to the assumed national models than their metropolitan counterparts*. This finding is particularly interesting because it indicates that the previously made assumption that the sampled traditional industrial regions would come closer to their respective national models because of having paralleled national industrial trajectories and being object to processes of *institutional isomorphism* is basically confirmed, at least for these cases. With that, also the assumption that the investigated *metropolitan regions vary stronger from their national models* has been affirmed, indicating that their more recently created institutional structures, connecting major urban agglomerations with their regional surroundings, deviate more from national economic trajectories and are less shaped by institutional isomorphisms imposed by the national level. Not least, the *tailoring* of these regional knowledge transfer systems to the particular structural requirements of metropolitan environments resulted to play a major role in this regard. In addition, this can also be read as underlining the formulated critique towards the VoC-approach, which relies too much on single paradigmatic regional cases to describe whole national market economy models, systematically underestimating structurally diverging regional economies within these national models, mostly using cases of traditional industrial regions to develop its assumptions, and not sufficiently taking into account metropolitan and other types of regional economies.

Apart from this, however, the way the metropolitan regions differ from their respective national environment, is *not fully consistent* across the cases. One major explanatory factor in this regard is that also within the sample of the investigated metropolitan regions certain structural differences exist, of which the most important is the fact that the MR HBGW is a *polycentric* region, while the Madrid Community and Greater London have a more *monocentric* character, requiring and allowing for distinct regional governance systems of knowledge transfer. This is manifested above all in the MR HBGW's strong sub-spatial coordination elements in the form of regional growth concepts, its fragmented corporatist structures, and its antagonistic actors' constellations, requiring a comparatively centralized policy approach through the interplay of regional ministries, the *Innovationszentrum Niedersachsen* and the *NBank*, as well as the central coordination of intermediary actors within the *Innovationsnetzwerk Niedersachsen*. The more centralized regional structures of the Madrid Community, on the other hand, with its consistent region-wide coordination of knowledge transfer activities and its coherent corporatist structures allow for a more self-regulating, cluster and network based regional policy approach, especially through *Madrid Network* and the *madri+d* system. The monocentric Greater London region, then, leaves these coordination efforts to the myriads of economic actors, giving room for a strongly national level dominated regional policy with a main focus on creating adequate framework conditions and providing an overview on the hard to oversee metropolitan knowledge transfer landscape through the regional policy arms of LDA and the GLA.

In addition, this refers again to the fact that it is especially the respective country's political system with *different degrees of federalism and regional policy autonomy* to create regionally specifically tailored innovation policies, which accounts for cross-national differences among the metropolitan regions. Therefore, it is not surprising that in the German case of the MR HBGW regional variation is the strongest among the sampled metropolitan regions, due to the high degree of freedom to create specific regional policies in the German *Länder* (even if the MR HBGW lacks an own regional government), and the low degree of centralism exercised by national government. Less pronounced is the variation of the Madrid Community in the Spanish context, even if the region has considerable autonomy rights. Here, the analysis shows clearly a homogenizing effect rooting in the strong role exercised by central government and its strong need to intervene and to compensate for cross-regionally existing structural deficits. In the Greater London case, which has the lowest degree of policy autonomy, this homogenizing effect can be clearly traced back to the UK's pointedly centralized political system.

Apart from these differences, one cross-case consistent feature resulting from the sample of metropolitan regions is that they are all characterized by a disequilibrium between their dense and variegated university landscapes, which then are not fully reflected in the regions' innovation and knowledge transfer activities. All investigated metropolitan regions had university structures with strong *foci on higher education and training*, but less on applied research and technological development, limiting their transfer activities and their cooperation propensity with the business sector. In addition, also the *matching of research focus and sectoral economic focus* in these regions resulted to be lower than in the historically grown traditional industrial regions, being a barrier to more intensive interaction between business and academia. Creating more congruent structures in this regard, as well as enhancing the efforts to connect the economic actors to the rich metropolitan academic knowledge bases is, therefore, to be seen as one major challenge for policy and other actors involved in the creation and further development of metropolitan regions as places for innovation, learning and knowledge transfer.

Altogether, this demonstrates that investigating regional varieties of innovation systems must take into account the *distinct types of regional economies* and the *different structural conditions* characterizing them. Focusing predominantly on established industrial regions and taking them as paradigmatic cases to describe national systems as a whole underestimates strongly the potential of regional institutional variation being inherent in distinct types of regions. Especially metropolitan agglomeration areas resulted to be shaped by markedly diverging knowledge transfer systems, accounting for regional variation in national systems. This shows, on the one hand, that the recent movement to create these metropolitan regions through connecting major urban agglomeration areas with their regional surroundings gives considerable freedom to economic and policy actors to create particular "metropolitan governance structures" in the fields of innovation and knowledge transfer, which can take into account the specific structural requirements of this type of region. On the other hand, it also reveals a necessity to focus more strongly than before on these metropolitan regions when analyzing innovation and knowledge transfer systems in general, not least because they represent innovative structures of connecting urban knowledge economies with regional industry. This makes their innovative policy and governance models as specific occurrences of the knowledge based economy interesting future research objects.

8.4 Sectoral Foci as Drivers of Regional Variation

This study also provided evidence that a region's *sectoral structure* is one central explanatory factor for the concrete shape of its governance system of knowledge transfer. Regions dominated by traditional industrial sectors, with medium and medium-high technology levels, synthetic and highly cumulative knowledge bases and predominantly incremental modes of innovation, as it would be Baden-Württemberg and Catalonia, and, even if to a minor extend, the MR HBGW and the Madrid Community, show a comparatively higher significance of network based mechanisms of knowledge transfer, especially with regard to vertical supply chain networks and institutionally stabilized networks. This holds true for associational moderation as well, while these regions make particularly low use of the market based mechanisms and require at least medium-high intervention by the state. Regions characterized by New Economy sectors with high-technology levels, analytical and lowly cumulative knowledge bases and predominantly radical innovation processes, on the other hand, like the East of England, rely much more on organization and market based mechanisms of knowledge transfer, with strong network links between business and academia, but low use of associational intermediation and non-market coordination.

Going beyond these general insights into the interplay of sectoral structures and regional knowledge transfer systems, this study also focused on two specific industries representing the two previously discussed types of sectors: Automotive and biotechnology. This analysis also allows to draw conclusions on how these sectors adapt to or deviate from their respective regional frameworks, and to give some insights into cross-regionally and cross-nationally consistent patterns in the structures and mechanisms of knowledge transfer applied by these two sectors. In this regard, the analysis showed that these sector specific structures fit best to their regional environments where the overall sectoral structure is dominated by the respective type of sector, as it would be the case for automotive in Baden-Württemberg and for biotechnology in the East of England. In all other cases, in which the overall sectoral structure is different or less pointedly developed, the analysis revealed that the sectors vary considerably from their regional frameworks, showing *sector specific deviations* at least to a certain degree. And even in those cases in which these deviations are not realized because of *region-specific structural barriers*, the absence of these adaptations is acknowledged as major shortcoming for the knowledge transfer activities and the overall competitiveness of the respective regional sectors, which has been shown most clearly in the case of Catalonia.

This refers to certain functional needs for successful knowledge transfer processes in the two investigated sectors: Dense, enduring and institutionally stabilized horizontal and vertical business networks, cooperative R&D-activities, efficient and qualified associational intermediation, recruitment of qualified staff and adequate support structures provided by the state constitute the main components of knowledge transfer processes in the mature sector of automotive (Figure 66). Biotechnology, on the other hand, relies strongly on dense network links between business and academia, intensive fluctuation of highly qualified personnel, high start-up and spin-off rates, strong policy support, especially in the initial phases of business activities, specialized external consultancy, as well as intensive trade of IPRs in its knowledge transfer activities (Figure 67). Both sectors in all six investigated regions show these tendencies at least to a certain degree, even if they are sometimes confronted with insuperable structural barriers. In these cases, as it would be automotive in the UK regions and biotechnology in the German and Spanish regions, however, to some extend the sectors also make use of and adapt to the available institutional frameworks and structural conditions in the respective regions. The dominant patterns of these adaptations are, above all, *sub-sector specialization*, as it is the case with the niche-sector orientation of the East of England's automotive industry and the focus on medical technology in the MR HBGW's biotech sector, or *strong support by the distinct levels of the state*, as in the case of biotechnology in Madrid and Catalonia, but also in Baden-Württemberg.

Fig. 66: Cross-Case Comparison of Knowledge Transfer Structures in Automotive

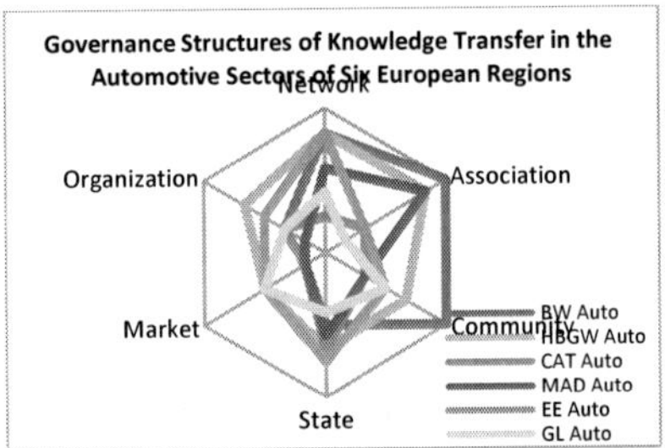

Source: Own Illustration

Fig. 67: Cross-Case Comparison of Knowledge Transfer Structures in Biotechnology

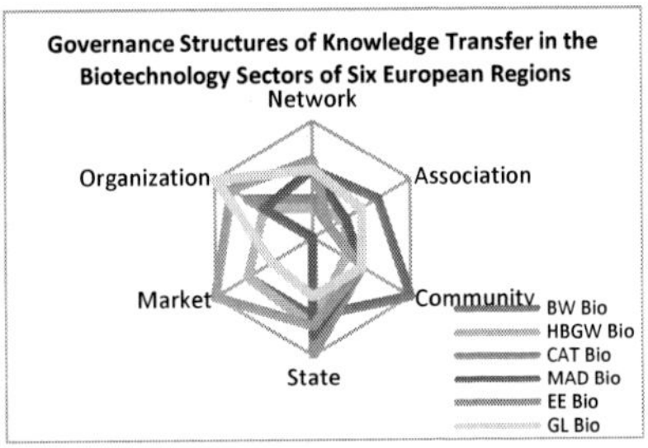

Source: Own Illustration

Overall, both tendencies have to be taken into account when looking at the sectoral dimension of knowledge transfer: *Functional requirements and distinct knowledge bases* in specific sectors account for regional variation, and these sectoral specialisms are, to a large degree, consistent across regions and across national economies. However, limiting the sectoral influence on regional variation, regional structural conditions and national institutional frameworks also exercise certain pressures on industrial sectors to adapt to their respective environment, to make use of the available institutional and support infrastructure, and to focus on business strategies being most compatible to their environments. *Strong state support* for specific "alien" sectors in inadequate environments to bypass these adaptation processes and to compensate for structural incompatibilities are, then, a third tendency accounting for regional variation. This is manifested in an increased importance of the state induced dimension of knowledge transfer in these sectors, especially in regional and national economies in which the state would otherwise play a less protagonist role, like, above all, in the German regions.

8.5 Towards a Varieties-Centered Research Agenda on Innovation and Knowledge Transfer

Apart from these antagonisms of traditional industrial regions versus metropolitan regions and of Old versus New Economy sectors, the analysis also provided further evidence for regional variation within national innovation and production systems, which also address some of the *conceptual shortcomings* identified for the *NIS- and the RIS-approach*. As the theoretical discussion in section 2.4.2 demonstrates, the *RIS-approach* indeed captures many of the central differentiating factors of innovation and knowledge transfer at the regional level in terms of *absolute* competitive advantage, although it still misses a clear focus on how these factors differ in a comparative perspective between regions within national systems and cross-nationally. This study, therefore, focused on the *comparative* institutional advantages of regional economies, confronting the main elements of absolute competitive advantage discussed in the RIS-approach with nationally and internationally comparative empirical evidence. The results show clearly that distinct *degrees of hierarchy or heterarchy*, as in the Spanish regions of Catalonia and Madrid, varying *degrees of dominance of one or more focal firms*, as in the German regions of the MR HBGW and Baden-Württemberg (or the absence of such firms like in Greater London), as well as the *strength of the cohesion of the commercial and technological community*, as in the UK regions of the East of England and Greater London are important explanatory factors of regional innovation systems varying from their respective national institutional frameworks. In addition, also differences in the ability of regions to offer *possibilities to establish informal and trust-based relationships*, and distinctly strong developed *regional cultural identities* through different historical development paths and experiences, resulted to be major drivers of regional variance.

However, the theoretical discussion also showed that both the RIS- and the NIS-approach are too *imprecise* when describing the processes of knowledge transfer and learning underlying innovation activities, not providing an encompassing and conclusive systematization of the institutional structures and concrete mechanisms through which knowledge is transferred in innovation processes. This study addressed these shortcomings in both approaches by developing a *heuristic model of knowledge transfer systems*, which has been proved to be highly valuable for *reconstructing single knowledge transfer systems*, as well as for *analyzing the modes of coordination* dominating these systems. In addition, this heuristic model makes the analysis of knowledge transfer systems compatible with other governance based institutionalist approaches like, above all, VoC, and allows with that for *within-national* as well as *cross-national comparisons* of knowledge transfer systems according to *clearly defined indicators and analytical dimensions.*

The suggestion arising from these results is, therefore, to include this heuristic model into the conceptual frameworks of the NIS- and the RIS-approach. With that, the concept of knowledge transfer in innovation processes would be lifted out of its implicitness and made more explicit with regard to the *concrete mechanisms and modes of coordination* shaping knowledge transfer processes at distinct geographical levels. This integration can be reached by looking at NIS and RIS as approaches defining innovation systems as *particular forms of social and institutional order.* As has been shown, up to now, the distinct dimensions of this institutional order are not sufficiently differentiated by these approaches. This is the particular contribution of the heuristic model: It *differentiates* knowledge transfer processes in innovation systems by the six ideal forms of social and institutional order, namely networks, associations, organizations, the state, the market and the community, and *ascribes* them concrete mechanisms following their particular *logic of coordination.* Therefore, as exercised in this study, also future studies on NIS and RIS should look at innovation systems in this differentiated way, not only when investigating knowledge transfer and innovation systems as a whole, but also when focusing on particular innovation related topics like sectoral innovation, financing, product development, institutional change, policy reform processes or innovation dilemmas, etc.

In addition, this integration would allow for the *weighting and differentiating of innovation systems* according to their respectively *dominant modes of coordination*, being an essential step for opening up especially the RIS-approach to a more *"varieties-centered" discussion* of innovation systems, and being with that a central building block of a *"Regional-Varieties-of-Capitalism-research-agenda"*, which this study demonstrates to be absolutely required not only for a further evolution of the NIS- and VoC-approaches, but also for a more precise understanding of economic, innovation and knowledge transfer processes per se. Up to now, this weighting is not sufficiently taken into account in the NIS- and the RIS-approaches, often disregarding regional and national variation across and within innovation systems, and *excluding the comparative perspective* from broad parts of the analyses. This is also true for both differentiations of RIS according to distinct knowledge bases and sectoral logics (ERIS and IRIS; see 2.4.2.5.2; Asheim/Gertler 2005; Cooke 2004; Cooke 2003), and attempts to create more governance oriented typologies (grassroots, network, *dirigiste;* see Cooke 2004: 10ff.; or territorially embedded, regionally networked or regionalized national innovation systems; see 2.4.2.5.1; Asheim/Gertler 2005: 300ff.; Cooke 2004: 10ff.; Asheim/Cooke 1998). Comparative studies basing on these approaches are hardly practicable and can be supposed to lead to imprecise results, since they *lack a clearly elaborated analytical model* and are not differentiated enough. Therefore, these approaches do not allow founded cross-case comparisons.

For the purpose of the integration of the heuristic model into the NIS- and RIS-approaches, thus, I suggest to apply a *concrete analysis scheme* when investigating national and regional systems of innovation (Figure 68), consisting of *six analytical steps*: First, the national or regional economy to be investigated must be *selected* basing on a *descriptive quantitative statistical portrait*, revealing its basic socio-economic indicators and structural specialisms with regard to the research interest. Second, the *central actors and institutional players* involved in knowledge transfer and innovation processes must be *identified* as units of analysis. Third, the investigated structures and mechanisms of innovation and knowledge transfer must be *classified* by the six analytical dimensions of institutional order constituting the *heuristic model.* Fourth, the particular national or regional institutional order of the innovation and knowledge transfer systems must be *reconstructed* as specific *configuration* of the six ideal types of institutional order. Fifth, the six distinct analytical dimensions of institutional order must be *weighted* according to their qualitative and/or quantitative relevance to detect the respectively *dominating modes of coordination* within the national and regional innovation systems. Sixth, national and regional innovation systems must be *compared* to investigate particular *comparative institutional advantages*, strengths and weaknesses, barriers, shortcomings and dilemmas as well as best practice models.

Fig. 68: Guide to the Comparative Analysis of Innovation and Knowledge Transfer Systems

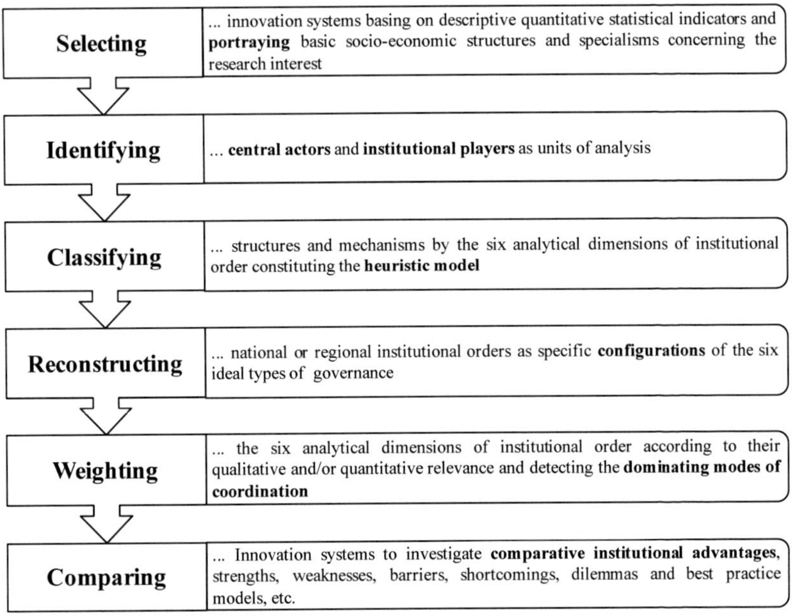

Source: Own Illustration

Using this analytical scheme, the heuristic model could work as the *consistent conceptual framework* which has been frequently demanded in the debate on NIS and RIS (e.g. Edquist 2005; Edquist 1997; Nelson 2008; Lundvall 2007; Audretsch/Lehmann 2005; Acs et al. 1999; Ortiz 2008), not only making distinct studies in the fields of innovation and knowledge transfer, but also both approaches and sectoral analyses *compatible* and *comparable*.

Altogether, the theoretical and empirical discussion in this study highlighted that national systems of innovation should by no means be conceptualized as monolithic blocks of institutions in the fields of innovation and knowledge transfer, solely offering *absolute* and nationally coherent competitive advantages to their economic actors. Rather, national innovation systems should be seen as *regionally variegated* sets of institutions and actors involved in the creation, storage, transfer and further development of knowledge, which are embedded in specific models of social order creating particular governance systems of innovation and knowledge transfer, from which also *comparative* institutional advantages arise. In this regard, also the approach of bringing together the NIS-concept and the VoC-approach and regarding national innovation systems as sub-systems of broader defined national production systems and political economies resulted to be well suited to expand the NIS-perspective to a more encompassing, systemized and varieties-centered perspective. Using the VoC-approach's institutionalist and governance centered theoretical implications for the analysis of national innovation systems came out to be helpful to specify the involved actors as well as the distinct institutional dimensions and coordination patterns of innovation and knowledge transfer, allowing with that to assign specific national innovation systems to particular varieties of market economy models and their respective modes of coordination.

8.6 Lessons for the VoC-Approach

This already refers to the central function of the VoC-approach when addressing the question of regional variation in national systems of innovation, knowledge transfer and production. Three major theoretical implications for the VoC-approach can be derived from the analyses conducted in this study: First, the critical discussion on this approach as well as the empirical case studies disclosed that it *overestimates its assumptions of institutional coherence and complementarity* strongly, not taking sufficiently into account factors driving regional variation within national models. National market economies and innovation systems are to be seen as *regionally variegated institutional models*, which allow for regionally diversified adaptations to national institutional configurations, and which *differ according to the degree they allow for this variation.*

Second, the VoC-approach differentiates national market economy models according to their respectively best supported mode of innovation, might it be incremental or radical or mixed. However, taking into account the central role exercised by innovation in modern knowledge based economies, most contributions to the VoC-approach remain *too vague* with regard to these processes. In this connection, the integrated discussion of the approaches of VoC and NIS in this study resulted to be *mutually fructifying* for both approaches, helping to overcome many of their respective conceptual shortcomings. Based on this, and complementing the already claimed necessity for an explicitly regional-VoC-centered research agenda, the need for the development of a "*national varieties of innovation systems*" perspective can be derived, focusing more strongly than before on the institutional determinants of innovation and the comparative institutional advantages arising from them.

Third, from this stronger focus on innovation also arises a need to further concentrate research in this field on the *underlying processes of knowledge transfer*. To the extend in which developed societies become knowledge societies, and market economies become knowledge based economies, and in which the innovative capacity of single market economies determines its international competitiveness strongly, an approach intending to describe the comparative economic advantages of nations must be strongly aware of the processes of knowledge transfer, which represent the basis of all innovation. Therefore, a stronger integration of the approaches of NIS/RIS and VoC, as exercised in this study, assigning the coordination modes and governance structures identified by VoC to national and regional knowledge transfer systems, and by including the detailed analysis of institutional innovation and knowledge transfer infrastructures into the VoC-framework, appears to be a valuable future research agenda to extend and deepen the understanding of capitalist diversity through addressing the fundaments of diverging comparative advantages, as well as their national and regional variation.

Overall, the empirical evidence provided by this study demonstrates clearly that the specific institutional configurations shaping the governance of national political economies also influence strongly the respectively dominating modes of knowledge transfer. These configurations create strong incentives and support structures to choose specific knowledge transfer channels, but also constrain the intensive use of others. Varieties of capitalism are, therefore, also to be seen as *varieties of knowledge transferring and learning economies*, and they are *regionally variegated sets of occurrences of national market economy models*. Contrasting many other positions, diagnosing and predicting a dissolution of these varieties after the fall of Communism and in the course of an accelerated globalization processes, this study presented extensive evidence for the persistence of these differences, as well as for the functional and structural requirements behind them. However, a general trend of liberalization of regional and national systems of innovation and production induced by a necessity to adapt to a globally dominant and outperforming liberal model, including a stronger orientation towards market and especially financial market based mechanisms, unstable labor markets or individualized actors' relations, could not be observed. Rather, a general trend towards a stronger academization and scientific knowledge orientation in innovation processes has been detected, indicating consistent, but variegated and, above all, distinctly successful, ways into the knowledge based economy. Especially for actors concerned with reform processes in the light of the current global economic and financial crisis this differentiated view on national and regional varieties of knowledge transfer systems can be highly valuable to find adequate solutions for securing sustainable competitiveness.

8.7 The Paths to Take—Policy Implications and Outlook

Which path can these reform processes take? And what can especially policy makers, but also other actors involved in innovation and knowledge transfer, learn from this analysis? Some major additional policy implications complementing those already formulated in section 8.1 and a brief outlook on possible future developments shall conclude this book.

Looking at the six investigated regions separately and starting with the German regions, the major policy challenge for Baden-Württemberg as a successful and efficient regional knowledge transfer system, building on historically grown industrial innovation and cooperation structures, is to further connect its abundant research and scientific knowledge base to its main industrial sectors. As shown, there already exists a large number and broad variety of cluster and cooperation initiatives involving academia, but still, all experts stated subjectively that there would be high potential for many further productive cooperations, which is, not least, underlined by the high percentage of both business and academic actors still not cooperating at all. One major task for the Baden-Württemberg innovation policy actors in this regard is certainly the reorganization of the technology transfer agencies at universities and research institutions. These should be given a more proactive and less administrative mission as central *marketplaces* for exchange processes between business and academia, and should with that be converted into central points of access to Baden-Württemberg's rich scientific knowledge base. In addition, also their knowledge and innovation management capacities should be professionalized, they should be better equiped with highly qualified and experienced staff, and their funding should be increased to create measures to actively approach science-averse actors like especially SMEs and to intensify their marketing efforts to become more visible. Also the stronger inclusion of these agencies into wider networks of intermediary actors to leverage their impact appears to be promising in this regard, as especially the policy practice models of the *Innovationsnetzwerk Niedersachsen* and the *madri+d* system showed.

Furthermore, also business start-up support should be readdressed intensively by regional policy actors: Many initiatives and programs do already exist, and they provide solid support for entrepreneurial activities in organizational and management issues. But they share one major problem with regional banks and other financial institutes: The qualified assessment of technological and high level service and process innovation is in most cases not given, leading to major bottlenecks in the provision of seed- and venture-capital and with that in business creation activities. Parts of the hesitant venture-capital provision and, thus, relatively low start-up-rates in the investigated German regions root in these deficient assessment capacities. The assessment services provided by banks and regional as well as national funding institutions must therefore be professionalized and upgraded not only to avoid misinvestments, but especially to avoid the wasting of good ideas and projects or their realization elsewhere. Regional actors should both jointly and individually reorganize their assessment services, and also consider to stronger coordinate and specialize their assessment and funding activities, even across private-public boundaries. Not least, this will require a joint effort to create and maintain a professional business angel network.

In sum, it is to be expected that Baden-Württemberg's regional knowledge transfer system will gradually evolve in the described and suggested direction and be able to overcome the few barriers currently still existing. The region has all the potential to preserve its advantages as an innovative and learning industrial region and even to expand them when making more use of its singular scientific knowledge base. One major challenge in this regard will be the further management of structural change in its core industrial sectors, resulting from the expectable further relocation of employment in production and manufacturing, and to a strong upgrading of industrial research and development as well as related services fields. With a stronger interaction of all parts of the business sector with academia building on diversified channels of knowledge transfer the region should be able to set itself on top in the competition of the most innovative industrial regions of the world. As can already be seen by the analysis at hand, these changes must not necessarily lead to major changes in the coordinated character of Baden-Württemberg's regional knowledge transfer system, but rather to slight adaptations towards a stronger role of academic knowledge in innovation processes.

Also for the MR HBGW, the aspects concerning business start-up support and the professionalization in the assessment of seed- and venture-capital provision constitute major challenges. But even more important for this region is a patient continuing of its region-building efforts creating more systemically condensed structures in the fields of innovation and knowledge transfer. In the long term the goal should be to convert the region into a space which strongly builds on its multipolar roots in the fields of innovation and knowledge transfer, but develops a strong awareness of being part of a common systemic context. The MR HBGW's efforts and programs to make regional actors and structures more visible, and to stronger connect them in joint interaction and transfer contexts as well as concrete cooperation projects should even be reinforced and be accompanied by encompassing marketing activities. Together with the efficient and positively evaluated centralized Lower Saxon innovation policy through the *Innovationsnetzwerk*, the *Innovationszentrum* and the *NBank*, this will allow the region to grow closer together and to make stronger use of its variegated industrial and academic structural potentials. This will require a continuation of the predominantly top-down approach realized by regional policy in the middle-term, which, however, should successively be supplemented by setting stronger incentives for self-coordination among the actors in the long term. One major substantial topic in this regard would be addressing the mismatch of the region's academic focus on life sciences, which is not even rudimentary reflected in its business landscape. Major policy programs and measures should be developed to give strong impulses to transfer this rich knowledge base in life sciences into industry and to the creation of new businesses-models from it. The *Gesundheitsregion*-program can only be a right start in this regard, and should be followed by many others.

Furthermore, regional policy should, together with actors from the business sector, develop new ideas to reorganize the Lower Saxon chambers of industry and commerce. Their current particularized and separated structure should be overcome, and reasons for antagonistic action from the past should be left behind to create future structures of intermediary support in innovation and knowledge transfer which are stronger coordinated at least across the Metropolitan Region, if not Lower-Saxony-wide. The analysis showed clearly that a more active role and coordination of the chambers in these fields is not only requested intensively by the practitioners, but would also be one important contribution to the overall functioning and efficiency of the regional knowledge transfer system, especially when taking into account the region's structural particularities.

One final recommendation that can be derived from the case study would be to intensify the current efforts to make the region a more attractive place to work and life for high potentials and talents from other parts of the country and the world to foster the talent driven channels of knowledge transfer. The perception of the MR HBGW as a pleasurable place to move is still far beyond its real qualities, not least because of insufficient self-marketing activities. One important step into the right direction would be the support of firm's recruiting activities by policy programs, for example through voucher schemes. Another would be continued marketing efforts addressing the promotion of the living conditions in the Metropolitan Region. The fostering and marketing of cultural activities by the Metropolitan Region highlighting its rich cultural life and heritage, as well as natural virtues, can be seen as one good start in this regard, as would also be the "Niedersachsen"-campaign by the Lower-Saxon government.

Therefore, the MR HBGW has to realize an ambitious agenda in the upcoming years. The process of constructing stable and sustainable regional structures in the multipolar Metropolitan Region will continue to be accompanied by problems, barriers and conflicts, and policy actors with patience and clear vision will continuously be required to bring this project forward. However, following the experts' perspective, this effort is worth it and for all subspaces of the Metropolitan Region the chances of further cooperation and growing together especially in the fields of innovation and knowledge transfer are extraordinarily high. The region has excellent potentials in both academia and its core industrial sectors, and only bundling and connecting these competencies will allow the region to gain further competition and market shares. Taking into account current strategical considerations in the region, one can expect that this will not only allow the region's competence in automotive and mechanical engineering to be further expanded, but also in making stronger industrial use of its excellence in the life sciences and other fields and diversifying with that its sectoral portfolio. The region's knowledge transfer system can, therefore, be expected to especially evolve in three directions: It will expand the role of academic knowledge transfers, it will construct further and denser network and intermediary structures, and, finally, it will have to continue to rely strongly on national and regional policy to manage these adaptation processes.

Turning to the Spanish regions, one major policy recommendation arising from the Catalan case study is to consequently continue the reorientation of the Catalan innovation and knowledge transfer policy towards initiating single technology oriented cooperation projects and networks (NUCLIS program) and towards making transfer oriented regional scientific institutes visible through labeling and supporting them, as well as to professionalize their interaction with the business sector within the framework of the TECNIO program. Given the (even for the Spanish case) extraordinarily pointed individualistic attitude predominating the Catalan business sector and the strong mental barriers between the business sector and the scientific world, these policy measures can be interpreted as adequate first steps in transmitting an innovation and collaboration culture to the regional actors and to motivate them to reinforce their knowledge transfer activities.

A second recommendation relates to the chambers and associations: Regional policy must find solutions together with the business sector to reorganize these corporatist agents in a way they can actively contribute as intermediaries to the management of knowledge transfer and innovation and respective policies in the region. This includes a stronger coordination between locally divided sub-units, the development of an explicit innovation and knowledge transfer agenda for all chambers and involved business associations, professional equipment with qualified staff and funding, as well as a stronger coordination with other intermediary agents and especially with regional policy. Taking into account Catalonia's structural deficiencies these corporatist actors could play a central role in transmitting a more transfer oriented culture and regional policy into their member organizations. In this point, they should be supported by the OTRIS, which shall be reorganized and upgraded to work in a more proactive manner as gateways between business and academia, and which should be included more strongly than now into the implementation of the NUCLIS and TECNIO policy measures mentioned above.

For the case of the Madrid Community, it can be recommended to further pursue the started reorganization of innovation and knowledge transfer policy with the *madri+d* system and *Madrid Network*. The taken focus on cluster and network-building, and, therefore, on establishing long term, rather than project oriented, cooperation structures, especially through the *madri+d* system as central interface, technological marketplace and assessing body for connecting firms and scientific actors, and *Madrid Network* as cluster management and coordination facility for the regional clusters and scientific parks, appears to be well suited to bring an advance for the region's knowledge transfer performance. However, it might be questioned if this Madrilenian cluster and network policy strategy might presently not be too optimistic and advanced for its strongly individualistic business sector. Regional policy actors and cluster and network managers should, therefore, strongly be aware of the implementation process of their policies, and ensure that the economic grass roots are sufficiently enabled to participate and to be included into these structures. For that goal, other, more basic measures might be necessary to supplement *madri+d* and *Madrid Network*, as it would be conferences, open days, support in proposal writing, exchange programs to bring graduate students into firms e.g. through regional voucher programs in addition to the expiring TORRES QUEVEDO program, assessment programs for SMEs to indentify where they could make use of scientific knowledge and staff, etc. Regional policy can and should rely in these basic work and implementation measures on the Madrid Chamber and CEIM, which appear to be powerful, competent and motivated partners and are strongly rooted into the region's business base.

Another recommendation arises from the Madrid Community's deficiencies as entrepreneurial breeding ground for business start-ups and spin-offs, and as a fluent job market for scientific and technical staff. One major reason for this might be that Madrid's rich metropolitan university and research landscape is mainly oriented towards basic research and, above all, higher education, and less on applied research and technology development. As a consequence, it not fully finds its expression in elevated innovation and transfer activities, which is, not least, reflected in the absence of any technological centers in the region. Therefore, regional policy should reconsider to still invest in technological centers as places of technology transfer and development, and to slightly reorient the region's academic focus in the long term more towards technology related fields and especially towards those which are or can be reflected in the region's business sector. Taking into account the dramatic unemployment among young Spanish graduates in the whole country, this also constitutes an overarching challenge for national and regional policy actors, not only with regard to innovation and knowledge transfer.

Overall, the success of these policy measures and recommendations in both Spanish regions will require a considerable shift of mentality as well as business and academic culture. The expectation is therefore, that both regions, but above all Catalonia might have major difficulties to overcome their barriers in the fields of innovation and knowledge transfer in the short term intrinsically, especially in times of the current deep economic and financial crisis. However, this dramatic crisis also opens up new opportunities for reform processes in this regard: On the one hand, it might lead to the insight that one important step towards a strengthening of the Catalan and the Madrilenian competitiveness is a reorientation of all regional actors towards learning and innovation. On the other hand, also reform pressures arising from compensation measures for financial support by other EU-countries might be able to extrinsically initiate substantial shifts in this regard. The expectable direction of these reform processes is towards the more successful European models providing this financial support under strict conditions, and, with that, especially towards the German coordinated market economy model. Not least, many interviewed experts in both Spanish regions also intrinsically recognized this need and also the opportunities in orienting themselves stronger towards the structures in the German regions, which are accepted as good practice models and reference points.

Focusing at the UK regions, for the case of the East of England one policy recommendation can clearly be formulated when looking at the analysis above: The path towards regional devolution policies to bring national policy closer to the English regions through the RDAs like EEDA, Government Offices for the Regions and Regional Assemblies has consistently been evaluated as positively and effective, and should, therefore, be continued. Nevertheless, the new national government and actual political developments in the UK are clearly moving towards a different direction. With the shift of government from *Labour* to the coalition of

Tories and *Liberal Democrats* in May 2010, the central pillars of regional devolution policy have been successively taken back, because of both political and financial rationales, giving room to a reinforced re-localization of policy. As a consequence, the *Government Offices* for the regions have been closed in March 2011, *Regional Assemblies* already in April 2010 and the RDAs including *EEDA* will stop their operations in March 2012 and will be presumably be closed in summer of that year. It can only be recommended to reconsider this decision, since this study showed clearly that regional actors in the East of England strongly wish to receive more autonomy to develop regional policy instruments specifically tailored to the particular needs of the region and to transform parts of central government support into more region specific tools taking into account the region's individual structural and economic characteristics. Regional actors indentified clearly the regional structural level to be able to bring the East of England's already successful knowledge transfer system forward. For the time before new attempts of regional devolution policy processes can be realized, regional actors should lobby national government for a stronger regional tailoring of national innovation and knowledge transfer policy, as well as for a stronger regional coordination of the respective local policies.

One interesting aspect in this regard is the national policy shift towards recovering parts of the UK's industrial sector as one consequence from the current economic and financial crisis. It is to be expected that also for the East of England this strategy could be a chance to reconstruct and further develop some of its existing but mostly incomplete and currently declining value added chains in traditional industrial sectors. The challenge will be to involve the region's strong research base in these fields into the reconstruction of these value added chains, and policy should accompany and manage this process within a clearly defined strategic framework which, in its regional dimension, is still to develop. However, this re-industrialization strategy can only be seen as a complementing factor to the regions focus on high technology, especially because of the fact that its innovation and knowledge transfer system is oriented clearly towards the latter and has singular competitive strengths which the East of England should continue to expand.

Turning to the Greater London case, also this region will be affected by the policy shifts described before, including the closure of the LDA in March 2012. However, since the intention is to fold most of the LDA's functions concerning innovation and knowledge transfer into the organizational structures of the Mayor of London and the Greater London Authority, it is to be expected that the impact of these changes on Greater London might be lower than on other UK regions, since at least one essential part of regional policy autonomy will be preserved. The recommendation to regional policy actors is in any case to continue their successful path of creating explicit regional innovation and knowledge transfer policies for Greater London. The existence of a regional government, represented by the Mayor of London, which, in dense collaboration with the LDA, creates a regional policy level with far reaching agenda and strategy setting autonomy, allowing to adapt more policy instruments to the specific needs of Greater London and its metropolitan structural conditions, has been evaluated throughout positively by the experts. The same holds true for the central role the LDA assumes in co-developing and, above all, implementing Greater London's policy agenda. Therefore, it will be to observe and be left to future research which consequences the new developments will have for the innovation and knowledge transfer policy in Greater London, and to which degree the LDA's successful work can be continued by the remaining institutions.

One major challenge in this regard will be the centralization and general reorganization of intermediaries and support institutions in the fields of innovation and knowledge transfer. Since the analysis above demonstrated clearly that the myriads of actors in these fields in Greater London are hard to oversee and that many barriers in the system do arise from demand and existing supply simply not finding together, the recommendation must be to stronger integrate and coordinate these activities within one common systemic context. The *Innovationsnetzwerk* in the MR HBGW could be a good practice model in this connection, and the Mayor of London/LDA should initiate the creation of such a "London Innovation Network." This will also require to establish a respective network management unit, which could also be located at the Mayor of London, or be delegated to an independent agency, as is the case with the *Innovationszentrum* in the MR HBGW.

Another challenge will be a reinforced opening process of Greater London's universities and research institutes for more applied research and technological development activities, to make a more efficient use of its impressive academic knowledge base. Due to its structural conditions as a metropolitan area described above, the recommendation cannot be to complete more of its only rudimentary developed industrial value added chains, since Greater London's future is not to be seen in a traditionally reindustrialized innovation and knowledge transfer system, but rather in innovative business and financial services as well as creative industries etc. Yet, both, academic and business actors should continue the path of cross-regional cooperations with other UK regions, in which Greater London actors can bring in their excellence, then complemented by the industrial structures, infrastructures and competences of other regions. The discussed examples of *One Nucleus* in the life sciences and the different automotive clusters are good practice models in this regard, which should be followed by other sectors and actors in Greater London as well.

Table Credits

Figure Credits

Bibliography

Abramson, H. Norman, José Encarnaçao, Proctor P. Reid and Ulrich Schmoch (eds.) (1997): *Technology Transfer Systems in the United States and Germany. Lessons and Perspectives.* Washington, D.C.: National Academy Press.

Abreu, Maria, Vadim Grinevich, Michael Kitson and Maria Savona (2008): *Absorptive Capacity and Regional Patterns of Innovation.* UK Department for Innovation, Universities and Skills (DIUS).

ACC1Ó (Agència de Suport a l'Empresa Catalana) (2010): *La Inversion en R+D de les 50 Empreses més Grans de Catalunya.* Barcelona: Departament d'Innovació, Universitats i Empresa: ACC1Ó.

Acs, Zoltan J., Felix R. Fitzroy and Ian Smith (1999): High Technology Employment, Wages and University R&D Spillovers. Evidence from US Cities. In: R. Henderson and A. Jaffe (eds.): *Economics of Innovation and New Technology 8*, Special Issue on R&D Spillovers: 57–78.

ADT (2011): Standorte. Bundesverband Deutscher Innovations-, Technologie- und Gründerzentren e. V. [Source: http://www.adt-online.de/standorte.html; last access: 01.02.2011].

AiF (Allianz Industrie Forschung) (2010): *AiF Jahresbericht 2009.* Köln: AiF.

Al-Laham, Andreas (2003): Organisationales Wissensmanagement. München: Vahlen.

Albert, Michel (1993): *Capitalism Against Capitalism. How America's Obsession with Individual Achievement and Short-Term Profit Has Led It to the Brink of Collapse.* New York: Four Walls Eight Windows.

Allen, John (2000): Power/Economic Knowledge. Symbolic and Spatial Formations. In: Bryson, John, Peter Daniels, Nick Henry (eds.): *Knowledge, Space, Economy.* London: Routledge: 15–33.

Allen, Matthew (2004): The Varieties of Capitalism Paradigm. Not Enough Variety? In: *Socio-Economic Review 2*: 87–108.

Amable, Bruno (2003): *The Diversity of Modern Capitalism.* Oxford: Oxford University Press.

— (2000): Institutional complementarity and diversity of social systems of innovation and production. In: *Review of International Political Economy 7*: 645–687.

Amable, Bruno, Ekkehard Ernst and Stefano Palombarini (2005): How do Financial Markets Affect Industrial Relations. An Institutional Complementarity Approach. In: *Socio-Economic Review 3*: 311–330.

Amable, Bruno and Stefano Palombarini (2009): A Neo-Realist Approach to Institutional Change and the Diversity of Capitalism. In: *Socio-Economic Review 7 (1)*: 123–143.

Amin, Ash and Nigel Thrift (1992): Neo-Marshallian Nodes in Global Networks. In: *International Journal of Urban and Regional Research 16*: 571–587.

Andersen, Esben Sloth and Lundvall, Bengt-Åke (1997): National Innovation Systems and the Dynamics of the Division of Labor. In: Edquist, Charles (ed.): Systems of Innovation. Technologies, Institutions and Organizations. London. Pinter: 242–265.

Archibugi, Daniele, Jeremy Howells and Jonathan Michie (1999): Innovation Systems in a Global Economy. In: *Technology Analysis & Strategic Management 11 (4)*: 527–539.

Archibugi, Daniele and Simona Iammarino (2010): The Globalization of Technology and National Policies. In: Archibugi, Daniele and Lundvall, Bengt-Åke (eds.): *The Globalizing Learning Economy*. Oxford University Press, Oxford: 111–126.

Arrow, K. J. (1962): The Economic Implications of Learning by doing. In: *Review of Economic Studies 29 (3)*: 155–173.

ASEBIO (Asociación Española de Bioempresas) (2010): *Informe Anual 2009*. Madrid: ASEBIO.

Asheim, Bjørn T. (2002): Localised Learning, Innovation and Regional Clusters. In: Mariussen, Åge (ed.): *Cluster Policies—Cluster Development? (Nordregio Report 2001: 2)*: Stockholm: 39–58.

— (1998): Territoriality and Economics. On the Substantial Contribution of Economic Geography. In: Jonsson O. and L.-O. Olander (eds.): *Economic Geography in Transition. The Swedish Geographical Yearbook 74*. Lund: 98–109.

Asheim, Bjørn T. and Lars Coenen (2006): Contextualizing Regional Innovation Systems in a Globalizing Learning Economy. On Knowledge Bases and Institutional Frameworks. In: *Journal of Technology Transfer 31*: 163–173.

Asheim, Bjørn T. and Philip Cooke (1998): Localised Innovation Networks in a Global Economy. A Comparative Analysis of Endogenous and Exogenous Regional Development Approaches. In: *Comparative Social Research 17*: 199–240.

Asheim, Bjørn T. and Meric S. Gertler (2005): The Geography of Innovation. Regional Innovation Systems. In: Fagerberg, Jan, David C. Mowery and Richard R. Nelson (eds.): *The Oxford Handbook of Innovation*. Oxford/New York: Oxford University Press: 291–317.

Asheim, Bjørn T. and Arne Isaksen (2002): Regional Innovation Systems. The Integration of Local 'Sticky' and Global 'Ubiquitous' Knowledge. In: *Journal of Technology Transfer 27*: 77–86.

Audretsch, David B. (2003): Globalization, Innovation and the Strategic Management of Places. In: Bröcker, Johannes, Dirk Dohse, Rüdiger Soltwedel (eds.): *Innovation Clusters and Interregional Competition.* Berlin, Heidelberg: Springer: 11–27.

Audretsch, David B. and Maryann P. Feldman, (1996): R&D-Spillovers and the Geography of Innovation and Production. In: *American Economic Review 86*: 630–640.

Audretsch, David B. and Erik E. Lehmann (2005): Does the Knowledge Spillover Theory of Entrepreneurship Hold for Regions? In: *Research Policy 34*: 1191–1202.

Audretsch, David B. and Paula E. Stephan (1996): Company-Scientist Locational Links. The Case of Biotechnology. In: *American Economic Review 86 (3)*: 641–652.

Audretsch, David B. and Roy Thurik (2001): What Is New About the New Economy? Sources of Growth in the Managed and Entrepreneurial Economies. In: *Industrial and Corporate Change 19*: 795–821.

AXESOR (2009): *Estudio sobre el Sector de Automoción.* Gabinete de Estudios Económicos. Madrid: Axesor.

Bacaria, Jordi, Susana Borràs Alomar and Andrea Fernàndez-Ribas (2004): The Changing Institutional Structure and Performance of the Catalan Innovation System. In: Cooke, Philip, Martin Heidenreich and Hans-Joachim Braczyk (eds.): *Regional Innovation Systems. The Role of Governance in a Globalized World (2nd edition).* London/New York: Routledge: 63–90.

Balconi, Margherita, Stefano Brusoni and Luigi Orsenigo (2010): *In Defence of the Linear Model. An Essay.* In: Research Policy 39: 1–13.

Bathelt, Harald (2003): Geographies of Production: Growth Regimes in Spatial Perspective 1—Innovation, Institutions and Social Systems. In: *Progress in Human Geography 27 (6)*: 763–778.

BBSR (2010): Regionales Monitoring 2010. *Daten und Karten zu den Europäischen Metropolregionen in Deutschland.* Bonn, Mannheim: Bundesinstitut für Bau-, Stadt- und Raumforschung (BBSR) im Bundesamt für Bauwesen und Raumordnung (BBR); Initiativkreis Europäische Metropolregionen in Deutschland (IKM).

Beckenbach, Frank (2007): Akteurberfragungen als Grundlage für die Erfassung von Innovationsprozessen—Möglichkeiten und Grenzen. In: Hof, Hagen and Ulrich Wengenroth: *Innovationsforschung. Ansätze, Methoden, Grenzen und Perspektiven.* Hamburg: LIT: 55–70.

Berger, Peter L. and Thomas Luckmann (1966): *The Social Construction of Reality.* Garden City: New York.

Berger, Suzanne and Ronald Philip Dore (1996): *National Diversity and Global Capitalism.* Cornell Studies in Political Economy. Ithaca, NY: Cornell University Press.

Bezirksregierung Braunschweig, Regionale Entwicklungsagentur für Südostniedersachsen e.V. and Zweckverband Großraum Braunschweig (eds.) (2001): *Regio-*

nales Entwicklungskonzept für Südostniedersachsen. Reson-Report Band 4. Braunschweig: Reson.

BioCat (Catalonia BioRegion Foundation) (2011): About BioCat. Driving the Catalan Biocluster. [Source: http://www.biocat.cat/en/about-biocat; last access: 25.10.2011].

BioCat (2009): *BioCat Report on the State of Biotechnology, Biomedicine and Medical Technology in Catalonia.* Barcelona: BioCat.

BIOPRO (2011): Das Biotechnologie und Life Sciences Portal Baden-Württemberg. BIOPRO Baden-Württemberg GmbH. [Source: http://www.bio-pro.de; last access: 01.02.2011].

— (2010): *Gründerleitfaden. Stationen auf dem Weg in die Selbstständigkeit.* Stuttgart: BIOPRO Baden-Württemberg GmbH.

BioRegioN (2010): *Life Science-Report 2009—Ergebnisse & Trends.* Hannover: Geschäftsstelle der Landesinitiative Life Sciences Niedersachsen—BioRegioN (c/o Corvay GmbH).

— (2009): *Life Science-Report Niedersachsen 2008.* Hannover: Geschäftsstelle der Landesinitiative Life Sciences Niedersachsen—BioRegioN (c/o Corvay GmbH).

Biotechnologie.de (2010): *The German Biotechnology Sector 2010.* Berlin: Biotechnologie.de.

— (2007): *Zahlen auf einen Blick. Biotechnologie in Deutschland.* Berlin: Biotechnologie.de.

BIS (Department for Business Innovation and Skills) (2011): Automotive Industry. Europe's Most Diverse and Dynamic Automotive Industry. [Source: http://www.bis.gov.uk/policies/business-sectors/automotive, last access: 15.11.2011].

— (2010a): Small and Medium-Sized Enterprise (SME). Statistics for the UK and Regions 2008. BIS. [Source: http://stats.bis.gov.uk/ed/sme; last access: 10.05.2011].

— (2010b): Manufacturing in the UK. An Economic Analysis of the Sector. BIS Economics Paper No. 10A. BIS. [Source: http://www.berr.gov.uk/Policies/economics-statistics/economics; last access: 10.05.2011].

— (2010c): Strength and Opportunity. The landscape of the medical technology, medical biotechnology and industrial biotechnology sectors in the UK. London: BIS. [Source: http://www.bis.gov.uk/assets/biscore/business-sectors/docs/s/10-p90-strength-and-opportunity-bioscience-and-health-technology-sectors.pdf; last access: 15.11.2011].

Blättel-Mink, Birgit (2006): *Kompendium der Innovationsforschung.* Wiesbaden: VS.

Block, Fred and Matthew R. Keller (2009): Where Do Innovations Come From? Transformations in the US Economy, 1970–2006. In: *Socio-Economic Review 7*: 459–483.

Blotevogel, Hans Heinrich (2002): Deutsche Metropolregionen in der Vernetzung. In: *Informationen zur Raumentwicklung 6/7*: 345–351.

BMBF (2010): *Deutschlands Spitzencluster—Germany's Leading-Edge Clusters.* Berlin: Bundesministerium für Bildung und Forschung.

— (2009): *Forschung und Innovation für Deutschland. Bilanz und Perspektive.* Berlin: Bundesministerium für Bildung und Forschung.

— (2005): *BioRegionen in Deutschland. Starke Impulse für die nat. Technologieentwicklung.* Berlin: Bundesministerium für Bildung und Forschung.

Bogner, Alexander and Wolfgang Menz (2005): Das theoriegenerierende Experteninterview. Erkenntnisinteresse, Wissensformen, Interaktion. In: Bogner, Alexander, Beate Littig and Wolfgang Menz (eds.): *Das Experteninterview. Theorie, Methode Anwendung.* Wiesbaden: VS: 33–70.

Boschma, Ron A. (2005): Proximity and Innovation. A Critical Assessment. In: *Regional Studies 39 (1)*: 61–74.

Bourdieu, Pierre (1971): Intellectual Field and Creative Project. In: Michael K. D. Young (ed.): *Knowledge and Control. New Directions for the Sociology of Education.* London: Collier Macmillan: 161–88.

Boyer, Robert (2005): How and Why Capitalisms Differ. *MPIfG Discussion Paper 05/4.* Cologne: Max Planck Institute for the Study of Societies.

Bozeman, Barry (2000): Technology Transfer and Public Policy. A Review of Research and Theory. In: *Research Policy 29*: 627–655.

Braczyk, Hans-Joachim, Philip Cooke and Martin Heidenreich (eds.) (1998): *Regional Innovation Systems.* London: UCL-Press.

Brandt, Arno, Stefan Krätke, Claudia Hahn and Renate Borst (2008): *Metropolregionen und Wissensvernetzung—Eine Netzwerkanalyse innovationsbezogener Kooperationen in der Metropolregion Hannover-Braunschweig-Göttingen.* LIT: Münster.

Brüsemeister, Thomas (2008): *Qualitative Forschung. Ein Überblick.* Wiesbaden: VS.

Buesa, Mikel (2006): El Sistema Nacional de Innovación en España. In: *Revista de Investigación en Gestión de la Innovación y Tecnología.* Madri+d.

— (2004): *El Sistema Regional de Innovación de la Comunidad de Madrid.* Madrid: Consejería de Educación de la Comunidad de Madrid.

Bundesagentur für Arbeit (2011): Beschäftigungsstatistik der Bundesagentur für Arbeit. [Source: http://statistik.arbeitsagentur.de/Navigation/Statistik/Statistik-nach-Themen/Arbeitsmarkt-im-Ueberblick/zu-den-Daten/zu-den-Daten-Nav.html; last access: 15.02.2011].

— (2010): Analytikreport der Statistik - Arbeitsmarkt in Deutschland. Zeitreihen bis 2009. [Source: http://statistik.arbeitsagentur.de; last access: 15.01.2011]

— (2009): Analytikreport der Statistik—Analyse des Arbeitsmarktes für Ausländer April 2009. [Source: http://www.pub.arbeitsamt.de/hst/services/statistik/200904/ama/auslaender_d.pdf; last access: 15.01.2011].

BW-International (2011): *Branchenüberblick Biotechnologie.* Stuttgart: Baden-Württemberg International.

BW-Invest (2011a): Aktuelle Angaben zur Bedeutung der Automobilindustrie in Baden-Württemberg. [Source: http://www.bw-invest.de/deu/index_deu_5019.aspx; last access: 01.02.2011].

— (2011b): Automobilzulieferer. [Source: http://www.bw-invest.de/deu/index_deu_265.aspx; last access: 01.02.2011].

— (2011c): Baden-Württemberg - Ein exzellenter Standort für Forschung und Entwicklung. [Source: http://www.bw-invest.de/deu/index_deu_6401.aspx; last access: 01.02.2011].

BW MoE (Baden-Württemberg Ministry of Economics) (2010a): *Technologie-Initiativen in Baden-Württemberg*. BW MoE.

— (2010b): *Regional Cluster Atlas Baden-Württemberg 2010. Overview of cluster-related networks and initiatives*. BW MoE.

— (2008): *Jahreswirtschaftsbericht 2007/2008*. BW MoE.

— (2007): *Vertragsforschung in Baden-Württemberg*. BW MoE.

Candil Gonzalo, Jesús (2010): *Peculiaridades de la Industria Española de Automoción y Planes de Actuación des Gobierno Relacionados con el Sector*. Madrid: Ministerio de Industria, Turismo y Comercio.

Casper, Steven and David Soskice (2004): Sectoral Systems of Innovation and Varieties of Capitalism. Explaining the Development of High-Technology Entrepreneurship in Europe. In: Malerba, Franco (ed.): *Sectoral Systems of Innovation. Concepts, Issues and Analyses of Six Major Sectors in Europe*. Cambridge: Cambridge University Press: 348–387.

CEIM (2009): *Memoria 2008*. Madrid: CEIM—Confederación Empresesarial de Madrid.

Cerny, Philip, Susanne Soederberg and George Menz (eds.) (2006): *Internalizing Globalization. The Rise of Neoliberalism and the Erosion of National Models of Capitalism*. London: Palgrave Macmillan.

Chandler, Alfred D. (1990): *Scale and Scope. The Dynamics of Industrial Capitalism*. Cambridge, MA: Harvard University Press.

Coates, David (2005): Paradigms of Explanation. In: David Coates (ed.): *Varieties of Capitalism, Varieties of Approaches*. Basingstoke: Palgrave Macmillan: 1–25.

Cohen, Wesley M. and Daniel A. Levinthal (1990): Absorptive Capacity. A New Perspective on Learning and Innovation. In: *Administrative Science Quarterly 35*: 128–152.

— Innovation and Learning. The two Faces of R&D. In: *The Economic Journal 99*: 569–596.

Cooke, Philip (2004): Introduction. Regional Innovation Systems—An Evolutionary Approach: In: Cooke, Philip, Martin Heidenreich and Hans-Joachim Braczyk (eds.): *Regional Innovation Systems (2nd edition)*. London; New York: Routledge: 1–18.

— (2003): Integrating Global Knowledge Flows for Generative Growth in Scotland. Life Sciences as a Knowledge Economy Exemplar. In: *Inward Investment, Entrepreneurship and Knowledge Flows in Scotland—International Comparisons*. Paris: OECD.

— (1998): Introduction. Origins of the Concept. In: Braczyk, Hans-Joachim, Philip Cooke and Martin Heidenreich (eds.): *Regional Innovation Systems.* London: UCL-Press: 2–25.

Cooke, Philip, Patrice Boekholt and Franz Tödtling (2000): *The Governance of Innovation in Europe.* London/New York: Pinter.

Cooke, Philip, Mikel Gomez Uranga and Goio Etxebarria (1997): Regional Innovation Systems. Institutional and Organisational Dimensions. In: *Research Policy 26*: 475–491.

Cooke, Philip, Martin Heidenreich and Hans-Joachim Braczyk (eds.) (2004): *Regional Innovation Systems (2nd edition).* London; New York: Routledge.

Cooke, Philip, Carla de Laurentis, Franz Tödtling and Michael Trippl (2007): *Regional Knowledge Economies. Markets, Clusters and Innovation.* Cheltenham, Northampton: Edward Elgar Publishing.

Cooke, Philip and Loet Leydesdorff (2006): Regional Development in the Knowledge-Based Economy. The Construction of Advantage. In: *The Journal of Technology Transfer 31*: 5–15.

Coriat, Benjamin and Olivier Weinstein (2004): National Institutional Frameworks, Institutional Complementarities and Sectoral Systems of Innovation. In: Franco Malerba (ed.): *Sectoral Systems of Innovation. Concepts, Issues and Analyses of Six Major Sectors in Europe.* Cambridge: Cambridge University Press: 325–347.

COTEC (2010): *Tecnología e Innovación en España Informe COTEC 2010.* Madrid: Fundación COTEC para la Innovación Tecnológica.

— (2004): *Libro Blanco de la Innovación en la Comunidad de Madrid.* Madrid: Fundación COTEC para la Innovación Tecnológica.

— (2003): *Libro Verde. Documento para el debate sobre el Sistema de Innovación de la Comunidad de Madrid.* Madrid: Fundación COTEC para la Innovación Tecnológica.

Crouch, Colin (2007): How to 'do' Post-Determinist Institutional Analysis. In: *Socio-Economic Review 5*: 527–567.

— (2005): *Capitalist Diversity and Change. Recombinant Governance and Institutional Entrepreneurs.* Oxford: Oxford University Press.

Crouch, Colin and Henry Farrell (2004): Breaking the Path of Institutional Development? Alternatives to the New Determinism. *Rationality and Society 16 (1)*: 5–43.

Colin Crouch, Patrick Le Galès, Carlo Trigilia and Helmut Voelzkow (eds.) (2004): *Changing Governance of Local Economies. Response of European Local Production Systems.* Oxford: Oxford University Press.

— (2001): *Local Production Systems in Europe. Rise or Demise?* Oxford: Oxford University Press.

Crouch, Colin and Wolfgang Streeck (1997): *Political Economy of Modern Capitalism. Mapping Convergence and Diversity.* London: Sage.

Crouch, Colin and Helmut Voelzkow (2009): Introduction. Local and Sectoral Diversity Within National Economic Systems. In: Crouch, Colin and Helmut

Voelzkow (eds.): *Innovation in Local Economies. Germany in Comparative Context.* Oxford: Oxford University Press: 1–21.

Crouch, Colin, Helmut Voelzkow and Martin Schröder (2009): Regional and Sectoral Varieties of Capitalism. In: *Economy and Society 38 (4)*: 654–678.

Deeg, Richard and Gregory Jackson (2007): Towards a More Dynamic Theory of Capitalist Variety. In: *Socio-Economic Review 5(1)*, 149–179.

Deeg, Richard (2005): Path Dependency, Institutional Complementarity, and Change in National Business Systems. In: Glenn Morgan et al. (eds.): *Changing Capitalisms? Internationalization, Institutional Change, and Systems of Economic Organization.* Oxford: Oxford University Press: 21–52.

Destatis (2011): [Source: http://www.destatis.de/jetspeed/portal/cms/Sites/destatis/Internet/DE/Navigation/Statistiken/Zeitreihen/Indikatoren/Strukturindikatoren__nk.psml; last access: 15.02.2011]

— (2010): *Bruttoinlandsprodukt 2009 für Deutschland.* Wiesbaden: Statistisches Bundesamt.

— (2009): *Statistisches Jahrbuch für die Bundesrepublik Deutschland.* Wiesbaden: Statistisches Bundesamt.

Deutsche Messe AG (2010): *Geschäftsbericht 2009.* Hannover: Deutsche Messe AG.

DiMaggio, Paul (1979): Review: On Pierre Bourdieu. In: *The American Journal of Sociology 84 (6)*: 1460–1474.

DiMaggio, Paul J. and Walter W. Powell (1983): The Iron Cage Revisited. Institutional Isomorphism and Collective Rationality in Organizational Fields. In: *American Sociological Review 48, (2)*: 147–160.

Döring, Thomas and Jan Schnellenbach (2006): What Do We Know about Geographical Knowledge Spillovers and Regional Growth? A Survey of the Literature. In: *Regional Studies 40 (3)*: 375–395.

Dosi, Giovanni (1996): The Contribution of Economic Theory to the Understanding of a Knowledge-Based Economy. In: OECD (ed.): *Employment and Growth in the Knowledge-Based Economy.* Paris: 81–92.

Dresing, Thorsten and Thorsten Pehl (2011): *Praxisbuch Transkription.* Regelsysteme, Software und praktische Anleitungen für qualitative ForscherInnen. Marburg: Eigenverlag. [Source: www.audiotranskription.de/praxisbuch; last access: 01.09.2011].

Drucker, Peter F. (1994): The Age of Social Transformation. In: *The Atlantic Monthly 274 (5)*: 53–80.

Ebbinghaus, Bernhard (2009): Vergleichende Politische Soziologie. Quantitative Analyse- oder qualitative Fallstudiendesigns? Vergleichende Politische Soziologie. In: Kaina, Viktoria and Andrea Römmele (eds.): *Politische Soziologie. Ein Studienbuch.* Wiesbaden: VS: 481–501.

— (2006): Qualitativer oder quantitativer Vergleich? Herausforderungen für die sozialwissenschaftliche Europaforschung. In: *Zeitschrift für Staats- und Europawissenschaften 4 (3)*: 388–404.

— (2005a): When Less is More. Selection Problems in Large-N and Small-N Cross-national Comparisons. In: *International Sociology 20 (2)*: 133–152.

— (2005b): Can Path Dependence Explain Institutional Change? Two Approaches Applied to Welfare State Reform. *MPIfG Discussion Paper 02/2005.*

— (1999): Does a European Social Model Exist and can it Survive? In: Gerhard Huemer, Michael Mesch and Franz Traxler (eds.): *The Role of Employer Associations and Labour Unions in the EMU. Institutional Requirements for European Economic Policies.* Aldershot: Ashgate: 1–26.

Ebner, Alexander (2009): Governance von Innovationssystemen und die Politische Ökonomie der Wettbewerbsfähigkeit. In: Blättel-Mink, Birgit and Alexander Ebner (eds.): *Innovationssysteme. Technologie, Institutionen und die Dynamik der Wettberwerbsfähigkeit.* Wiesbaden: VS: 119–142.

Edquist, Charles (2005): Systems of Innovation. Perspectives and Challenges. In: Fagerberg, Jan, David C. Mowery and Richard R. Nelson (eds.): *The Oxford Handbook of Innovation.* Oxford/New York: Oxford University Press.

— (1997): Systems of Innovation Approaches—Their Emergence and Characteristics. In: Edquist, Charles (ed.): *Systems of Innovation. Technologies, Institutions and Organizations.* London: Pinter: 1–35.

Edquist, Charles and Björn Johnson (1997): Institutions and Organizations in Systems of Innovation. In: Edquist, Charles (ed.): *Systems of Innovation, Technologies, Institutions and Organisations.* London: Pinter: 41–63.

EEDA (East of England Development Agency) (2011): Life Sciences. [Source: http://www.eeda.org.uk/5140.asp; last access: 15.11.2011].

— (2009): EEDA's contribution to the strategic development of the East of England. Final Report. At PA Consulting Group; SQW consulting. [Source: http://www.eeda.org.uk/files/EEDAs_contribution_to_the_strategic_developmen t_of_the_East_of_England.pdf; last access: 15.11.2011].

Egetemeyr, Christian and Joachim Werner (2008): Forschung und Entwicklung in Baden-Württemberg. Ein Spitzenplatz im internationalen Vergleich. In: *Statistisches Monatsheft Baden-Württemberg 6/2008.* Statistisches Landesamt Baden-Württemberg: 17–25.

Enright, Michael J. (1996): Regional Clusters and Economic Development. A Research Agenda. In: Staber, Udo H., Norbert V. Schaefer, Basu Sharma (eds.): *Business Networks, Prospects for Regional Development.* De Gruyter: Berlin New York: 190–214.

EofE (2011a): East of England. Space for Ideas. Key Sectors. [Source: http://www.eastofengland.uk.com/business/key-sectors; last access: 15.11.2011].

— (2011b): East of England—Driving automotive innovation. EEDA and UK Trade and Investment [Source: http://www.eastofengland.uk.com/wp-content/uploads/EofE-Automotive.pdf; last access: 15.11.2011].

— (2011c): East of England—A centre of aerospace excellence. EEDA and UK Trade and Investment [Source: http://www.eastofengland.uk.com/wp-content/uploads/EofE-Aerospace.pdf; last access: 15.11.2011].

— (2011d): East of England—Opportunities in drug discovery. EEDA and UK Trade and Investment [Source: http://www.eastofengland.uk.com/wp-content/uploads/EofE-Drug-discovery.pdf; last access: 15.11.2011].

Ernst & Young (2010): *Neue Spielregeln. Deutscher Biotechnologie-Report 2010.* Mannheim: Ernst and Young.

— (2009): *Fallstrick Finanzierung. Deutscher Biotechnologie-Report 2009.* Mannheim: Ernst and Young.

Etzkowitz, Henry and Loet Leydesdorff (2000): The Dynamics of Innovation. From National Systems and "Mode 2" to a Triple Helix of University-Industry-Government relations. In: *Research Policy 29*: 109–123.

Etzkowitz, Henry, Andrew Webster, Christiane Gebhardt, Branca Regina Cantisano Terra (2000): The Future of the University and the University of the Future. Evolution of Ivory Tower to Entrepreneurial Paradigm. In: *Research Policy 29*: 313–330.

Eurostat (2011a): [Source: http://epp.eurostat.ec.europa.eu/portal/page/portal/structural_indicators/indicators/innovation_and_research; last access: 15.02.2011].

— (2011b): [Source: http://epp.eurostat.ec.europa.eu/portal/page/portal/region_cities/regional_statistics; last access: 15.05.2011].

— (2010): *Europe in Figures. Eurostat Yearbook 2010.* Luxembourg: Publications Office of the European Union.

— (2004a): Portrait of the Regions—United Kingdom—The East of England. Geography and history. [Source: http://circa.europa.eu/irc/dsis/regportraits/info/data/ukh_geo.htm; last access: 15.11.2011].

— (2004b): Portrait of the Regions—United Kingdom—The East of England. Economy. [Source: http://circa.europa.eu/irc/dsis/regportraits/info/data/ukh_eco.htm; last access: 15.11.2011].

Fagerberg, Jan (2005): Innovation. A Guide to the Literature. In: Fagerberg, Jan, David C. Mowery and Richard R. Nelson (eds.): *The Oxford Handbook of Innovation.* Oxford: Oxford University Press: 1–26.

Fira de Barcelona (2009): El Sector de la Automoción en España. Departament d'Investigació i Estrategia de Mercats. [Source: http://www.salonautomovil.com/portal/ShowProperty?nodeId=/BEA%20Repository/542002//document; last access: 15.11.2011].

Fischer, Manfred M. and Josef Fröhlich (2001): Knowledge, Complexity and Innovation Systems. Prologue. In: Fischer, Manfred M. and Josef Fröhlich (eds.): *Knowledge, Complexity and Innovation Systems.* Heidelberg, Berlin: Springer: 1–20.

Fischer, Manfred M., Javier Revilla Diez and Folke Snickars (2001): *Metropolitan Innovation Systems. Theory and Evidence from Three Metropolitan regions in Europe.* Berlin, Heidelberg: Springer.

Flick, Uwe (2009): *Qualitative Sozialforschung. Eine Einführung. Vollständig überarbeitete und erw. Neuausgabe.* Reinbek: Rowohlt-Taschenbuch-Verlag.

— (2008): Design und Prozess Qualitativer Forschung. In: Flick, Uwe, Ernst von Kardorff and Ines Steincke (eds.): *Qualitative Forschung. Ein Handbuch.* Reinbeck: Rowohlt: 252–265.

Fligstein, Neil (2001): *The Architecture of Markets. An Economic Sociology of Twenty-First-Century Capitalist Societies.* Princeton, NJ: Princeton University Press.

Foray, Dominique (1998): The Economics of Knowledge Openness. Emergence, Persistence and Change of Conventions in the Knowledge Systems. In: Lazaric N. and E. Lorenz (eds.): *Trust and Economic Learning.* Cheltanham: 162–189.

Foray, Dominique and Bengt-Åke Lundvall (1996): The Knowledge-Based Economy. From the Economics of Knowledge to the learning Economy. In: OECD (ed.): *Employment and Growth in the Knowledge-Based Economy.* Paris: 11–32.

Franzoni, Chiara and Francesco Lissoni (2009): Academic Entrepreneurs. Critical Issues and Lessons for Europe. In: Varga, Attila (ed.): *Universities, Knowledge Transfer and Regional Development. Geography, Entrepreneurship and Policy.* Cheltenham, Northampton: Edward Elgar: 163–190.

Freeman, Chris (1995): The "National System of Innovation" in Historical Perspective. In: *Cambridge Journal of Economics 19*: 5–24.

— (1988): Japan. A New National System of Innovation? In: Dosi, G., C. Freeman, R. Nelson, G. Silverberg and L. Soete (eds.): *Technical Change and Economic Theory.* London: Pinter.

— (1987): *Technology Policy and Economic Performance—Lessons from Japan.* London: Pinter.

Freeman, Chris and Luc Soete (2009): Developing Science, Technology and Innovation Indicators. What we can learn from the Past. In: *Research Policy 38*: 583–589.

Friedberg, Erhard (1995): *Ordnung und Macht.* Frankfurt a.M.; New York: Campus.

Fromhold-Eisebith, Martina and Günter Eisebith (2005): How to institutionalize Innovative Clusters? Comparing explicit Top-down and implicit Bottom-up Approaches. In: *Research Policy 34*: 1250–1268.

Fuchs, Gerhard and Sandra Wassermann (2004): *The Regional Innovation System of Baden-Württemberg. Lock-In or Breakthrough?* Stuttgarter Beiträge zur Risiko- und Nachhaltigkeitsforschung. Stuttgart: Institut für Sozialwissenschaften; Abt. für Technik- und Umweltsoziologie; Universität Stuttgart.

Fukuyama, Francis (1992): *Das Ende der Geschichte. Wo stehen wir?* München: Kindler.

GenCat (Generalitat de Catalunya) (2010): *Informe Anual Sobre la Indústria a Catalunya 2009.* Barcelona: Generalitat de Catalunya: Departament d'Innovació, Universitats i Empresa.

Genoma España (Fundación Española para el Desarrollo de la Investigación en Genómica y Proteómica) (2009): *Relevancia de la Biotechnología en España 2009.* Madrid: Genoma España.

Ghoshal, Sumantra and Christopher A. Bartlett (1990): The Multinational Corporation as an Interorganizational Network. In: *Academy of Management Review 15*: 603–625.

Giddens, Anthony (1990): *The Consequences of Modernity*. Stanford: Stanford University Press.

GLA (2010): *Economic Evidence Base to support the London Plan, the Transport Strategy and the Economic Development Strategy*. London: Greater London Authority.

Gläser, Jochen and Grit Laudel (2009): *Experteninterviews und Qualitative Inhaltsanalyse*. 3., überarbeitete Auflage. Wiesbaden: VS.

Godin, Benoît (2006): The Knowledge-Based Economy: Conceptual Framework or Buzzword? In: *Journal of Technology Transfer 31*: 17–30

Granovetter, Mark (1985): Economic Action and Social Structure. The Problem of Embeddedness. In: *American Journal of Sociology 91 (3)*: 481–510.

Gresse, Christopher (2010): *Wissensmanagement im Technologietransfer. Einfluss der Wissensmarkmale in F&E-Kooperationen*. Wiesbaden: Gabler.

Grimpe, Christoph and Heide Fier (2010): Informal University Technology Transfer. A Comparison between the United States and Germany. In: *The Journal of Technology Transfer 35 (6)*: 637–650.

Grimpe, Christoph and Katrin Hussinger (2008): Formal and Informal Technology Transfer from Academia to Industry. Complementarity Effects and Innovation Performance. *ZEW Discussion Papers 08–080*. ZEW—Center for European Economic Research.

Gunasekera, Chrys (2006): Reframing the Role of Universities in the Development of Regional Innovation Systems. In: *Journal of Technology Transfer 31*: 101–113.

Hall, Peter A. (2006): Stabilität und Wandel in den Spielarten des Kapitalismus. In: Beckert, Jens, Bernhard Ebbinghaus, Anke Hassel, and Philip Manow (eds.): *Transformationen des Kapitalismus*. Frankfurt: Campus: 181–204.

Hall, Peter A. and Daniel W. Gingerich (2009): Varieties of Capitalism and Institutional Complementarities in the Macroeconomy. An Empirical Analysis. In: *British Journal of Political Science 39*: 449–482.

Hall, Peter A. and Kathleen Thelen (2009): Institutional Change in Varieties of Capitalism. In: *Socio-Economic Review 7 (1)*: 7–34.

Hall, Peter A. and David Soskice (eds.) (2001a): *Varieties of Capitalism. The Institutional Foundations of Comparative Advantage*. Oxford: Oxford University Press.

— (2001b): An Introduction to Varieties of Capitalism. In: Hall, Peter A. and David Soskice (eds.): *Varieties of Capitalism. The Institutional Foundations of Comparative Advantage*. Oxford: Oxford University Press: 1–68.

Hancké, Bob and Michel Goyer (2005): Degrees of Freedom. Rethinking the Institutional Analysis of Economic Change. In: Glenn Morgan et al. (eds.): *Changing Capitalisms? Internationalization, Institutional Change, and Systems of Economic Organization*. Oxford: Oxford University Press: 53–77.

Hancké, Bob, Martin Rhodes and Mark Thatcher (eds.) (2007a): *Beyond Varieties of Capitalism. Conflict, Contradictions, and Complementarities in the European Economy*. Oxford: Oxford University Press.

— (2007b): Introduction. Beyond Varieties of Capitalism. In: Hancké, Bob, Martin Rhodes and Mark Thatcher (eds.): *Beyond Varieties of Capitalism. Conflict, Con-*

tradictions, and Complementarities in the European Economy. Oxford: Oxford University Press: 3–38.

Hay, Colin (2005): Two Can Play at That Game … Or Can They? In: David Coates (ed.): *Varieties of Capitalism, Varieties of Approaches.* Basingstoke: Palgrave: 106–121.

Heidenreich, Martin (2011a): Regionale Netzwerke. In: Johannes Weyer (ed.): *Soziale Netzwerke.* München: Oldenbourg: 167–188.

Heidenreich, Martin, with Jannika Mattes (2011b): Innovation in Multinational Enterprises. Between Organizational Control and Social Embeddedness. In: Martin Heidenreich, Christoph Barmeyer and Knut Koschatzky (eds.): *Multinational Enterprises and Innovation. Regional Learning in Networks.* London, New York: Routledge.

Heidenreich, Martin (2006): Die Organisationen der Wissensgesellschaft. Zwischen regionalem und grenzüberschreitendem Lernen. In: Heinrich-Böll-Stiftung (ed.): *Die Verfasstheit der Wissensgesellschaft.* Münster: Westfälisches Dampfboot: 43–56.

— (2004b): The Dilemmas of Regional Innovation Systems. In: Cooke, Philip, Martin Heidenreich and Hans-Joachim Braczyk (eds.): *Regional Innovation Systems. The Role of Governance in a Globalized World (2nd edition).* London/New York: Routledge: 363–389.

— (2004a): Regional Capabilities and the European Employment Strategy. In: Salais, Robert and Robert Villeneuve (eds.): *Europe and the Politics of Capabilities.* Cambridge, Cambridge University Press: 54–72.

— (2003a): Die Debatte um die Wissensgesellschaft. In: Böschen, Stefan and Ingo Schulz-Schaeffer (eds.): *Wissenschaft in der Wissensgesellschaft.* Opladen: Westdeutscher Verlag: 25–51.

— (2003b): Regional Inequalities in the Enlarged Europe. In: *Journal of European Social Policy 13 (4)*: 313–333.

— (2002): Merkmale der Wissensgesellschaft. In: Bund-Länder-Kommission für Bildungsplanung und Forschungsförderung et al. (ed.): Lernen in der Wissensgesellschaft. Innsbruck a.o.: Studienverlag: 334–363. [Source: http://www.uni-oldenburg.de/sozialstruktur/dokumente/blk.pdf; last access: 27.11.2007].

— (2001a): Regionale Innovationssysteme. Zwischen Wandel und Beharrung. In: Fuchs, Gerhard and Karin Töpsch (eds.): *Baden-Württemberg—Erneuerung einer Industrieregion.* Stuttgart: Akademie für Technikfolgenabschätzung: 87–106.

— (2001b): Regionale Netzwerke. Ein Weg zur Stärkung der Innovationstätigkeit ostdeutscher Betriebe? In: Wolf, Jürgen, Jens Rannenberg, Harald Mattfeldt and Heiner Giebel (eds.): *Jahrbuch für Politik und Gesellschaft in Sachsen-Anhalt 2001.* Halle (Saale): Mitteldeutscher Verlag: 222–243.

— (2000): Regionale Netzwerke in der globalen Wissensgesellschaft. In: Weyer, Johannes (ed.): *Soziale Netzwerke. Konzepte und Methoden der sozialwissenschaftlichen Netzwerkforschung.* München/Wien: Oldenbourg: 87–110.

Heidenreich, Martin, Christoph Barmeyer and Knut Koschatzky (eds.) (2011): *Multinational Enterprises and Innovation. Regional Learning in Networks.* London, New York: Routledge.

Heidenreich, Martin and Gerhard Krauss (2004): The Production and Innovation Regime of Baden-Württemberg. Between Past Successes and New Challenges. In: Cooke, Philip, Martin Heidenreich and Hans-Joachim Braczyk (eds.): *Regional Innovation Systems. The Role of Governance in a Globalized World (2nd edition).* London/New York: Routledge: 186–213.

Heidling, Eckhard (2000): Strategische Netzwerke. Koordination und Kooperation in asymmetrisch strukturierten Unternehmensnetzwerken. In: Weyer, Johannes (ed.): *Soziale Netzwerke. Konzepte und Methoden der sozialwissenschaftlichen Netzwerkforschung.* München/Wien: Oldenbourg: 63–86.

Held, David, Anthony McGrew, David Goldblatt and Jonathan Perraton (1999): *Global Transformations.* Cambridge: Polity Press.

Hess, Martin (2004): 'Spatial' Relationships? Towards a Reconceptualization of Embeddedness. In: *Progress in Human Geography, 28 (2)*: 165–186.

Hirst, Paul and Grahame Thompson (1996): *Globalization in Question. The International Economy and the Possibility of Governance.* Cambridge: Polity Press.

Höpner, Martin (2005): What Connects Industrial Relations and Corporate Governance? Explaining Institutional Complementarity. In: *Socio-Economic Review 3(2)*: 331–358.

Hollingsworth, J. Rogers and Robert Boyer (1997): *Contemporary Capitalism. The Embeddedness of Institutions.* Cambridge Studies in Comparative Politics. Cambridge: Cambridge University Press.

Hollingsworth, J. Rogers, Philippe C. Schmitter and Wolfgang Streeck (1994): *Governing Capitalist Economies.* Oxford: Oxford University Press.

Holweg, Matthias, with Yung Tran, Philip Davies and Stephan Schramm (2010): *Growing the Automotive Supply Chain. The Road Forward.* The Automotive Council UK. Buckingham: PICSIE Books.

Holweg, Matthias, with Philip Davies and Dmitry Podpolny (2009): *The competitive status of the UK automotive industry.* Buckingham: PICSIE Books.

Howell, Chris (2003): Varieties of Capitalism. And Then There Was One? In: *Comparative Politics 36 (3)*: 103–124.

Howells, Jeremy (1999): Research and Technology Outsourcing and Innovation Systems. An Exploratory Analysis. In: *Industry and Innovation 6 (1)*: 111–129.

Howlett, Robert J. (2010): Knowledge Transfer between UK Universities and Business. In: Howlett, Robert J. (ed.): *Innovation through Knowledge Transfer.* Berlin, Heidelberg: Springer: 1–15.

Hudson, Ray (1999): The Learning Economy, the Learning Firm and the Learning Region. In: *European Urban and Regional Studies, 6 (1)*: 59–72.

Hübner, Kurt (2009): Innovationssysteme und „Varieties of Capitalism" unter Bedingungen ökonomischer Globalisierung. In: Blättel-Mink, Birgit and Alexander Ebner (eds.): *Innovationssysteme. Technologie, Institutionen und die Dynamik der Wettberwerbsfähigkeit.* Wiesbaden: VS: 143–158.

Iammarino, Simona (2005): An Evolutionary Integrated View of Regional Systems of Innovation. Concepts, Measures and Historical Perspectives. In: *European Planning Studies 13*: 497–519.

IAW (Institut für Angewandte Wirtschaftsforschung e.V.) (2008): *Innovationstätigkeit und Innovationshemmnisse Baden-Württembergischer Betriebe Ergebnisse der Auswertungen des IAB-Betriebspanels Baden-Württemberg.* IAW-Kurzbericht 7/2008. Tübingen: IAW.

ICO (Instituto de Crédito Oficial) (2009): *Situación Económica de Comunidad de Madrid.* Madrid: ICO.

IDESCAT (Institut d' Estadística de Catalunya) (2011): Indicadors d' Estructura Económica. IDESCAT. [Source: http://www.idescat.cat/cat/economia/eco-indust.html, last Access: 15.11.2011].

IESTADIS (Instituto de Estadística de la Comunidad de Madrid) (2011a): Contabilidad Regional de la Comunidad de Madrid. IESTADIS [Source: http://www.madrid.org/iestadis/fijas/estructu/economicas/contabilidad/estructucr.htm; last access: 15.11.2011].

— (2011b): Colectivo Empresarial de la Comunidad de Madrid. IESTADIS. [Source: http://www.madrid.org/iestadis/fijas/estructu/economicas/otros/estructucolectivo.htm; last access: 15.11.2011].

— (2011c): Estadística de I+D e Innovación Tecnológica. IESTADIS. [Source: http://www.madrid.org/iestadis/fijas/estructu/economicas/otros/estructuid.htm; last access: 15.11.2011].

IMI (2007): *Automotivate—The Sector Skills Agreement for the Downstream Motor Industry. Sector Skills Agreement. Stage 5—Action Plans. East of England.* Fanshaws, Brickendon, Hertford: Institute of the Motor Industry.

INE (Instituto Nacional de Estadística) (2010a): *Panorámica de la Industria.* Madrid: INE.

— (2010b): *España en cifras 2010.* Madrid: INE.

Innovationszentrum Niedersachsen (2005): *Technologietransfer in Niedersachsen. Akteure und Zusammenhänge.* Broschüre des Innovationszentrums Niedersachsen. Hannover.

IW (2010): *Deutschland in Zahlen. Ausgabe 2010.* Köln: Institut der Deutschen Wirtschaft Köln.

Jackson, Gregory and Richard Deeg (2006): How Many Varieties of Capitalism? Comparing the Comparative Institutional Analyses of Capitalist Diversity. *MPIfG Discussion Paper 06/2.* Cologne: Max Planck Institute for the Study of Societies.

Jacobson, Nora, Dale Butterill and Paula Goering (2005): Consulting as a Strategy for Knowledge Transfer. In: *The Milbank Quarterly 83 (2)*: 299–321.

Jaffe, Adam B. (1989): Real Effects of Academic Research. In: *American Economic Review 79 (5)*: 957–970.

Jepperson, Ronald L. (1991): Institutions, Institutional Effects, and Institutionalism. In: Powell, Walter W. and Paul J. DiMaggio (eds.): *The New Institutionalism in Organizational Analysis.* Chicago: University of Chicago Press: 143–163.

Johnson, Björn (1992): Institutional Learning. In: Lundvall, Bengt- Åke (ed.): *National Systems of Innovation. Towards a Theory of Innovation and Interactive Learning.* London: Pinter: 23–44.

Kaiser, Robert and Heiko Prange (2004): The Reconfiguration of National Innovation Systems. The Example of German Biotechnology. In: *Research Policy 33*: 395–408.

Karlsson, Charlie and Agostino Manduchi (2001): Knowledge Spillovers in a Spatial Context—A Critical Review and Assessment. In: Fischer, Manfred M. and Josef Fröhlich (eds.): *Knowledge, Complexity and Innovation Systems.* Heidelberg, Berlin: Springer: 101–123.

Keeble, David (1998): Local Industrial Development and Dynamics. The Anglian Case. *ESRC Centre for Business Research, University of Cambridge: Working Paper No. 96.* Cambridge.

Kenworthy, Lane (2009): Institutional Coherence and Macroeconomic Performance. In Bob Hancké (ed.): *Debating Varieties of Capitalism. A Reader.* Oxford: Oxford University Press: 180–199.

Kitschelt, Herbert et al. (1999): *Continuity and Change in Contemporary Capitalism.* Cambridge: Cambridge University Press.

Kline, Stephen J. and Nathan Rosenberg (1986): An Overview of Innovation. In: Landau, Ralph and Nathan Rosenberg (eds.): *The Positive Sum Strategy. Harnessing Technology for Economic Growth.* Washington D.C.: National Academy Press: 275–304.

Koschatzky, Knut (2009): The Uncertainty in Regional Innovation Policy: Some Rationales and Tools for Learning in Policy Making. *Working Papers Firms and Region No. R6/2009.* Karlsruhe: Fraunhofer ISI.

— (2001a): *Räumliche Aspekte im Innovationsprozess. Ein Beitrag zur neuen Wirtschaftsgeographie aus Sicht der regionalen Innovationsforschung.* Münster: LIT.

— (2001b): Networks in Innovation Research and Innovation Policy—An Introduction. In: Koschatzky, Knut, Marianne Kulicke and Andrea Zenker (eds.): *Innovation Networks. Concepts and Challenges in the European Perspective.* Heidelberg/New York: Physika: 3–23.

Koschatzky, Knut and Joachim Hemer (2009): Firm Formation and Economic Development. What Drives Academic Spin-offs to Success or Failure? In: Varga, Attila (ed.): *Universities, Knowledge Transfer and Regional Development. Geography, Entrepreneurship and Policy.* Cheltenham, Northampton: Edward Elgar: 191–218.

Koschatzky, Knut and Henning Kroll (2007): Which Side of the Coin? The Regional Governance of Science and Innovation. In: *Regional Studies 41(8)*: 1115–1127.

Krätke, Stefan (2009): Die Metropolisierung des europäischen Stadtsystems. In: Basten, L. (ed.): *Metropolregionen—Restrukturierung und Governance. Deutsche und internationale Fallstudien.* Dortmund: Rohn Verlag.

Krauss, Gerhard (2009): Baden-Württemberg als Prototyp eines regionalen Innovationssystems. In: Blättel-Mink, Birgit and Alexander Ebner: *Innovationssysteme. Technologie, Institutionen und die Dynamik der Wettbewerbsfähigkeit.* Wiesbaden: VS.

— (2005): Regionale und soziale Einbettung von Wissensökonomien in Metropolregionen. In: Kujath, Hans-Joachim (ed.): *Knoten im Netz. Zur neuen Rolle der Metropolregionen in der Dienstleistungswirtschaft und Wissensökonomie.* Münster: LIT.

Krauss, Gerhard and Thomas Stahlecker (2001): New Biotechnology Firms in Germany. Heidelberg and the BioRegion Rhine-Neckar Triangle. In: *Small Business Economics 17*: 143–153.

Krumbein, Wolfgang (2003): Südniedersachsen. Kompetenzregion oder Problemregion? *Schriftenreihe des Instituts für Regionalforschung e. V. an der Universität Göttingen, 15/2003.* Göttingen.

Kruse, Jan and Hildegard Wenzler-Cremer (2008): Interviews auf Computer überspielen und transkribieren. Ein Manual für die Aufnahme und Transkription von Interviews mit einfachen EDV-basierten Lösungen. [Source: http:// www 2.soziologie.uni-freiburg.de/Personen/kruse/Manual.pdf; last access: 15.11. 2011].

Kujath, Hans-Joachim (2005): Die neue Rolle der Metropolregionen in der Wissensökonomie. In: Kujath, Hans-Joachim (ed.): *Knoten im Netz. Zur neuen Rolle der Metropolregionen in der Dienstleistungswirtschaft und Wissensökonomie.* Münster: LIT: 23–64.

Lagendijk, Arnoud (2001): Scaling Knowledge Production. How Significant is the Region? In: Fischer, Manfred M. and Josef Fröhlich (eds.): *Knowledge, Complexity and Innovation Systems.* Heidelberg, Berlin: Springer: 79–100.

Lagendijk, Arnoud and David Charles (1999): Clustering as a New Groth Strategy for Regional Economies? A Discussion of New Forms of Regional Industrial Policy in the United Kingdom. In: OECD (ed.): *Boosting Innovation. The Cluster Approach.* Paris: OECD: 127–154.

Lambooy, Jan (2002): Knowledge and Urban Economic Development. An Evolutionary Perspective. In: *Urban Studies 39 (5–6)*: 1019–1035.

Lamnek, Siegfried (2010): *Qualitative Sozialforschung. Lehrbuch. 5. Überarbeitete Auflage.* Weinheim, Basel: Beltz.

— (1995): *Qualitative Sozialforschung. Band 1. Methodologie. 3. korrigierte Auflage.* Weinheim: Beltz: PsychologieVerlagsUnion.

Lane, Christel and Reinhard Bachmann (1996): The Social Constitution of Trust. Supplier Relations in Britain and Germany. In: *Organization Studies 17 (3)*: 365–395.

Le Galès, Patrick and Helmut Voelzkow (2001): Introduction. The Governance of Local Economies. In: Crouch, Collin, Patrick Le Galès, Carlo Trigilia and Helmut Voelzkow (eds.): *Local Production Systems in Europe. Rise or Demise?* Oxford: Oxford University Press: 1–24.

Legler, Harald (2004): Innovationsstandort Niedersachsen. In: Niedersächsisches Institut für Wirtschaftsforschung (NIW) (Autorengemeinschaft): Positionsbestimmung Niedersachsen. Ausgewählte Befunde und Schlussfolgerungen für die Wirtschaftspolitik. *Forschungsberichte des NIW, 32.* Hannover: 17–54.

Legler, Harald, Birgit Gehrke, Olaf Krawczyk, Ulrich Schasse, Christian Rammer, Nina Leheyda and Wolfgang Sofka (2009): *Die Bedeutung der Automobilindustrie für die deutsche Volkswirtschaft im europäischen Kontext.* Hannover, Mannheim: NIW, ZEW.

Legler, Harald and Ulrich Schasse (2010): *Forschungs- und Entwicklungsaktivitäten der Deutschen Wirtschaft—Eine strukturelle Langfristbetrachtung.* Studie zum Deutschen Innovationssystem Nr. 2-2010. Berlin: EFI.

— (2004): Wirtschaftlicher Strukturwandel in Niedersachsen. In: *Neues Archiv für Niedersachsen. Zeitschrift für Landesforschung 2/2004*: 3–24.

Lehrer, Mark, Phillip Nell and Lisa Gärber (2009): A national systems view of university entrepreneurialism. Inferences from comparison of the German and US experience. In: *Research Policy 38*: 268–280.

Liebold, Renate and Rainer Trinczek (2009): Experteninterview. In: Kühl, Stefan, Petra Strodtholz and Andreas Taffertshofer: (eds.): *Handbuch Methoden der Organisationsforschung. Quantitative und Qualitative Methoden.* Wiesbaden: VS: 32–56.

— (2002): Experteninterview. In: Kühl, Stefan and Petra Strodtholz (Ed.): *Methoden der Organisationsforschung - Ein Handbuch.* Reinbek bei Hamburg: Rowohlt Taschenbuch Verlag: 33–71.

Löw, Martina (2001): *Raumsoziologie.* Frankfurt/Main: Suhrkamp.

LSKN (Landesbetrieb für Statistik und Kommunikationstechnologie Niedersachsen) (2010): *Niedersachsen-Monitor 2010.* Hannover: LSKN.

Lundvall, Bengt-Åke (2007): National Innovation Systems—Analytical Concept and Development Tool. In: *Industry and Innovation 14 (1)*: 95–119.

Lundvall, Bengt-Åke (ed.) (1992): *National Systems of Innovation. Towards a Theory of Innovation and Interactive Learning.* London: Pinter.

Lundvall, Bengt-Åke (1988): Innovation as an Interactive Process. From User-Producer Interaction to the National Systems of Innovation. In: Giovanni Dosi, Chris Freeman, Richard Nelson, Gerald Silverberg, and Luc Soete (ed.): *Technical Change and Economic Theory.* London/New York: Pinter: 221–238.

Lundvall, Bengt-Åke and Björn Johnson (1994): The Learning Economy. In: *Journal of Industry Studies 1 (2)*: 23–42.

MacKinnon, Danny, Andrew Cumbers and Keith Chapman (2002): Learning, Innovation and Regional Development. A Critical Appraisal of Recent Debates. In: *Progress in Human Geography 26 (3)*: 293–311.

Madri+d (2008): La Fundación madri+d—Un Espacio Para la Ciencia y la Tecnología. [Source: http://www.madrimasd.org/queesmadrimasd/estrategia/doc/Espacio_ciencia_tecnologia.pdf; last access: 15.11.2011].

Madrid Emprende (2007): *Biotechnology in Madrid*. Madrid: Madrid Emprende.

Madrid Network (2011): Cluster Automoción. El Sector en Madrid. [Source: http://www.madridnetwork.org/red/automoción/SectorEnMadrid; last access: 15.11.2011].

Mahoney, James (2007): Qualitative Methodology and Comparative Politics. In: *Comparative Political Studies 40 (2)*: 122–144.

Mahoney, James and Kathleen Thelen (eds.) (2010): *Explaining Institutional Change. Ambiguity, Agency, and Power.* Cambridge: Cambridge University Press.

Malecki, Edward J. and Pääivi Oinas (eds.) (1999): *Making Connections. Technological Learning and Regional Economic Change.* Aldershot: Ashgate.

Malerba, Franco (2004): Sectoral Systems of Innovation. Basic Concepts. In: Franco Malerba (ed.): *Sectoral Systems of Innovation. Concepts, Issues and Analyses of Six Major Sectors in Europe.* Cambridge: Cambridge University Press: 9–41.

Markusen, Ann (1999): Fuzzy Concepts, Scanty Evidence, Policy Distance. The Case for Rigour and Policy Relevance in Critical Regional Studies. In: *Regional Studies 33 (9)*: 869–884.

— (1994): Studying Regions by Studying Firms. In: *The Professional Geographer 46 (4)*: 477–490.

Martin, Ron and Peter Sunley (2003): Deconstructing Clusters. Chaotic Concept or Policy Panacea? In: *Journal of Economic Geography 3*: 5–35.

Maskell, Peter and Anders Malmberg (1999a): The Competitiveness of Firms and Regions. Ubiquitification and the Importance of Localized Learning. In: *European Urban and Regional Studies 6 (1)*: 9–25.

— (1999b): Localised Learning and Industrial Competitiveness. In: *Cambridge Journal of Economics 23*: 167–185.

Mason, Geoff and Karin Wagner (1999): Knowledge Transfer and Innovation in Britain and Germany. "Intermediate Institution" Models of Knowledge Transfer Under Strain? *Industry and Innovation 6*: 85–110.

Mattes, Jannika (2010): *Innovation in Multinational Companies. Organisational, International and Regional Dilemmas.* Frankfurt, London: Peter Lang.

Mayer, Otto (2008): *Interview und schriftliche Befragung. Entwicklung, Durchführung und Auswertung.* München, Wien: Oldenbourg.

McMichael, Philip (2005): Globalization. In: Janoski, Thomas, Robert Alford, Alexander Hicks and Mildred Schwartz (eds.): *The Handbook of Political Sociology. States, Civil Societies and Globalization.* Cambridge: Cambridge University Press.

Meinefeld, Werner (2008): Hypothesen und Vorwissen in der Qualitativen Forschung. In: Flick, Uwe, Ernst von Kardorff and Ines Steincke (eds.): *Qualitative Forschung. Ein Handbuch.* Reinbeck, Rowohlt: 265–275.

Metcalfe, Stan (1995): The Economic Foundations of Technology Policy. In: Stoneman, Paul (ed.): *Handbook of the Economics of Innovation and Technological Change.* Oxford/Cambridge (Mass.): Blackwell: 409–512.

Metropolregion (2010): MAP 2011. Metropolregion Arbeitsprogramm/Herbst 09 ► Sommer 2011. Hannover: Metropolregion Hannover Braunschweig Göttingen Wolfsburg GmbH. [Source: http://www.metropolregion.de/meta_ downloads/8882/map_2011.pdf; last access: 15.11.2011].

Meuser, Michael and Ulrike Nagel (2009): Experteninterview und der Wandel der Wissensproduktion. In: Bogner, Alexander, Beate Littig and Wolfgang Menz (eds.): *Das Experteninterview. Theorien, Methoden, Anwendungsfelder*. Wiesbaden: VS: 35–60.

— (2005): ExpertInneninterviews—Vielfach erprobt, wenig bedacht. Ein Beitrag zur Qualitativen Methodendiskussion. In: Bogner, Alexander, Beate Littig and Wolfgang Menz (eds.): *Das Experteninterview. Theorie, Methode, Anwendung*. Wiesbaden: VS: 71–94.

Meyer, John W. and Brian Rowan (1991): Institutionalized Organizations. Formal Structure as Myth and Ceremony. In: Powell, Walter W. And Paul J. DiMaggio (eds.): *The New Institutionalism in Organizational Analysis*. Chicago, London: The University of Chicago Press: 41–62.

— (1977): Institutional Organizations. Formal Structure as Myth and Ceremony. In: *American Journal of Sociology, 83 (2)*: 340–363.

MITC (Ministerio de Industria, Turismo y Comercio) (2010): *Tecnologías de la Información y las Comunicaciones en la Microempresa Española*. Madrid: MITC. Gobierno de España.

Molina, Oscár and Martin Rhodes (2007): The Political Economy of Adjustment in Mixed Market Economies. A Study of Spain and Italy. In: Hancké, Bob, Martin Rhodes and Mark Thatcher: *Beyond Varieties of Capitalism. Conflict, Contradictions, and Complementarities in the European Economy*. Oxford: Oxford University Press: 223–252.

Mora Valentín, Eva María (2002): Co-operative Relationships. A Theoretical Review of Co-operative Relationships between Firms and Universities. In: *Science and Public Policy 29 (1)*: 37–46.

Morgan, Glenn, Richard Whitley and Eli Moen (eds.) (2005): *Changing Capitalisms? Internationalization, Institutional Change, and Systems of Economic Organization*. Oxford: Oxford University Press.

Morgan, Kevin (2007): The Learning Region. Institutions, Innovation and Regional Renewal. In: *Regional Studies 41(1)*: 147–159.

Morgan, Robert P. (1998): University Research Contributions to Industry. The Faculty View. In: Blair, Peter D. and Robert A. Frosch (eds.): *Trends in Industrial Innovation. Industry Perspectives & Policy Implications*. Research Triangle Park: Sigma Xi, Scientific Research Society of North America, Incorporated: 163–170.

MRA (Motorsport Research Associates) (2004): A Study into the UK Motorsport and Performance Engineering Cluster. Final Report Prepared for the Department of Trade and Industry (Automotive Unit). MRA. [Source: http://www.northamptonshirehpeandmotorsport.com/resources/UK_Motorsport_and_Performance_Engineering_Study.pdf; last access: 15.11.2011].

Mughan Terry, Lester Lloyd-Reason and Carsten Zimmerman (2004): Management Consulting and International Business Support for SMEs. Need and Obstacles. In: Education & Training 46 (8/9): 424–432.

Mühlfeld, Claus and Paul Windolf (1981): Auswertungsprobleme Offener Interviews. In: *Soziale Welt 32*: 325–354.

MW NDS (Niedersächsisches Ministerium für Wirtschaft, Arbeit und Verkehr) (2011a): Automobilindustrie in Niedersachsen. MW NDS. [Source: http://www.mw.niedersachsen.de/live/live.php?navigation_id=5583&article_id=1600; last access: 01.03.2011].

— (2011b): Landesinitiative Gesundheitswirtschaft—Life Sciences Niedersachsen. MW NDS. [Source: http://www.mw.niedersachsen.de/live/live.php? Navigation_id=5580&article_id=1590; last access: 01.03.2011].

Nauwelaers, Claire (2001): Path-Dependency and the Role of Institutions in Cluster Policy Generation. In: Mariussen, Åge (ed.): *Cluster Policies—Cluster Development? (Nordregio Report 2001:2)*. Stockholm.

NGlobal (2009a): Kompetenz in Automotive. Niedersachsen Global GmbH. [Source: http://www.nglobal.de/fileadmin/media/docs/Automotive_Folder_dt. pdf; last access: 01.11.2011].

— (2009b): Life Sciences Expertise. Niedersachsen Global GmbH. [Source: http://www.nglobal.de/fileadmin/media/docs/wirtschaftsraum_niedersachsen/schwerpunktbranchen/WEB-ng2847_Folder_LifeSciences_eng.pdf; last access: 01.11.2011].

Nee, Victor (2005): The New Institutionalisms in Economics and Sociology. In: Smelser, Neil J. and Richard Swedberg (eds.): *The Handbook of Economic Sociology*, 2nd edition. Princeton/Oxford: Princeton University Press: 49–74.

Nelson, Andrew J. (2009): Measuring Knowledge Spillovers. What Patents, Licenses and Publications Reveal about Innovation Diffusion. In: *Research Policy 38*: 994–1005.

Nelson, Richard and Nathan Rosenberg (2009): Technische Innovation and National Systeme. In: Blättel-Mink, Birgit and Alexander Ebner (eds.): *Innovationssysteme. Technologie, Institutionen und die Dynamik der Wettbewerbsfähigkeit.* Wiesbaden: VS: 51–68.

Nelson, Richard R. (2008): What Enables Rapid Economic Progress. What are the Needed Institutions? In: *Research Policy 37*: 1–11.

Nelson, Richard R. (ed.): (1993): *National Systems of Innovation. A Comparative Analysis.* Oxford: Oxford University Press.

NIW (Niedersächsisches Institut für Wirtschaftsforschung) (2008): *Forschung, Experimentelle Entwicklung und Innovationen in der Niedersächsischen Wirtschaft.* Hannover: NIW.

NIW/Nord/LB (2005): Wirtschaftsstandort Region Hannover. Regionaler Entwicklungsbericht 2005. Auf dem Weg zur europäischen Metropolregion. Studie im Auftrag der Region Hannover. *Beiträge zur regionalen Entwicklung, Heft-Nr. 104*. Hannover.

NMWK (Niedersächsisches Ministerium für Wissenschaft und Kultur) (2011): NGlobal. Standortinformationen Wissensschaft [Source: http://www. nglobal. de/standortinformationen/wissenschaft/; last access: 15.11.2011].

— (2010): Hochschulen in Niedersachsen. Zahlen. Daten. Fakten. Broschüre des Niedersächsischen Ministeriums für Wissenschaft und Kultur. Hannover. [Source: http://www.mwk.niedersachsen.de/download/51108; last access: 15. 11.2011].

Nonaka, Ikojiro and Hirotaka Takeuchi (1997): *Die Organisation des Wissens. Wie japanische Unternehmen eine brachliegende Ressource nutzbar machen.* Frankfurt/Main, New York: Campus.

Nord/LB (Norddeutsche Landesbank) (2010): *Wirtschaft Niedersachsen. Die 100 größten Unternehmen in Niedersachsen.* Analysen und Kommentare Dezember 2010.

— (2007): *Wissensvernetzung in der Metropolregion Hannover-Braunschweig-Göttingen. Studie im Auftrag der Metropolregion Hannover-Braunschweig-Göttingen.* Hannover.

North, Douglass C. (1990): *Institutions, Institutional Change and Economic Performance.* Cambridge: Cambridge University Press.

OECD (2011a): OECD Key Biotech Indicators. Percentage of Biotechnology Firms with Fewer than 50 Employees. OECD Biotechnology Statistics Database. [Source: http://www.oecd.org/dataoecd/6/58/47023351.xls; last access: 15.02.2011].

— (2011b): OECD Key Biotech Indicators. Percentage of biotechnology R&D performed by small biotechnology R&D firms. OECD Biotechnology Statistics Database. [Source: http://www.oecd.org/dataoecd/6/41/47025109.xls; last access: 15.02.2011].

— (2010a): *The Autonomous Region of Catalonia, Spain. Self-Evaluation Report.* OECD Reviews of Higher Education in Regional and City Development. IMHE.

— (2010b): Country Statistical Profiles 2010: Spain. [Source: http://stats.oecd. org/Index.aspx?DataSetCode=CSP2010, last access: 10.04.2011].

— (2010c): *OECD Reviews of Regional Innovation. Catalonia, Spain.* Paris: OECD.

— (2010d): Higher Education in Regional and City Development. Catalonia, Spain. OECD Publishing. [Source: http://dx.doi.org/10.1787/9789264089006 -en, last access: 15.11.2011].

— (2009a): OECD Science, Technology and Industry Scoreboard 2009. Paris: OECD. [Source: http://www.oecd-ilibrary.org/content/book/sti_scoreboard-2009-en, last access 15.11.2011].

— (2009b): OECD Biotechnology Statistics 2009. Paris: OECD. [Source: http:// www.oecd.org/dataoecd/4/23/42833898.pdf; last access: 15.11.2011].

— (2007a): *R&D and Innovation in Spain. Improving the Policy Mix.* Paris: OECD; Madrid: Fundación Española para la Ciencia y la Tecnología (FECYT).

— (2007b): *OECD Territorial Reviews. Madrid, Spain.* Paris: OECD.

— (2001): *OECD Science, Technology and Industry Scoreboard. Towards a Knowledge-Based Economy.* Paris: OECD.

Ohmae, Kenichi (1993): The Rise of the Region State. *Foreign Affairs 72 (2)*: 78–87.

One Nucleus (2011): One Nucleus. The Business Catalyst for Life Sciences. [Source: http://www.onenucleus.com/; last access: 15.11.2011].

ONS (Office for National Statistics) (2011a): Population. ONS. [Source: http://www.statistics.gov.uk/hub/population/index.html; last access: 15.11.2011].

— (2011b): *Statistical Bulletin. Regional, Sub-Regional and Local Gross Value Added 2009.*

— (2011c): *Statistical Bulletin. UK Business Enterprise Research and Development 2008.* ONS.

— (2011d): *Statistical Bulletin. UK Gross Domestic Expenditure on Research and Development 2009.* ONS.

— (2011e): Business Enterprise Research and Development, 2008. ONS. [Source: http://www.ons.gov.uk/ons/rel/rdit1/bus-ent-res-and-dev/2008/index.html; last access: 15.11.2011].

— (2011f): UK Regional Economic Performance Indicators. ONS. [Source: http://stats.bis.gov.uk/sd/repi/REPI_2010_Compendium_corrected%2020_10_2010.xls; last access: 15.11.2011].

— (2011g): Regional Trends—Economy. ONS. [Source: http://www.ons.gov.uk/ons/rel/regional-trends/regional-trends-online-tables/regional-statistics---economy/regional-statistics---economy-.xls; last access: 15.11.2011].

— (2009): *UK Business. Activity, Size and Location—2009.* Newport: ONS.

— (2003): UK Government Office Regions as at August 1998. ONS. [Source: http://www.ons.gov.uk/ons/guide-method/geography/beginner-s-guide/maps/government-office-regions--gors--as-at-august-1998.pdf, last access: 15.11.2011].

Ortiz, Michael (2008): *Die Rolle regionaler Institutionen in Innovationsprozessen. Das Beispiel der Metropolregion Hannover-Braunschweig-Göttingen.* Saarbrücken: VDM.

Padmore, Tim, Hans Schuetze and Hervey Gibson (1998): Modeling Systems of Innovation. An Enterprise-Centered View. In: *Research Policy 26*: 605–624.

Peck, Jamie and Nik Theodore (2007): Variegated Capitalism. In: *Progress in Human Geography 31 (6)*: 731–772.

Perry, Beth (2007): The Multi-Level Governance of Science Policy in England. *Regional Studies 41*: 1051–1067.

Perry, Beth and Tim May (2007): Governance, Science Policy and Regions. An Introduction. In: *Regional Studies 41*: 1039–1050.

Pierson, Paul (2007): The Costs of Marginalization Qualitative Methods in the Study of American Politics. *Comparative Political Studies 40 (2)*: 145–169.

PIMEC (Petita i Mitjana Empresa de Catalunya) (2010): *Anuari de la PIME Catalana.* Barcelona: PIMEC.

Polanyi, Michael (1985): *Implizites Wissen.* Frankfurt/Main: Suhrkamp.

Polt, Wolfgang, Martin Berger, Patries Boekholt, Katrin Cremers, Jürgen Egeln, Helmut Gassler, Reinhold Hofer and Christian Rammer (2010): *Das deutsche Forschungs- und Innovationssystem. Ein internationaler Systemvergleich zur Rolle von Wis-*

senschaft, Interaktionen und Governance für die technologische Leistungsfähigkeit. Studien zum deutschen Innovationssystem Nr. 11-2010. Berlin: EFI.

Pontusson, Jonas (2005): Varieties and Commonalities of Capitalism. In: David Coates (ed.): *Varieties of Capitalism, Varieties of Approaches.* Basingstoke: Palgrave Macmillan: 163–188.

Porter, Michael E. (1998a): The Competitive Advantage of Nations. In: Porter, Michael E.: *On Competition.* Boston, Harvard Business School Press: 155–196.

— (1998b): Clusters and the New Economics of Competition. In: *Harvard Business Review 76*: 77–90.

— (1990): *The Competitive Advantage of Nations.* New York: Free Press.

Prätorius, Gerhard (2003a): Die Region und das Unternehmen als Entwicklungspartnerschaft. [http://www.reson-online.de/servlet/PB/show/1017802/ent wicklungsparterschaft.pdf; last access: 13.12.2010].

— (2003b): Politische Steuerung der Regionalentwicklung. Der niedersächsische Weg. In: *Neues Archiv für Niedersachsen. Zeitschrift für Landesforschung 1/2003*: 19–32.

Priebs, Axel (2004): Raum Hannover-Braunschweig-Göttingen auf dem Weg zur Europäischen Metropolregion. In: *Neues Archiv für Niedersachsen. Zeitschrift für Landesforschung 2/2004*: 101–112.

Prognos (2009): *Analytische und konzeptionelle Grundlagen zur Clusterpolitik in Baden-Württemberg. Bericht im Auftrag des Wirtschaftsministeriums des Landes Baden-Württemberg.* Basel: Prognos AG.

Przyborski, Aglaja and Monika Wohlrab-Sahr (2009): *Qualitative Sozialforschung. Ein Arbeitsbuch.* München: Oldenbourg.

Przeworski, Adam and Henry Teune (1970): *The Logic of Comparative Social Inquiry.* New York: Wiley.

Ragin, Charles C. (2007): Qualitative Comparative Analysis Using Fuzzy Sets (fsQCA). In Rihoux, Benoit and Charles Ragin (eds.): *Configurational Comparative Analysis.* Thousand Oaks/London: Sage.

— (2000): *Fuzzy-Set Social Science.* Chicago: University of Chicago Press.

— (1987): *The Comparative Method. Moving Beyond Qualitative and Quantitative Strategies.* Berkeley: University of California Press.

Rammer, Christian and Britta Weißenfeld (2008): *Innovationsverhalten der Unternehmen in Deutschland 2006. Aktuelle Entwicklungen und ein internationaler Vergleich. Studien zum deutschen Innovationssystem Nr. 04-2008.* Berlin: BMBF.

Revilla Diez, Javier (2004): Regionale Wachstumskonzepte in Niedersachsen—Eine Bewertung aus regionalökonomischer Sicht. In: *Neues Archiv für Niedersachsen. Zeitschrift für Landesforschung 2/2004*: 65–82.

— (2002): Metropolitan Innovation Systems. A Comparison between Barcelona, Stockholm and Vienna. In: *International Regional Science Review: 25 (1)*: 63–85.

Robertson, Roland (2001): Globalization Theory 2000+. Major Problematics. In: Ritzer, George and Barry Smart (eds.): *Handbook of Social Theory.* London: Sage.

Rosenfeld, Stuart A. (1997): Bringing Business Clusters into the Mainstream of Economic Development', *European Planning Studies 5 (1)*: 3–23.

Royo, Sebastián (2007): Varieties of Capitalism in Spain. Business and the Politics of Cooperation. In: *European Journal of Industrial Relations 13 (1)*: 47–65.

SÄdL (Statistische Ämter des Bundes und der Länder) (2011): SÄdL. [Source: http://www.statistikportal.de/Statistik-Portal/; last access: 01.02.2011].

Sassen, Saskia (2001): *The Global City*. New York, London, Tokyo (2nd edition): Princeton: Princeton University Press.

Schäfer, Rudolf, Johanna Stackelberg and Floria Stellmacher (2004): *Bildung einer Metropolregion Hannover-Braunschweig-Göttingen von europäischer Bedeutung*. Konzeptstudie im Auftrag des Städtenetzes EXPO-Region. o. O.

Scharpf, Fritz W. (2011): Monetary Union, Fiscal Crisis and the Preemption of Democracy. *MPIfG Discussion Paper 11/11*. Max Planck Institute for the Study of Societies, Cologne.

Schmidt, Vivian. A. (2006): Bringing the State Back into Varieties of Capitalism and Discourse Back into the Explanation of Change. Draft Paper. Paper Prepared for Presentation for the Panel. 7-5 Explaining Institutional Change in Different Varieties of Capitalism of the Annual Meetings of the American Political Science Association (Philadelphia, PA, Aug. 31-Sept. 3, 2006).

— (2002): *The Futures of European Capitalism*. Oxford: Oxford University Press.

Schmoch, Ulrich (2003): *Hochschulforschung und Industrieforschung. Perspektiven der Interaktion*. Frankfurt/Main, New York: Campus.

Schneider, Carsten Q. and Claudius Wagemann (2007): *Qualitative Comparative Analysis and Fuzzi Sets. Ein Lehrbuch für Anwender und jene, die es werden wollen*. Opladen, Farmington Hills: Budrich.

Schulze, Gerhard (2004): *Einführung in die Methoden der empirischen Sozialforschung. Bamberger Beiträge zur empirischen Sozialforschung, Band 1, 2*. Auflage. Bamberg.

Schumpeter, Josef A. (1939): *Business Cycles. A Theoretical, Historical and Statistical Analysis of the Capitalist Process. 2 Vols*. New York: McGraw Hill.

Scott, W. Richard (1994): Conceptualizing Organizational Fields. Linking Organizations and Societal Systems. In: Derlien, Hans-Ulrich, Uta Gerhardt and Fritz W. Scharpf (eds.): *Systemrationalität und Partialinteresse. Festschrift für Renate Mayntz*. Baden-Baden: Nomos: 203–221.

— (2008): Institutional Processes and Organizational Fields. In: W. Richard Scott: *Institutions and Organizations. Ideas and Interests*. Los Angeles, New York: Sage: 181–210.

SEMTA (Sector Skills Council for Science, Engineering and Manufacturing Technologies) (2011): Automotive Manufacture. SEMTA. [Source: http://www.Semta.org.uk/pdf/IAGLMI_Automotive_March2011.pdf; last access: 15.11. 2011].

Shaw, Julian (2011): Business Population Estimates for the UK and Regions—Introducing improved statistics on the UK enterprise population. Economic &

Labour Market Review April 2011. [Source: http://www.bis.gov.uk/analysis/statistics/business-population-estimates, last access: 15.11.2011].

Siebenhüner, Bernd (2007): Methoden und Methodenprobleme der Innovationsforschung. In: Hof, Hagen and Ulrich Wengenroth: *Innovationsforschung. Ansätze, Methoden, Grenzen und Perspektiven.* Hamburg: LIT: 103–115.

Simmie, James and James Sennett (1999): Innovative Clusters. Global or Local Linkages? *National Institute Economic Review 170*: 87–98.

SLA BW (Statistisches Landesamt Baden-Württemberg) (2011): [Source: http://www.statistik-bw.de; last access: 20.10.2011].

— (2009): *Wirtschafts- und Sozialentwicklung in Baden-Württemberg 2008/ 2009.* SLA BW.

Staehler, Tanja, Dirk Dohse and Philip Cooke (2007): *Evaluation der Fördermaßnahmen BioRegio und BioProfile.* Berlin: BMBF.

Stehr, Nico (1994): *Arbeit, Eigentum und Wissen. Zur Theorie von Wissensgesellschaften.* Frankfurt/Main: Suhrkamp.

Sternberg, Rolf (2002): The Regional Impact of Innovation Networks. In: Schätzl, Ludwig and Javier Revilla Diez (eds.): *Technological Change and Regional Development in Europe.* Heidelberg: Physica: 135–155.

Strambach, Simone (2004): Wissensökonomie, organisatorischer Wandel und wissensbasierte Regionalentwicklung. Herausforderungen für die Wirtschaftsgeographie. In: *Zeitschrift für Wirtschaftsgeographie 48 (1)*: 1–18.

Strauss, Anselm (1991): *Grundlagen Qualitativer Sozialforschung. Datenanalyse und Theoriebildung in der empirischen Soziologischen Forschung.* München: Wilhelm Fink.

Streeck, Wolfgang (2010): E Pluribus Unum? Varieties and Commonalities of Capitalism. *MPIfG Discussion Paper 10/12.* Max Planck Institute for the Study of Societies: Cologne.

— (2009): *Re-Forming Capitalism. Institutional Change in the German Political Economy.* Oxford: Oxford University Press.

— (2004): Globalisierung. Mythos und Wirklichkeit. *MPIfG Working Paper 04/4.* Max Planck Institute for the Study of Societies: Cologne.

— (1997): German Capitalism. Does It Exist? Can It Survive? In: *New Political Economy 2 (2)*: 237–256.

— (1991): On the Institutional Conditions of Diversified Quality Production. In: Egon Matzner and Wolfgang Streeck (eds.): *Beyond Keynesianism. The Socio-Economics of Production and Employment.* London: Edward Elgar: 21–61.

Streeck, Wolfgang and Kathleen Thelen (2005): Introduction. Institutional Change in Advanced Political Economies. In: Streeck, Wolfgang and Kathleen Thelen (eds.): *Beyond Continuity. Institutional Change in Advanced Political Economies.* Oxford: Oxford University Press: 1–39.

Strodtholz, Petra and Stefan Kühl (2002): Qualitative Methoden der Organisationsforschung—Ein Überblick. In: Kühl, Stefan and Petra Strodtholz (eds.): *Methoden der Organisationsforschung. Ein Handbuch.* Reinbek bei Hamburg: Rowohlt Taschenbuch Verlag: 11–29.

Stühring, Karsten (2006): Die Europäische Metropolregion Hannover-Braunschweig-Göttingen. Ein Raum zwischen Kontinuität und Wandel. In: *Neues Archiv für Niedersachsen. Zeitschrift für Landesforschung 1/2006*: 82–93.

Swann, Peter and Martha Prevezer (1996): A Comparison of the Dynamics of Industrial Clustering in Computing and Biotechnology. In: *Research Policy 25*: 1139–1157.

Sydow, Jörg (2010): Management von Netzwerkorganisationen. Zum Stand der Forschung. In: Sydow, Jörg (ed.): *Management von Netzwerkorganisationen*. 5. Aufl. Wiesbaden: Gabler: 373–470.

Tappeiner, Gottfried, Christoph Hauser and Janette Walde (2008): Regional Knowledge Spillovers. Fact or Artifact? In: *Research Policy 37*: 861–874.

The Economist (2010): Older and Wiser. A Special Report on Germany. *The Economist March 13th 2010.*

Thelen, Kathleen (2010): Beyond Comparative Statics. Historical Institutional Approaches to Stability and Change in the Political Economy of Labor. In: John Campbell et al. (eds.), *Handbook of Comparative Institutional Analysis*. Oxford: Oxford University Press: 41–61.

— (2003): How Institutions Evolve. Insights from Comparative-Historical Analysis. In: Mahoney, James and Dietrich Rueschemeyer (eds.): *Comparative Historical Analysis in the Social Sciences*. Cambridge: Cambridge University Press: 208–240.

Thomi, Walter and Robert Werner (2001): Regionale Innovationssysteme. Zur territorialen Dimension von Wissen und Innovation. In: *Zeitschrift für Wirtschaftsgeographie 45 (3–4)*: 201–218.

TLB (2007): *Ideen suchen Produzenten—Produzenten finden Ideen*. Karlsruhe: Technologie-Lizenz-Büro (TLB) der Baden-Württembergischen Hochschulen GmbH.

Tödtling, Franz and Michaela Trippl (2005): One Size fits All? Towards a Differentiated Regional Innovation Policy Approach. In: *Research Policy 34*: 1203–1219.

Toffler, Alvin (1990): *Machtebenen. Wissen, Wohlstand und Macht im 21. Jahrhundert*. Düsseldorf: Econ.

Trigilia, Carlo (2004): High-Tech Districts. In: Colin Crouch, Patrick Le Galès, Carlo Trigilia and Helmut Voelzkow: *Changing Governance of Local Economies. Response of European Local Production Systems*. Oxford: Oxford University Press: 229–236.

Tushman, Michael L. and Richard R. Nelson (1990): Introduction. Technology, Organizations, and Innovation. In: *Administrative Science Quarterly 35*: 1–8.

UKTI (UK Trade and Investment) (2007): Motorsport in the UK. An industry with a winning formula. UKTI. [Source: http://www.ukti.gov.uk/download/109421_101037/Motorsport%20in%20the%20UK.html; last access: 15.11.2011].

Van den Berg, Leo, Erik Braun and Willem van Winden (2001): Growth Clusters in European Cities—An Integral Approach. In: *Urban Studies 38 (1)*: 185–205.

Varga, Attila (ed.) (2009): *Universities, Knowledge Transfer and Regional Development. Geography, Entrepreneurship and Policy*. Cheltenham, Northampton: Edward Elgar.

VDA (Verband der Automobilindustrie e.V.) (2011): *Niedersachsen—Zentrum der Automobilindustrie.* Berlin: VDA.

— (2010): *Jahresbericht 2010.* Berlin: VDA.

Voelzkow, Helmut, with Elbing, Sabine and Martin Schröder (2007): *Jenseits nationaler Produktionsmodelle? Die Governance regionaler Wirtschaftscluster. International vergleichende Analysen.* Marburg: Metropolis.

Vogel, Friedrich (2000): *Beschreibende und Schließende Statistik.* München, Wien: Oldenbourg.

Walter, Achim (2003): *Technologietransfer zwischen Wissenschaft und Wirtschaft. Voraussetzungen für den Erfolg.* Wiesbaden: Deutscher Universitäts-Verlag.

Watson, Matthew (2003): Ricardian Political Economy and the "Varieties of Capitalism" Approach. Specialization, Trade and Comparative Institutional Advantage. In: *Comparative European Politics 1(2)*: 227–240.

Weingart, Peter (2001): *Die Stunde der Wahrheit? Zum Verhältnis der Wissenschaft zu Politik, Wirtschaft und Medien in der Wissensgesellschaft.* Weilerswist: Velbrück Wissenschaft.

Wentland, Alexander, Andreas Knie and Dagmar Simon (2011): Warum aus Forschern keine Erfinder werden. Innovationshemmnisse im deutschen Wissenschaftssystem am Beispiel der Biotechnologie. *WZBrief Bildung 17.* Berlin: Wissenschaftszentrum Berlin für Sozialforschung.

Werr, Andreas and Torbjörn Stjernberg (2003): Exploring Management Consulting Firms as Knowledge Systems. In: *Organization Studies 24 (6)*: 881–908.

Weyer, Johannes (2000): Einleitung. Zum Stand der Netzwerkforschung in den Sozialwissenschaften. In: Weyer, Johannes (ed.): *Soziale Netzwerke. Konzepte und Methoden der Sozialwissenschaftlichen Netzwerkforschung.* München/ Wien: Oldenbourg: 1–34.

Whitford, Josh (2001): The Decline of a Model? Challenge and Response in the Italian Industrial Districts. In: *Economy and Society 30 (1)*: 38–65.

Whitley, Richard (1999): *Divergent Capitalisms. The Social Structuring and Change of Business Systems.* Oxford: Oxford University Press.

Widuckel, Werner (2004): AutoVision—Wachstum durch Strategie und Qualität. In: *Neues Archiv für Niedersachsen. Zeitschrift für Landesforschung 2/2004*: 60–64.

Willke, Helmut (2001): *Systemisches Wissensmanagement.* Stuttgart: Lucius & Lucius.

Wood, Steward (2001): Business, Government and Patterns of Labor Market Policy in Britain, and the Federal Republic of Germany. In: Hall, Peter A. and David Soskice (eds.): *Varieties of Capitalism. The Institutional Foundations of Comparative Advantage.* Oxford: Oxford University Press: 247–274.

Yin, Robert K. (2009): *Case Study Research. Design and Methods.* 4th Edition. Applied Social Research Methods Series, Vol. 5. Thousand Oaks: SAGE.

Acknowledgements

This dissertation deals with the governance of knowledge transfer and innovation. What holds true for the actors investigated in this study, also applies to the researcher studying them: Without the feedback, support and advice of supervisors, colleagues, friends and family members realizing an innovative research project would not have been possible.

Therefore, first and foremost, I am grateful to my supervisor Bernhard Ebbinghaus for always supporting my dissertation project and especially for creating the organizational framework conditions in which this project could be realized. Furthermore, I am particularly indebted to Knut Koschatzky for his willingness to co-supervise my dissertation project, his helpful and highly instructive comments as well as his constructive suggestions on my research.

Critical comments are essential to further develop and improve research ideas, hypotheses and empirical models. I profited from many of them coming from professors, colleagues and friends as a doctoral student and candidate in the excellent and stimulating research environment of the University of Mannheim's Graduate School of Economic and Social Sciences (GESS) and its Center for Doctoral Studies in Social and Behavioral Sciences (CDSS). I am proud and grateful having received financial support and scholarships during this time granted by the German Research Foundation (DFG), the Landesgraduiertenförderung by the Land of Baden-Württemberg, as well as by the Karin-Islinger-Stiftung.

My strong gratitude shall be expressed towards the friendly and generous funding I was awarded by the Axel Springer Stiftung and the Johanna and Fritz Buch Gedächtnis-Stiftung in Hamburg. This financial support has been essential for allowing a young researcher to realize the publication process of this encompassing project.

Moreover, working for many years on an ambitious research project is impossible without a supportive family in the background. Therefore, I am heartly grateful to Nathalie for all her support, patience and love over the last years. Furthermore, I am especially indebted to my brother André, who inspired my work with his sharp feedbacks, his brilliant analytical comments and his deep understanding of the ups and downs in the course of writing a dissertation. Above all, I am deeply grateful to him for being the person I could always rely on. Finally, I want to thank my parents, who always supported all stages of my life and career in so many ways, not least with their love and their neverending believe in me. This book is dedicated to them.

Index